Acid Soils Their Chemistry and Management

NEW INDIA PUBLISHING AGENCY
New Delhi-110 034

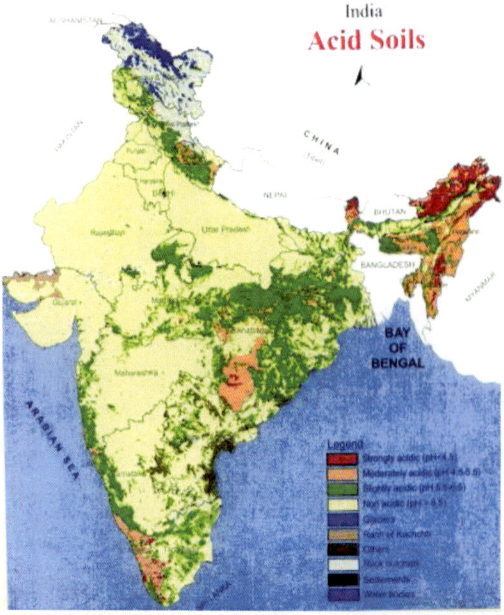

Strongly Acidic (pH<4.5)
6.19 mha

Moderately Acidic
(pH 4.5-5.5)
24.81 mha

Slightly Acidic (pH 5.5-6.5)
58.94 mha

Total Area under Acid Soils in the Country – 89.94 m ha

Acid Soils
Their Chemistry and Management

Edited by

A.K. Sarkar

Formerly Professor & Head (Soil Science) and
Dean, Faculty of Agriculture
Birsa Agricultural University
Ranchi – 834 006
Jharkhand

NEW INDIA PUBLISHING AGENCY
New Delhi-110 034

NEW INDIA PUBLISHING AGENCY
101,103, Vikas Surya Plaza, CU Block
L.S.C. Market, Pitam Pura, New Delhi-110 034
Ph : +91 11 27341616, 27341717, 27341718
E-mail: newindiapublishingagency@gmail.com
www: www.nipabooks.com

For customer assistance, please contact
Phone: + 91-11-27 34 17 17
Fax: + 91-11- 27 34 16 16
E-Mail: feedbacks@nipabooks.com

ISBN: 978-93-81450-38-3

Composed and Designed by NIPA®.

Dr. M. VELAYUTHAM, Ph. D (U.K.)
Former Director
National Bureau of Soil Survey & Land Use Planning
(ICAR) and
Former Executive Director
M.S. Swaminathan Research Foundation

Foreword

Agriculture is the primary industry,source of livelihood and way of life for more than 60 per cent of the population in India.There is an increasing pressure on the land resource and good quality land available for Agriculture, due to alternate demands on the requirement for urbanization,industrialisation and infra-structure development. In addition, the "Productive potential" of the existing cultivable lands is decreasing due to its improper management and use (including its abuse and misuse), the impact of various man-induced land degradation processes and the lack of "Public awareness" on the effect of these actions and processes on the "Quality of Land".

Soil Acidity is one of the major intrinsic soil-related constraints limiting the production potential of these soils for crop growth. About 30 million hectares of cultivated land in the country is highly acidic with soil pH ranging from 4.5 to 5.5.Crop diversity, cropping intensity,crop productivity & profitability in these areas is very low, resulting in livelihood insecurity, poverty, and slow pace of development. Major part (about 80 per cent) of the Eastern and North-eastern India are of greater concern in this regard.

The Chemistry of soil acidity has been one of the much researched domains in Soil Science. The investigations on "Clay acids or Acid clays" in the early decades of the last century and the publication of the paper in the mid-60's by M.L.Jackson on "Soil Acidity-a-merry-go-round" provides an interesting reading on the subject to the students of Soil Science. What has been lacking is the practical application of these principles into the Best Management Practices in the cultivation of these soils for diverse crops and cropping systems and plantation crops.

I congratulate Dr.A.K.Sarkar, Former Dean,College of Agriculture, Birsa Agricultural University, Ranchi and Former Project Co-ordinator, ICAR Network Project on Acid Soils (2000-2005), for his creative thinking and painstaking efforts in compiling and editing the recent advances made in the Management of acid soils by the National Agricultural Research system, in this book. Dr. Sarkar is a distinguished Scientist of national repute, who followed the school of thought on the studies on Acid Soils of Chotanagpur plateau, created by late Professor S.C. Mandal.

This multi-authored book with contributions from several leading scientists provide valuable information for all stake-holders engaged in the management of Acid soils and sustainable development of these regions.

This book, I hope,will serve as a milestone in Acid Soil Management Programme through a synergy between Science,technology and Policy support for addressing the Food and Livelihood concerns of the population inhabiting these areas and regions.

M. Velayutham

M. Velayutham

Chennai
31-05-2013

Preface

The book on **Acid Soils: Their Chemistry and Management** is the result of a novel exercise to highlight the technological, management and policy issues involved in increasing the agricultural out put on low productive acid soils. During the last 10 to 15 years, excellent work has been done in this field after taking note of the pioneering research on liming acid soils by eminent scientists. It was thought proper to collate and consolidate the work for effective use by scientists, research scholars, post-graduate students, and extension personnels.

Our knowledge on soil acidity and its effect on soil health and crop productivity has expanded in recent years. Management of knowledge and its fast dissemination for better life is of great value. Though basic chemistry of acid soils has remained similar, but the land use, cropping patterns, methodology and sources of lime application have changed over the years. This has made the planners to realize that the untapped potential of acid soil regions must be taken care of by a technology based development.

The chapters for the book have been written by scientists, who are engaged in this work for the past two to three decades. I express sincere appreciation and gratitude to the authors for their hard work, cooperation and adherence to a time frame.

I am grateful to NIPA, New Delhi for helping us immensely by accepting to publish this book, which, I am sure, will act as an important reference material on the subject.

Farmers, in general, and of the eastern and north-east India in particular, ignoring all the hardships, distress, poverty, and un-sustainable lifestyles, are toiling hard to provide food to all of us. Let us pledge to help them in this gigantic task by empowering them with profitable and low cost technologies for mutual benefits.

To our readers, in the words of Swami Vivekananda,"Books are infinite in number and time is short. The secret of knowledge is to take what is essential. Take that and try to live up to it."

A.K. Sarkar
Editor

Place: Gurgaon, Haryana
Date: May, 2013

Contributors

A.K. Sarkar
Birsa Agricultural University
Ranchi – 834 006
Jharkhand

A.K. Ghosh
Department of Soil Science and Agricultural Chemistry
Institute of Agricultural Sciences
Banaras Hindu University
Varanasi, Uttar Pradesh – 221 005

A.K. Sahoo
National Bureau of Soil Survey and Land Use Planning, Regional Centre
Kolkata
West Bengal

B. Mandal
Bidhan Chandra Krishi Viswa-Vidyalaya
Kalyani
West Bengal

Dipak Sarkar
National Bureau of Soil Survey and Land Use Planning
Nagpur, Maharashtra
India

D. Jena
Department of Soil Science & Agricultural Chemistry
Odisha University of Agriculture & Technology
Bhubaneswar – 751 003
Odisha

Mausumi Raychaudhuri
Directorate of Water Management
Chandrasekharpur
Bhubaneswar – 751 023
Odisha

Patiram
ICAR Complex for NEH Region
Barapani

Sushanta Kumar Pattanayak
Professor (Soil Science)
OdishaUniversity of Agriculture and Technology
Bhubaneswar – 751 003
Odisha

S.P. Sharma
Himachal Pradesh Krishi Viswavidyalaya
Palampur
Himachal Pradesh

S.C. Talashilkar
Konkan Krishi Vidyapeeth
Dapoli
Maharashtra

S. Raychaudhuri
Directorate of Water Management
Chandrasekharpur
Bhubaneswar –751 023, Odisha

Surendra Singh
Department of Soil Science and Agricultural Chemistry
Institute of Agricultural Sciences
Banaras Hindu University
Varanasi – 221 005, Uttar Pradesh

T.C.Baruah
Assam Agricultural University
Jorhat
Assam

V.K.Venugopal
Kerala Agricutural University
Vellayani
Kerala

Contents

Foreword v

Preface vii

Contributors ix

1. Introduction 1
A.K. Sarkar

2. Soil Acidity – Concepts and Applications 5
A.K. Sarkar

3. Chemistry of Acid Soils 21
Sushanta Kumar Pattanayak

4. Genesis and Classification of Acid Soils of India 49
A.K.Sahoo and Dipak Sarkar

5. Field Studies on Acid Soils of India 105
A.K. Sarkar, S.P. Sharma, T.C. Baruah, Patiram, B. Mandal, S.K. Pattanayak, V.K. Venugopal, and S.C. Talashilkar

6. Acid Soils of North-Eastern India 133
Mausumi Raychaudhuri and S. Raychaudhuri

7. Acid Soils of Jharkhand 165
A.K. Sarkar

8. Acid Soils of Odisha 197
D. Jena

9. Secondary and Micronutrients in Acid Soils 257
Surendra Singh and A.K. Ghosh

10. Paper Mill Sludge as an Acid Soil Ameliorant 289
Sushanta Kumar Pattanayak

11. Way Forward 311
A.K. Sarkar

1

Introduction

A.K. Sarkar

Soil acidity is a major constraint for crop production. The production of acid in soils is a natural process, especially in the high rainfall areas. The removal of basic cations especially Ca and Mg, by leaching and erosion results in their replacement by acidic cations, H, Al and Fe on cation exchange sites and in the soil solution, which are in dynamic equilibrium with each other. Soils, that develop from weathered granite are likely to be more acidic than those developed from shale or limestone. Organic matter decay produces H^+ in soils, which also contribute to soil acidity. Intensive cropping and use of hybrid & high yielding crop varieties remove large quantities of basic elements essential for their growth i.e. Ca, Mg and K & deplete the soils. Rate of acidification is also accelerated by faulty fertilizer (especially Nitrogen) management practices.

Acid soils occupy approximately 30 per cent or 3950 Mha of the land area of the world. 67 per cent of the acid soils support forests and woodlands and about 18 per cent are covered by savanna, prairie and steppe vegetation. Only 4.5 per cent (179 Mha) of the acid soils area across the globe is used for arable crops. A further 33 M ha is utilized for perennial tropical crops. The value of annual production in these areas is approximately US$129 Billion. Value of products from forests, woodlands and permanent pastures on acid soils is difficult to evaluate (Uexkull and Mutert, 1995).

In India, about 90 M ha of the total geographical area comprise of acid soils. Out of this, about 30 M ha of cultivated soils with pH below 5.5 are critically degraded with very poor physical, chemical and biological characteristics ,and thus are of poor soil health. These soils are restricted mainly to the Himalayan region and red and lateritic zones (comprising both the southern, eastern and north eastern hilly regions) and some pockets in the greater plains of the country. Major crops grown in these regions are rice, jute,maize wheat, millets, potato, pulses, oilseeds & vegetables.

Soil acidity has both direct and indirect effects on crops. The direct effect is due to low pH of soils. This is due to the increase in the concentration of H^{+}, Al^{3+}, Fe^{3+}, Mn^{2+} which are often toxic to plants and soil micro organisms. Indirect effect of soil acidity on crops is related to the effect of soil pH on nutrient availability and organic matter decomposition. Besides, the actual soil acidity (solution + exchangeable acidity), buffering capacity of soil (resilience) of a soil to respond to changes in soil pH is important and vary with soil texture, clay (amount as well as nature of clay) and soil organic matter content.

Efficient soil amelioration and nutrient management methods are necessary for crops susceptible to soil acidity. For this, locally available low cost liming materials need to be promoted. Soils of eastern and north eastern India pose major problems of soil moisture deficiency in winter months. Appropriate and situation specific water management practices need to be practised. Basically, two-pronged strategy for acid soil management is to be adopted- soil amelioration with balanced use of plant nutrients for crops and using crops & varieties of the same crop, which are relatively tolerant to Al^{3+} toxicity. In fact, planting relatively Al^{3+}-tolerant species or varieties can considerably reduce the requirement of lime for soil amelioration. Future advances in the use of biotechnological tools and related genetic manipulation has great potential for inducing Al^{3+} tolerance in crops.

Liming acidic soils is an ancient agricultural practice.Particle size of liming materials such as calcite,dolomite, basic slag and papermill sludges influence their speed of reaction with soil.When lime is added to acid soils,the particles begin to dissolve.The dissolved $CaCO_3$ or $MgCO_3$ react with H+ ions to produce Ca-soil,$Ca(HCO_3)2$ and H_2CO_3.The H_2CO_3 equilibrates with H_2O and CO_2 and the excess CO_2 is lost by volatilization. Hence, the reaction is complete and the exchangeable H^+ and Al^{3+} are replaced by Ca^{2+}.Dissolution of limestone particles is a slow process .Diffusion of Ca depends on the fineness of the liming material and the distance it has to travel in soil to reach plant roots.Thus,localised placement of liming materials seems appropriate.

Earlier attempts of liming by broadcast application did not prove economical for crops grown in acid soils , because of the large quantity of lime to be transported to farmers fields and cost of the material to be borne by farmers at one time. The focus, therefore, shifted towards furrow method of application of lime. This method requires only one-tenth of lime required for broadcast application and is more farmer friendly and scientific. The technology of furrow application of lime can benefit oilseeds and pulses crops in yield improvement and greater economic return to farmers.This needs to be operationalized by creating requisite infrastructure of marketing and distribution network for quality liming materials in the acid soil regions.

The present book provides credence to the technology of lime application in furrows and the immense benefits one can derive from its implementation in farmers fields. Basic concepts of soil acidity, its chemistry, genesis and classification have been outlined to provide an insight to the problem of their occurrence and mitigation options. Challenges that have to be faced over years in acid soil management are: small holdings, poor farmers, rainfed areas, non availability of lime sources at site, absence of proper soils testing facilities for farmers,and lack of skills and empowerment of farmers. Emerging options such as integrated farming systems, organic farming, site-specific nutrient management, conservation agriculture, in-situ moisture retention practices need to be integrated with the acidity management programme for higher crop productivity, profitability from input use and sustainable soil health.

References

Von Uexkull, H.R. and Mutert, E. (1995). Global extent,development and economic impact of acid soils. *Plant and Soil*, 171:1-15.

2

Soil Acidity – Concepts and Applications

A.K. Sarkar

Soil pH

Danish scientist Sorensen (1909) first used the term pH. The H+ concentration is usually small and is expressed in terms of pH, defined as the logarithm of the reciprocal of the H^+ concentration, in moles per litre. The pH scale ranges from 0 to 14, with 7 being neutral. Soil acidity is the concentration of H^+ in the soil solution. This is active acidity. Besides, there is a potential acidity, that includes the H^+ adsorbed on the colloidal fraction of soil. The pH scale is derived from the dissociation constant of water, Kw:

$$Kw = [H^+][OH^-] = 10^{-14}$$

$$Or, -(\log[H^+]) + \log[OH^-] = 14$$

$$pH = -\log[H^+]$$

So that

$$pH = 14 + \log[OH^-]$$

In most acid soils with pH <4, free strong acid exists. In the pH range of 5.5 to 7.0, H^+ in the soil solution is supplied mainly by hydroxyl-Aluminium hydrolysis and between pH 4.0 to 5.5, the H^+ of the soil solution comes from exchangeable Aluminium hydrolysis. Organic matter rich soils have significant amount of H^+. Intensively weathered soils tend to produce Al toxicity and deficiencies of Mg and Ca in plants.

Studies have shown that the actual concentrations of H^+ or OH^- are not very important. But, the associated chemical and biological environment at a certain pH is very important. The greatest general influence of pH on plant growth is its effect on the availability of nutrients to plants. Soil pH is a master variable providing indication of soil properties.

Most nitrogen fixing legume bacteria are in–active in low pH soils. Organic matter decomposition by Bacteria suffers in acidic soils. Fungi usually tolerate acidity better than other micro organisms. Earthworms are inhibited by high soil acidity. Agricultural crops in mineral soils grow best at pH 6.5, while that in organic soils, at pH 5.5. Increasing acidity is associated with higher Fe and Al in soil solution with the formation of relatively insoluble iron and aluminium phosphates. This restricts availability of phosphorus to plants grown in low pH soils with increase in soil acidity. K, Ca & Mg becomes less available because of decreasing CEC and lower exchangeable basic cations. Boron, copper and zinc are leachable and often deficient in leached acid soils. Molybdenum is commonly deficient in low pH soils owing to its reaction with iron to form an insoluble compound.

Soil as a Buffer

The buffering capacity of soil is its ability of ions associated with the solid phase to buffer changes in ion concentration in the solution phase. The buffering capacity increases as the CEC increases. If calcium in solution is precipitated, additional exchangeable calcium will be out of exchange sites into solution, so that only a slight change in soluble calcium will occur. Also, adding calcium salts to soil will lead to its higher concentration in exchange complex, lowering the soil solution calcium. Buffering action should be effective in controlling soluble concentrations of ions such as H^+, Al^{3+}, Ca^{2+}, Mg^{2+}, K^+, Na^+ etc.

Havlin et al. (2010) have clearly outlined how a soil behaves like a weak acid that will buffer pH. In acid soils, adsorbed Al^{3+} is in equilibrium with Al^{3+} in soil solution. Al^{3+} hydrolyses to produce H^+. If H^+ is neutralized by liming (addition of $CaCO_3$), Al^{3+} in soil solution precipitates as $Al(OH)_3$, causing exchangeable Al^{3+} to desorb to re-supply solution Al^{3+}. Thus, soil pH remains the same or is buffered. As more $CaCO_3$ is added, the reaction continues with greater replacement of Ca^{2+} for adsorbed or exchangeable Al^{3+}. As a result, the soil pH gradually increases.

In acid soils buffering refers to the ability of the exchangeable Al^{3+}, H^+ and hydroxy aluminium, to maintain a certain concentration of H^+ in the soil solution. The amount of soil solution H^+ with a pH of 6.0, for example,is very small compared to adsorbed H^+ or Al^{+++} (which can hydrolyze to produce H^+). Neutralization of solution H^+ by addition of bases (liming materials) results in rapid replacement of H^+ from the relatively large amount of H^+ associated with the solid phase.

Soil containing higher amounts of clay and organic matter, exhibit high buffering capacity & require higher amounts of liming materials to increase the soil pH than soils with lower buffering capacity. Also, soils with dominance of 1:1 clay minerals are less buffered compared to that with 2:1 clay minerals.

Hydrolysis of Fe & Al in soil

The primary source of exchangeable Al^{3+} in acid soils is the acidic environment of weathering of primary as well as secondary aluminosilicate minerals. As soil pH decreases, the concentration of Al^{3+} in the soil solution increases because of dissolution of $Al(OH)_3$

$$Al(OH)_3 + 3H^+ \rightleftharpoons Al^{3+} + 3H_2O$$

The equilibrium reaction shows that, with decrease in pH (increase in H^+ concentration), the equilibrium shifts to the right, where $Al(OH)_3$ dissolves to produce Al^{3+}, which can then be adsorbed to the exchange complex. Depending on pH, Al^{3+} will hydrolyze as :

$$Al^{3+} + H_2O \rightleftharpoons Al(OH)^{+}_{2} + H^+ \text{ (Below pH 4.0)}$$

$$A1(OH)^{2+} + H_2O \rightleftharpoons Al(OH)^{+}_{2} + H^+ \text{ (Below pH 4.6 to 6.5)}$$

$$A1(OH)_2^{+} + H_2O \rightleftharpoons Al(OH)^{0}_{3} + H^+ \text{ (Below pH 6.5 to 8.0)}$$

$$Al(OH)^{0}_{3} + H_2O \rightleftharpoons Al\,OH)^{-}_{4} + H^+ \text{ (Below pH 8.0 to 11.0)}$$

At lower pH, more of the Al hydrolysis species are positively charged, which enables stronger bonding of Al in the exchange complex. Hydrolysis of Al^{3+} produces H^+, which lowers the pH.

$$Al^{3+} + H_2O \rightleftharpoons Al(OH)^{2+} + H^+$$

$$\text{Lattice Al} \rightleftharpoons \text{Al (OH)}^{2+} + H^+ \text{ Hydroxy-aluminium}$$

When lime is added to acid soils, H^+ in the soil solution is neutralized first with additional lime added, Al^{3+} hydrolyses with the production of H^+. Thus, Al^{3+} hydrolysis buffers the increase in solution pH. A pH of 5.5 is typically cited on the reference point above which no exchangeable Al^{3+} remains associated with the soil clays.

Fe hydrolysis is similar to that of Al

$$Fe^{3+} + H_2O \rightleftharpoons Fe\ (OH)^{+2}$$

$$Fe(OH)^{2+} + H_2O \rightleftharpoons Fe\ (OH)^+_2 + H^+$$

$$Fe(OH)_2^+ + H_2O \rightleftharpoons Fe\ (OH)^0_3 + H^+$$

$$Fe\ (OH)^0_3 + H_2O \rightleftharpoons Fe\ OH)^-_4 + H^+$$

This reaction is more acidic than Al hydrolysis, but the acidity is buffered by Al hydrolysis reactions. Thus, Fe hydrolysis does not affect soil pH till most of soil Al has reacted.

Hydrolysis of Fe is specially important, when oxidation reactions follow the drainage of wetlands. Drainage of wetlands can generate substantial amount of acidity. Drainage of soil results in the conversion of Fe^{2+} to Fe^{3+} ions. The resulting Fe^{3+} ions undergo hydrolysis and release H^+ ions to the soil solution. In acid-sulphate soils, the reaction taking place due to presence of Pyrites (FeS_2) is the oxidation of both Iron and Sulphide,which may be indicated as follows:

$$2FeS_2+7H_2O+7.5\ O_2 = 4SO_4{}^{2-}+ 8H^+ +2Fe(OH)_3$$

The above reaction thus, can produce substantial amounts of H+ that make liming impractical for managing such soils. It is better to keep the soil environment an– aerobic by flooding to prevent oxidation of Pyrites.

Degree and Forms of Soil Acidity

The extent of soil acidity depends on different factors, such as rainfall, slopy terrain, soil type, fertilizers, crop removal of basic cations,and organic matter status . The interaction of these factors influence the amount and speciation of nutrient ions in the soil solution and ultimately determine their availability for plant growth. The degree of soil acidity is measured in terms of soil pH in a soil testing laboratory (Table 1).

Table 1: Degree of soil acidity and the pH range*

Sl. No.	Soil	pH range
1.	Extremely acidic	<4.5
2.	Very strongly acidic	4.5-5.0
3.	Strongly acid	5.1-5.5
4.	Moderately acid	5.6-6.0
5.	Slightly acid	6.1-6.5
6.	Neutral	6.6-7.3

*Soil Science Society of America (1987)

Soil pH below 5.5 needs major emphasis for amelioration of soil acidity, because major portion of exchangeable Al is present in soils with pH below 5.5, which adversely affects plant growth and yield. Toxic levels of both soluble and exchangeable Al is eliminated by raising the soil pH to 5.5.

Acid soils have two components of acidity i.e. acidity exchangeable with neutral salt and that which is titratable but non-exchangeable. In other words, Aluminium hydroxy ions and H^+ and Al^{+++} ions present in non-exchangeable forms with organic matter and clays account for the potential or reserve acidity of a soil. The potential acidity is much higher as compared to exchange acidity. Soil pH do not measure the potential acidity. H^+ and Al^{3+} concentration of soil solution (active acidity) and exchange complex is referred to as exchangeable acidity. It is the acidity developed due to adsorbed H^+ and Al^{3+} ions on the soil colloids, which are in equilibrium with the soil solution. Exchangeable acidity refers to the acidic cations (primarily Fe^{3+} and Al^{3+}), that can be replaced from the ion- exchange complex by an unbuffered salt solution, such as, KCl or NaCl (Gregorich, 2001). Measurable amounts of exchangeable H^+ may also be present due to the acid functional groups associated with soil organic matter (Stevenson, 1982).

The amount of potential (residual) acidity is different to quantify, because the rate of its reaction with base is generally slower than the exchangeable acidity. Source of residual acidity include Al^{3+} and Fe^{3+} tightly bound by SOM and the acid functional groups located on the edges of layer silicate minerals and the surface of Fe and Al oxides and oxyhydroxides (Sumner et al. 1991).

Soil Acidity measurements

The common quantitative procedure to determine the exchange

acidity is extraction of soil with a neutral solution of KCl (Thomas, 1996). The procedure employs the titration of an aliquot of the extracting solution to the phenolphthalein endpoint with a standardized base. The equivalents of base added corrected to the volume of the extracting solution, is the estimate of total exchangeable acidity. This is followed with addition of NaF to release OH^- ions added with base, that are bound to the Al^{3+} in the original extract. Back titration with standardized acid provides the quantitative estimate of exchangeable Al^{3+}. Difference between the estimates of total exchangeable acidity and exchangeable Al^{3+} is a measure of the exchangeable H^+ present in the soil.

Total potential acidity is determined in acid soils using the $BaCl_2$-TEA procedure (Thomas, 1982). Potential acidity refers to exchangeable+ residual acidity contained in a soil referenced to a given pH value. The $BaCl_2$-TEA (pH 8-2) buffer is allowed to equilibrate with a soil sample. After friltration, the filtrate is titrated with standardized acid to the methyl red-bromocreasol green end point. The equivalents of acid required are taken as a measure of the total potential acidity of soil. The pH of 8.2 selected for TEA buffer solution corresponds to the H^+ activity maintained in solution by $CaCO_3$ in equilibrium, with the partial presence of CO_2 in the atmosphere.

Methods to determine lime requirement are based on the fact that the initial soil pH is to be increased to an optimum soil pH suited for plant growth. The approach to determine lime requirement is to equilibrate the soil with a buffer solution.

The buffer solution itself is usually a mixture of two or more buffers (Table 2), such that there is a linear change in pH with incremental additions of acid (H^+) over a range of several pH units. This linear change in pH with incremental addition of acid is in turn calibrated to the lime requirement for a particular soil group, the latter being determined from lime incubation studies or titration with base.

The SMP method is commonly used for soils with high lime requirements, pH<5.8, high organic matter and high soluble Al^{3+}. On the other hand, the Mehlich buffer method is predominantly used for highly weathered and leached soils.

Table 2: Composition of Buffer soilutions used to determine lime requirements

Sl. No.	Methods	Buffer reagent
1.	Shoemaker Mclean and Pratt (SMP)*	P-nitropohenol Potassium Chromate Triethanol amine
2.	Mehlich	Triethanol amine, Acetic acid
3.	Woodruff	P-nitrophenol, Mg, Calcium acetate
4.	Adams and Evans	P-nitrophenol, boric acid
5.	Nommik	Imidazole, malic acid, sod. or Pot. acetate.
6.	Yuan	Imidazole, Pyridine tris**, pot, chromate Hydroxy methyl-amino methane

**tris Hydroxy methyl-amino methane

**Source*: Yuan, 1974; McLean, 1978; Adams, 1984; Van Lierop, 1990

Aluminium Toxicity in Soil

Aluminium toxicity is a major growth limiting factor for plants in strongly acidic soils particularly at pH <5.0 to 5.5. Al toxicity limits the microbial breakdown of organic matter. Excess Al has been reported to interfere with cell division in root tips and lateral roots, increase cell wall rigidity by cross linking pectins, reduce DNA replication by increasing the rigidity of DNA double helix, fix soil P in less available forms, decrease root respiration and interfere with the uptake, transport and use of several essential nutrient elements e.g. Ca, Mg. K, P, Fe & Mo. The greatest benefit of liming acid soils of pH < 5.5, is the reduction in the solubility of Al^{3+}. Liming precipitates & decreases solution Al^{3+} (and also exchangeable Al^{3+}) to $Al(OH)_3$, which improves the chemical environment of acid soils.

Al^{3+} toxicity in plants is difficult to identify. In some plants, foliar symptoms resemble those of P deficiency (stunting, dark green leaves, purpling of stem, leaf). In others, Al toxicity appears as Ca deficiency (curling of young leaves). Aluminium injured roots are characteristically stubby and brittle. Root tips and lateral roots become thickened and may turn brown (Adams, 1984). Young seedlings are generally more susceptible to Al^{++} toxicity than older plants. One short term management strategy for reducing Al^{3+} toxicity to seedlings, is band application of P fertilizers. Al^{3+} toxicity can be observed in soils with pH value < 5.0, while that of Mn^{+2} toxicity is observed in soils between pH 5.0 and 5.5.

Crops and varieties of the same crop differ widely in their susceptibility to Al^{3+} toxicity or lime requirement (Table 3). In general, acid soil toxicity affects the growth of different plants through different physiological and biochemical pathways, depending on the genetic inheritance of an individual plant species (Table- 4).

Table 3: Relative lime requirement of selected plants

High lime requirement	Medium lime requirement	Low lime requirement	Very low lime requirement
Alfalfa	Black berry	Oats	Coffee
Barley	Cabbage	Potato	Napier grass
Field Beans	Corn	Pea nut	Pine apple
Cotton	Sorghum	Rasp berry	Tea
Pea	Grasses	Rice	Stylosanthes
Red clover	Lettuce	Rye	
Soybean	Peanut	Straw berry	
Spinach	Sweet Potato	Water melon	
Sugar beet	Wheat		
Sun flower			

Table 4: Major constraints and responses of plant growth under Acid Soil Conditions.

Sl. No.	Constraint	Response	Affected part
1.	> H^+ conc.	H+ toxicity	Root
2.	> Al^{3+} conc.	Al toxicity	Root
3.	> Mn^{2+} conc.	Mn toxicity	Shoot
4.	< Basic cations	Ca,K,Mg deficiency	Whole plant
5.	< P and Mo solubility	P, Mo deficiency	Whole plant
6.	Inhibition of root growth	Nutrient and water deficiency	Whole pliant
7.	Increase in leaching	Nutrient deficiency	Whole plant

*Source : Adams, 1984; Marschner, 1991

Subsoil Acidity

In some acid soils (e.g. ultisols), the acidity extends down into the subsoil. The increase in subsoil acidity depends on the leaching of H^+ and Al^{3+} ions from the surface horizons as counter ions of anions. Root growth is often confined to the surface horizons in these soils because of subsoil acidity.

Lime application to neutralise surface soil acidity in these soils has little effect on subsoil acidity, because bulk of the Ca^{++} ions remains in the surface horizon (Thomas and Hargrove, 1984).

$$2\ Al^{3+}\text{-Soil} + 3\ CaCO_3 + 3H_2O = 3\ Ca^{2+}\text{ - Soil} + 2\ Al(OH)_3 + 3\ CO_2$$

Deep placement of lime in such soils has proved to be un economical and not feasible due to high cost. McCray and Summer (1990) suggested amelioration of subsoil acidity with surface applications of $CaSO_4 . 2H_2O$.

Limited retention of the SO_4^{2-} anion in the surface horizon provides the mobile anion for movement of Ca^{2+} ion into the subsoil. The net result is an increase in exchangeable Ca^{2+} at the expense of each Al^{3+}.

Since, direct incorporation of lime into subsoil is essentially impractical, judicious application of soil amendments on the surface soil on a regular basis, may offer a feasible alternative. Ultisols, by definition, have low base saturated subsoil horizons and thus, there exists a potential for acid subsoil limiting crop yields in a wide geographical area. The restricted root development & spread in the subsurface layers of soil due to high acidity results in poor utilization of soil water, and inability of crops to survive in drought prone areas.

Lime and its reaction in Acid Soils

Liming materials are carbonate, oxides and hydroxides of calcium and magnesium. Gypsum and other neutral salts can not neutralize H^+ ion :

$$CaSO_4 . 2H_2O + 2H^+ \rightleftharpoons Ca^{2+} + 2H^+ + SO_4^{2-} + 2H_2O.$$

The Ca^{2+} replaces adsorbed $Al^{3+,}$ that increases solution Al^{3+} and decreases soil pH. Most of the agricultural lime is carbonate of Ca (calcite, stomatolitic limestone) and carbonate of Ca and Mg (dolomite). Quick lime (CaO) is burned limestone & its reaction in soil is faster than those of carbonates.

$$CaCO_3 + heat \rightarrow CaO + CO_2 \uparrow$$

$$Ca, Mg\ (CO_3)_2 + heat \rightarrow CaO + MgO + 2CO_2 \uparrow$$

Addition of water to burnt lime makes hydroxide or hydrated lime (slaked lime), which is more caustic than burnt lime.

$$CaO + H_2O \rightarrow Ca\ (OH)_2$$

$$MgO + H_2O \rightarrow Mg\ (OH)_2$$

Industrial wastes, such as basic slag, lime sludge from paper mills, pressmud, phosphogypsum are also good liming materials.

The value of a liming material depends on the amount of acid that the material will neutralize. Pure $CaCO_3$ with neutralizing value of 100 is compared to other lime sources to judge their efficiency. One modecute of CaO, MgO or Ca $(OH)_2$ neutralizes the same amount of acidity as does one molecule of $CaCO_3$.

So, $CaCO_3$ equivalent of CaO

$$= \frac{CaCO_3}{CaO} = \frac{100}{56} = 1.786$$

This molecular weight ratio, when multiplied by the weight of pure $CaCO_3$ gives $CaCO_3$ equivalent.

i.e. 100 x 1.786 = 178.6

The same procedure is used to calculate the neutralizing value of other liming materials (Table- 5). Dolomite is unique, in that there are 2 CO_3^{2-} in each Ca, Mg $(CO_3)_2$, so, ½ the molecular weight is used to determine $CaCO_3$ equivalent.

Table 5: Properties of some liming materials[1]

Liming materials	Molecular wt (g/mole)	Equivalent wt. (g/equiv.)	CCE* %	Ca content %
CaO	56	28	179	71
$Ca(OH)_2$	74	36	135	54
Ca $Mg(CO_3)_2$	184	46	109	22(13% Mg)
$CaCO_3$	100	50	100	40
$CaSiO_3$	116	58	86	46

*CCE – Calcium carbonate equivalent 1. Adapted from Havlin *et al.* 2010

Liming materials, on addition to acid soils, react in the following manner:

$$CaO + H_2O + 2\ CO_2 \rightarrow Ca\ (HCO_3)_2$$

$$Ca(OH)_2 + 2\ CO_2 \rightarrow Ca\ (HCO_3)_2$$

$$CaCO_3 + H_2O + CO_2 \rightarrow Ca\ (HCO_3)_2$$

On reaction with soil colloids, these materials replace H^+ and Al^{3+} ions from colloidal phase to soil solution.

$$\text{Clay} \begin{cases} H+ \\ H+ \end{cases} + \text{Ca}\,(\text{OH})_2 = \text{C} \begin{cases} \text{Ca}^{2+} \text{ - clay} \\ + 2\ \text{H2O} \end{cases}$$

$$\text{Clay} \begin{cases} H+ \\ H+ \end{cases} + \text{Ca}\,(\text{HCO}_3)_2 \;=\; \text{Ca}^{\,2+} \text{——} \text{Clay} + 2\text{H}_2\text{O} + 2\text{CO}_2$$

$$\text{Clay} \begin{cases} H+ \\ H+ \end{cases} + 2\text{Ca CO}_3 \;=\; \text{Ca}^{\,2+} \text{——} \text{Clay} + \text{H}_2\text{O} + \text{CO}_2$$

Lime requirement of acid soils depend mainly on four factors :

i) Required changes to be brought on soil pH

ii) Buffer capacity of soil

iii) Chemical composition of liming material

iv) Fineness of liming material

Lime required for ameliorating acidity is higher in a fine textured soil than that for a coarse textured soil. Similarly, it is higher for soil with high buffering capacity (organic matter and pH). Soils rich in simectite clays will require more lime compared to a soil rich in Kaolinite.

Fertilizer use and Soil Acidity

Fertiliser use alters the pH of the soil. Nitrate sources carrying a basic cation are less acid forming than NH_4^+ sources. Compared with P fertilizers, those containing or forming NH_4^+ exhibit greater effect on soil pH. The acidity produced is greater when S and P sources are combined with NH_4^+ sources. Among the phosphatic fertilizers such as TSP or MAP, the phosphoric acid released from their application in soil can acidify temporarily at the **zone** of application (pH 1.5 in case of TSP and pH 3.5 in case of MAP). But, the quantity of H^+ produced is small and has little effect on soil pH (Havlin et al. 2010).

This has been explained by Adams (1984) in soils of Southern US. As an exampile, $NH_4\ NO_3$ oxidizes as:

$$NH_4NO_3 + 2O_2 \rightarrow 2NO_3^- + 2H^+ + H_2O \qquad (1)$$

Reaction (1) shows that each 28 g of N from $NH_4\ NO_3$ produce acidity equivalent to 100g of $CaCO_3$, or 3.57 of $CaCO_3$/kg N. In the nitrification of Urea/Anhydrous ammonia.

$$(NH_2)\ CO + 4O_2 \rightarrow 2NO_3^- + 2H^+ + CO_2 + H_2O \qquad (2)$$

$$NH_3 + 2O_2 \rightarrow NO_3^- + 2H^+ + H_2O \qquad (3)$$

The method to determine the $CaCO_3$ equivalent for urea is as follows:

2 mole of H^+ produced / 2 mole of N applied.

or 2 equivalent weight of H^+ produced/ 2 equivalent weight of N

Thus,

2 equivalent weights of $CaCO_3$ are needed to neutralize 2 equivalent weight of H^+ produced/ 2 equivalent weight of N

$$\frac{2x50\ CaCO_3/eq.}{2x14\ g\ N/eq.} = \frac{100\ kg\ CaCO_3}{28\ kg\ N} = 3.60\ kg\ CaCO_3/kg\ N$$

In each of the above examples, only about one half of the predicted acidity will actually develop or 1.8 mg $CaCO_3$ per kg of N (Pierre, 1934).

In case of N fertilizers containing acidic anions other than NO_3^-, more than half of the predicted acidity is produced. For example, in case of Ammonium Sulphate oxidation :

$$(NH_4)_2SO_4 + 4O_2 \rightarrow 2NO_3^- + SO_4^{2-} + 4H^+ + 2H_2O \qquad (4)$$

It is estimated that 3/4th of the predicted value i.e. 5.36 kg of $CaCO_3$/ kg N instead of 7.14; and in DAP oxidation.

$$(NH_4)_2HPO_4 + 4O_2 \rightarrow 2NO_3^- + H_3PO_4^- + 3H^+ + 2H_2O \qquad (5)$$

It is estimated to give 2/3 rd of the predicted value i.e. 3.57 kg $CaCO_3$ per kg N instead of 5.36 (Table 6).

Table 6: Soil acidity produced by majors fertilizers

N Source	Nitrification reaction	Residual soil acidity				
		Maximum		Minimum		Official value *
		Acid residue	$CaCO_3$ equiv.	Acid residue	$CaCO_3$ equiv.	
Anhydrous Ammonia	$NH_3(g)+2O_2$ $H^+{-}NO_3^- + H_2O$	$H^+NO_3^-$	50/14=3.6	None	0	1.8
Urea	$(NH_2)_2CO + 4O_2$ $2H^+ + 2NO_3^- + CO + H_2O$	$2H^+2NO_3^-$	100/28=3.6	None	0	1.8
Ammonium Nitrate	$NH_4NO_3 + 2O_2$ $2H^+ + 2NO_3^- + H_2O$	$2H^+2NO_3^-$	100/28=3.6	None	0	1.8
Ammonium Sulfate	$(NH_4)_2SO_4 + 4O_2$ $4H^+ + 2NO_3^-SO_4^{2-} + 2H_2O$	$4H^+2NO_3^-SO_4^{-2}$	200/28=7.2	$2H^+SO_4^{2-}$	100/28=3.6	5.4
Monoammonium Phosphate	$NH_4H_2PO_4 + O_2$ $2H^+ + NO_3^-H_2PO_4 + H_2O$	$2H^+NO_3^-H_2PO_4^-$	100/14=7.2	$H^+H_2PO_4^-$	50/14=3.6	5.4
Diammonium Phosphate	$(NH_4)_2HPO_4 + O_2$ $3H^+ + 2NO_3^- H_2PO_4^- + H_2O$	$3H^+2NO_3^-H_2PO_4^-$	150/28=5.4	$H^+H_2PO_4^-$	50/28=1.8	3.6

*Value adopted by Association of Official Analytical Chemists (Pierre, 1934).

The theoretical equivalents are usually an overestimate of the $CaCO_3$ required to neutralize the acidity produced from fertilizer application. Actually, root absorption of fertilizer anions (NO_3^{-1}, SO_4^{-2} or $H_2PO_4^{-1}$) would neutralize some of the acidity produced from nitrification of NH_4^+ or oxidation of sulphur. When anion uptake effects are considered, the $CaCO_3$ equivalent is often reduced by 50% (Table-6).

Soil Acidity and Soil Properties

Soil acidity affects plant growth partly through the effect of pH as such on root growth and partly through its effect on soil properties. Toxic effect of different species of Al is a major concern in acid soils. Many micronutrients, with the exception of Mo can exist at a higher concentration in soil solution of acid soils. P availability is reduced as solution P precipitates as insoluble Fe/Al-P compounds at low pH. Symbiotic and non-symbiotic N_2 fixation is favoured by increasing the soil pH by liming to about 5.5 . Decomposition of plant residues and breakdown of soil organic matter is also faster at higher pH of soil. Aluminium x Ca interactions are important in acid soil. Below pH of 5.5, Al x Ca antagonism is the most important factor affecting Ca uptake by plants.

Soil pH influences the occurrence and activity of soil micro-organisms. Mycorrhizal fungi play an important role in phosphate uptake in P deficient acid soils. Nitrosomonas and Nitrobacter, that nitrify NH_4 –N and NO_2 –N, respectively, prefer more neutral soil conditions as do free living and symbiotic bacteria that fix N_2.

Raising the soil pH, improves the structure of fine textured soils as a result of increased soil organic matter content and enhanced flocculation of Ca- saturated clay.

Reducing soil acidity may have a significant role in the control of certain plant pathogens.

Growth- limiting factors in acid soils thus, are inter-dependent on direct and indirect effects of soil pH on the toxicities of some metal ions, deficiencies and unavailability of mainly Ca, Mg, P and Mo and decreased soil biological activity (Rhizobia, mycorrrhizae and other micro organisms). Nutrient uptake by plants is also dependent on soil pH. Ecosystem with abundant vegetation have a greater potential for acid neutralization by uptake of strong acid anions (SO_4^{2-}, NO_3^-).

References

Adams, F. (1984) Soil Acidity and liming. 2nd edition. American Society of Agronomy; 373 pp.

Chesworth, W. (2008) Encyclopedia of Soil Science. 'Acidity' pp. 10-20- Publ. Springer. The Netherlands.

Gregorich, E.G. (2001) Soil and Environmental Science Dictionary. Boca Raton. FL : CRC Press, 577pp.

Havlin John L., Beaton James D., Tisdale Samuel L., Nelson Werner L. (2010) Soil Fertility and Fertilisers – An Introduction to Nutrient Management, 7th ed. (Eastern Economy Edition) PHI Learning Pvt. Ltd., New Delhi.

Marschner, H. (1991) Mechanisms of adaptation of plants to acid soils. *Plant and Soil*, 134; 1-20.

Mc Cray,J.M. and Sumner, M.E. (1990). Assessing and modifying Ca and Al levels in acid subsoils. *Advances in Soil Science*. 14; 45-75.

Mc Lean, E.O. (1978) Principles underlying the practice of determining lime requirements of acid soils by use of buffer methods. *Communication in Soil Science and Plant Analysis*. 9; 699-715.

Pierre, W.H. (1934) The equivalent acidity and basicity of fertilizers as determined by a newly proposed method. *Association of Official Agricultural Chemists Journal* 17: 101-107.

Sorensen, S.P.L. (1909) Enzyme studies II. *On maalingen Og betydningen of brinton-koncentration en ved enzymatiske processor. Middelelser fra Carlsberg laboratories* 8; 1- 153.

Stevenson, F.J. (1982) Humus Chemistry. Genesis, Composition, Reaction. New York : John Wiley, p. 443

Sumner, M.E., Fey, M.V. and Noble, A.D. (1991) Nutrient Status and toxicity problems in acid soils. In Ulrich, B. and Summer, M.E. eds. Soil Acidity, Berlin: Springer – Verlag pp 149-182.

Thomas, G.W. (1982) Exchangeable cations. In page, A.L., Chair ed. Comm., Methods of Soil Analysis, Part 2. Chemical and Microbiological Properties. 2nd Edn. Madison, WI : *Soil Science Society of America*, pp 159-165.

Thomas, G.W. and Hargrove, W.L. (1984). The Chemistry of Soil Acidity. In Adams ed. Soil Acidity and Liming, Madison, WI : *American Society of Agronomy*, pp 3-56.

Thomas, G.W. (1996) Soil pH and Soil Acidity. In Sparks D.L. ed., Methods of Soil Analysis. Part 3. Chemical Methods Madison, WI: Soil Science Society of America, American Society of Agronomy, pp 475-490.

Van Lierop, W. (1990) Soil pH and lime requirement determination. In westerman R.L. ed, Soil Testing and Plant Analysis. Madison, WI : *Soil Science Society of America* pp. 73-126.

Yuan, T.B. (1974). A double buffer method for the determination of lime requirement of acid soil. *Soil Science Society of America Journal*, 38:437-440.

3

Chemistry of Acid Soils

Sushanta Kumar Pattanayak

Soil is an interface of inorganic and organic componentsof the planet earth, which combines carbon and nitrogen from the atmosphere with different mineral elements of the lithosphere through millions of microorganisms, intercepting energy from the sun and moisture from the hydrosphere and atmosphere. It is a source as well as sink to adsorb, desorb, fix or release mineral elements and gases, decompose organic residues, simultaneously and support the growth of plants. All these natural processes are accompanied by runoff, percolation and evapotranspiration processes of the hydrologic cycle. Thus, soil is a dynamic body that acquires properties in accordance with the forces, which act upon it. Many soil properties emanate from physical, chemical and biologically mediated reactions, where the factors like climate, vegetation, topography and time influences the processes. Soil reaction is one of the most important chemical characteristics of the soil solution,which has both direct and indirect effects on plant nutrition.

The soil acidity can be defined as the soil system's proton (H^+ ions) donating capacity during its transition from a given state to a reference state.

Sources of Soil Acidity

The acidity in soils arises from different sources. The sources are :

i) Rocks and minerals

ii) Humus

iii) Clay mineral make of soil

iv) Hydrous oxides of iron and aluminium

v) Exchangeable aluminium

vi) Soluble salts (oxidation of nitrogen and sulphur) and

vii) Carbon dioxide

i) Rocks and minerals

Soils originating from acidic parent materials lead to the development of acidic soils. These parent materials are over saturated with free quartz (SiO_2>66 %). The major minerals in soils are quartz, feldspar, and oxides of Al and Fe . The soils developed from these parent rocks and minerals contain low amount of basic elements e.g. Ca^{2+} and Mg^{2+}.

ii) Humus

Humus materials in soils formed as a result of microbiological decomposition of organic matter, contain different functional groups like carboxylic (– COOH), phenolic (-OH), amino and similar type groups , which are capable of attracting and dissociating hydrogen (H^+) ions. The strength of the acid produced depend upon the nature of the predominant groups present.

The contribution of acidity from humus can be presented through the following diagram:

$$R-\overset{O}{\overset{\|}{C}}-OH \xrightarrow[(-H_2O)]{OH^-} R-\overset{O}{\overset{\|}{C}}-O^- + H^+ \quad pKa = 3 \text{ to } 5$$

(Carboxylic acid)

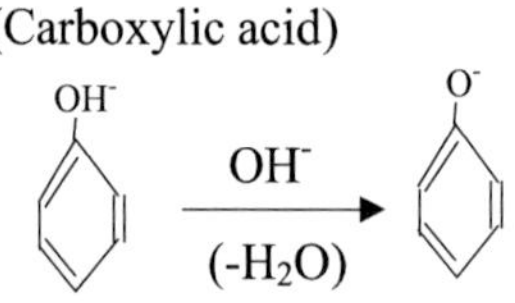

(Phenol) (Dissociate at higher pH level)

Al-Organic matter complexes

Complexation of aluminium (Al) in soil solution by decomposition products of organic materials, particularly low-molecular weight organic acids has been suggested by Hue *et al.* (1986).

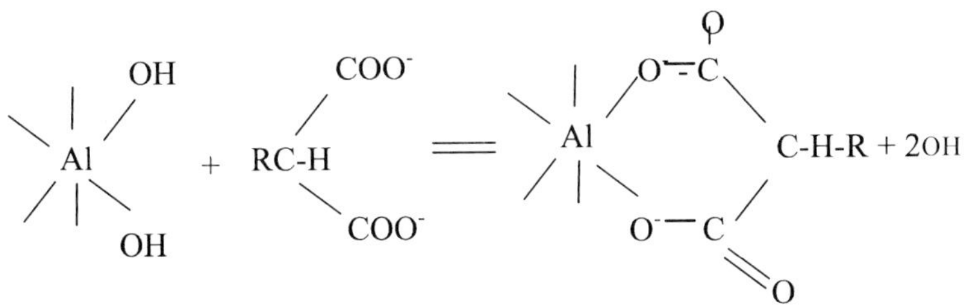

Adsorption of soluble Al on the surface of organic material has also been postulated as a mechanism of soil acidity reduction by plant materials (Asghar and Kanehiro, 1980). The mechanism that is best explained the neutralization reactions was found to be microbial decarboxylation of Ca–organic matter complex leading to the release and subsequent hydrolysis of Ca^{2+} ions. The hydroxyl ions released in the hydrolytic reaction then react with both exchangeable H^+ in and aluminium ions to form water and insoluble aluminium hydroxide $[Al(OH)_3]$, respectively (Ano and Ubochi, 2007)

iii) Aluminosilicate clay minerals and sesquioxides

The aluminosilicate clay minerals are considered tobe layer minerals, two or three layer silicate clays typified by Kaolinite (1:1) and montmorillonite (2:1), which contribute to soil acidity by way of holding on to H^+ (and Al^{3+}) ions to neutralize the negative charge on them. The negative charge arises from (i) isomorphous substitution within Al octahedral and Si tetrahedral layers of higher valence cations with lower valence ones of identical ionic size, and from (ii) the dissociation of protons from the exposed hydroxyl groups of bound water of constitution, both of which are structural components of the crystal lattice leading to the development of the negative charges, which naturally depends on pH of the surrounding soil solution.

The sequioxide (hydrous oxides of Fe and Al (and Mn), particles of clay sized dimensions may also contribute to the pool of pH dependent negative charges in soil. The following equations depict how silicate clays and sesquioxides give rise to pH dependent charges in soil.

i) On hydrous oxides of Fe and Al (pH dependent negative charge)

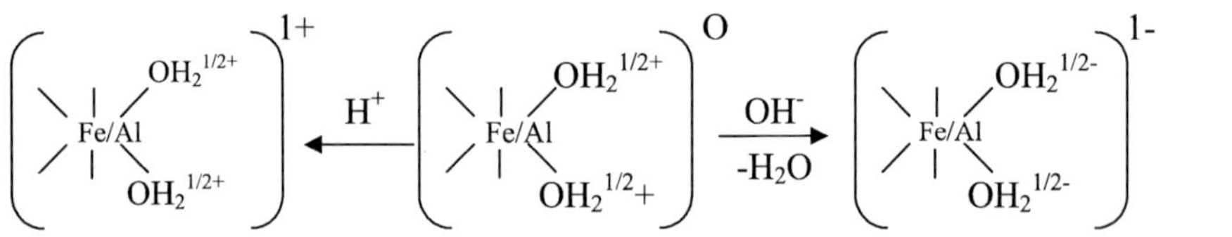

Acidic (at pH < ZPC contributes towards positive charge and anion exchange) | Neutral (at pH of ZPC) | Alikaline (at pH > the ZPC, contributes towards negative charge and cation exchange)

ii) On clay minerals (pH dependent negative charge)

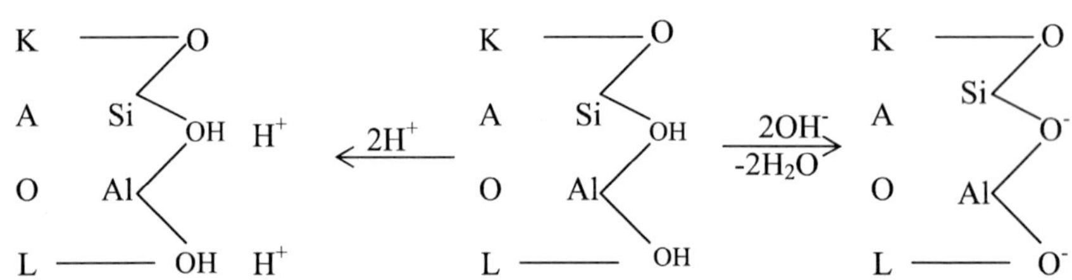

Acidic | Neutral | Alkaline

iii) On clay minerals (permanent negative charge)

The Mg^{2+}, Fe^{2+} replacing Al^{3+}, in octahedral layer

The Al^{3+} replacing Si^{4+} in tetrahedral layer

The Al and Fe polymers

The Al^{3+} ions displaced from clay minerals by cations are hydrolyzed to monomeric and polymeric hydroxyaluminium complexes. Hydrolysis of the monomeric forms proceeds stepwise and in each case liberates H^+ and lower the pH unless neutralized by OH^- ion present in the system (Tisdale *et al.*, 1990).

1. $Al^{3+} + H_2O \rightarrow Al(OH)^{2+} + H^+$
2. $Al(OH)^{2+} + H_2O \rightarrow Al(OH)_2^+ + H^+$
3. $Al(OH)_2^+ + H_2O \rightarrow Al(OH)_3^0 + H^+$

iv) Hydrolysis of hexahydrated aluminium

At low pH values, most of the aluminium is present as hexahydratedAl $(H_2O)_6^{3+}$ ion. At pH values above 5.0, hydrated hydroxyaluminium ions probably exist in exchangeable form. These reactions are shown as :

	pH levels
$Al(H_2O)_6^{3+} + H_2O \rightarrow Al(OH)(H_2O)_5^{2+} + H_3O^+$	< 4.7
$Al(OH)(H_2O)_5^{2+} + H_2O \rightarrow Al(OH)_2(H_2O)_4^+ + H_3O^+$	4.7-6.5
$Al(OH)_2(H_2O)_4^+ + H_2O \rightarrow Al(OH)_3(H_2O)^o_3 + H_3O^+$	6.5-8.0
$Al(OH)_3(H_2O)_3^o + H_2O \rightarrow Al(OH)_4(H_2O)^-_2 + H_3O^+$	>8.0

Each successive step occurs at higher pH. These aluminium hydrolysis products can be re-adsorbed by the clay minerals causing further hydrolysis. The H^+ ions resulting from the hydrolysis of aluminium and iron compounds react with and dissolve or decompose soil minerals. Further, the addition of salts, such as those contained in fertilizers, to sesquioxide coated interlayer minerals increase the hydrolysis of non-exchangeable iron and aluminium, resulting in an increase in the H^+ ion concentrations of the soil solution and hence a lower pH. The insoluble $Al(OH)_3$ will be precipitated throughout the pH range, whenever its solubility product is exceeded.

Iron hydrolysis is similar to that of aluminium. Although the reaction is more acidic (pK of 1[st]dissociation constant is 3.0) than the corresponding hydolysis of aluminium (pK=5.0), the acidity that develop is buffered by the reaction of aluminium. Thus, iron hydrolysis has little effect on soil pH until most of the reserves of aluminium in soil are reacted (Tisdale *et al.*, 1990).

$Fe(H_2O)_6^{3+} + H_2O \rightarrow Fe(OH)(H_2O)_5^{2+} + H_3O^+$

$Fe(H_2O)_3 (H_2O)^o + H_2O \rightarrow Fe(OH)_4 (H_2O)^-_2 + H_3O^+$

Polymerization process

The monomeric positively charged species can polymerize to form both large and small positively charged polynuclear complexes. In the polymerization process, hydroxyl groups are shared by adjacent Al^{3+} ions to produce polymers of general formula : [Al (OH) x $(H_2O)_{6-x}$ $(3\text{-}X)^+$]n, where 'n' is a large number.

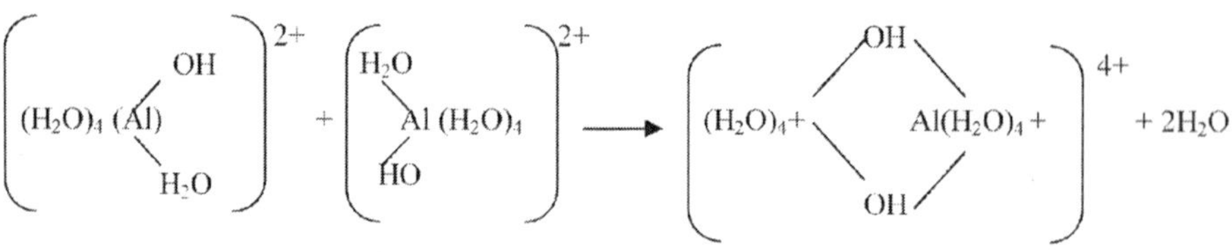

Single chain $[Al_6(OH)_{12}(H_2O)_{12}]^{6+}$

Double chain$[Al_{10}(OH)_{22}(H_2O)_{18}]^{8+}$

These species often aging, release H^+ ions (H_3O^+) or polymerise further and become firmly bound to colloidal surface. These polymers are positively charged and are non-exchangeable.

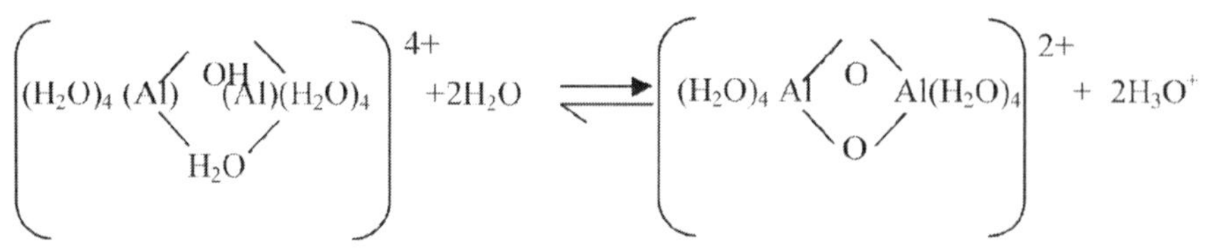

The CEC of soil colloids can be affected by the formation of these positive charged polymers on their surfaces. At high pH values, the polymeric positive charges become less and can even become negative as a result of the dissociation of H^+ ions or protons. As a result, CEC of the soil colloid polymer complex will increase. Decreasing pH in soils with large amounts of these Al polymers closely associated with soil colloids will lower CEC by increasing the positive charge on the polymers.

Under strongly acidic soils, the adsorbed aluminium is in equilibrium with Al^{3+} ions in the soil solution. The Al^{3+} ions have a strong tendency to hydrolyze, splitting water molecules into H^+andOH^-ions. The aluminium combines with the OH^- ions, leaving the $H^+$to lower the pH of the soil solution. For this reason, Al^{3+} ion thus release up to three H^+ ions as the following reversible reaction series proceeds to the right in stepwise manner :

$$\boxed{\text{Clay}}\ \text{Al} \rightleftharpoons Al^{3+}$$

Adsorbed aluminium Soil solution aluminium

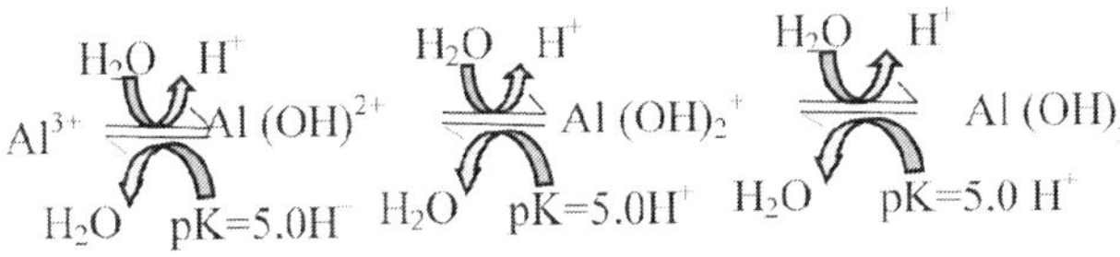

Gibbsite/amorphous (solid)

Most of the hydroxy aluminium ions $[Al(OH)_x^{y+}]$ formed as the pH increases are strongly adsorbed to clay surfaces or complexed with organic matter. Often the hydroxy aluminium ions together, forming large polymers with many positve charges. When tightly bound to negative charge sites of the colloids, these polymers are not exchangeable and so mask much of the colloids potential CEC. As the pH is raised and more of the hydroxyl aluminium ions precipitate as uncharged $Al(OH)_3^o$, the negative sites on the coloids become available for cation exchange. This is one reason for the increase in soil CEC as the pH is raised from 4.5 to 7.0 [at which pH virtually all the aluminium cations precipitate as $Al(OH)_3$].

v) Addition of soluble salts (Fertilizers)

Oxidation reactions generally produce H^+ ions as one of their products. Reduction reactions on the other hand, tend to consume H^+ ions and raise soil pH. Addition of N and P fertilizers contribute significantly to the formation of acid soils. The ammonium N is a major factor for acidification of sandy soils with low buffering capacity. Upon addition (from organic matter or from fertilizers) this NH_4^+ -N undergo nitrification (microbial oxidation) to nitrate (NO_3^-),releasing two H^+ ions. The proton also replaces exchangeable cations (Ca^{2+}, Mg^{2+} which are lost through leaching. About 7.2 kg $CaCO_3$ is required per kg N applied to replace lost Ca and neutralize the acidity produced. The acid equivalent of ammonium sulphate is 110.

$$(NH_4)_2SO_4 \rightleftharpoons 2NH_4^+ + SO_4^{2-}$$

$$2NH_4^+ + Ca \leftarrow \boxed{Clay} \quad CaSO_4\downarrow + \text{Leached out} \qquad \begin{matrix} NH_4 \\ NH_4 \end{matrix}\boxed{Clay}$$

$$\begin{matrix} NH_4 \\ NH_4 \end{matrix} \Big> \text{Clay} + 3O_2 \text{ Nitrification} \rightleftharpoons \begin{matrix} H \\ H \end{matrix}\boxed{Clay} + 2\,HNO_3$$

Acidic soil

$$(NH_4)_2\,SO_4 + 4O_2 \rightleftharpoons 2NO_3^- + 4H^+ + 2H_2O + SO_4^{2-}$$

Urea, upon application to soil under go enzymatic (urease) hydrolysis forming NH_4^+ ions from NH_2^- ions.The reaction is less acidifying than ammonium sulphate. About 3.0 kg $CaCO_3$is required per kg N applied as urea to neutralize acidity. Acid equivalent of urea is 80.

$$NH_2 - CO\text{-}NH_2\ (\text{Urea}) + 2H^+ + 2H_2O = 2NH_4^+ + H_2CO_3$$

Use of phosphatic fertilizers also has an acidifying effect on soils. The monocalcium phosphate component [Ca $(H_2PO_4)_2$] of the fertilizer by reaction with water in soil form dicalcium phosphate and phosphoric acid (H_3PO_4). The micro-site pH around fertilizer is as low as 1.5.

$$Ca\,(H_2PO_4)_2 + H_2O = CaHPO_4 + H_3PO_4$$

$$H_3PO_4 = H^+ + H_2PO_4^- = 2H^+ + HPO_4^= = 3H^+ + PO_4^{\equiv}$$

Such lowering of soil pH is marked, when the fertilizer is band placed (Lindsay and Stephenson, 1959).

The phosphate compound when supplied as mono/diammonium phosphate,the ammonium component will also contribute to soil acidity (as shown previously).

Even though potassic fertilizers (KCl/K_2SO_4) do not create acidity directly in soil, but the associated anions ($Cl^-/SO_4^=$) carry away Ca^{2+} and Mg^{2+} through leaching particularly in sandy coarse textured upland soils. Such loss is equivalent to 1.1 kg $CaCO_3$ from 1 kg K or 1.2 kg K_2O in the fertilizer.

Oxidation of sulphur

The decomposition of plant residues commonly involves the oxidation of organic –SH groups to yield sulphuric acid (H_2SO_4). Another important source of this strong acid, is the oxidation of reduced sulphur in minerals such as pyrite (FeS_2).

$$FeS_2 + 3\frac{1}{2}O_2 + H_2O \rightleftharpoons FeSO_4 + 2H^+ + SO_4^{2-}$$

Pyrite — Ferrous sulphate — Dissociated sulphuric acid

vi) Carbon dioxide

Carbon dioxide in soil is produced from many sources. Decaying of organic matter, root and microbial respiration produce CO_2. This CO_2 by reacting with water in soil produce weak acid called carbonic acid. This is the same acid that develops,when CO_2 in the atmosphere reacts with rain water in the form of acid rain. Organic matter decomposition also produces several other organic weak acids to make the soil acidic slowly over a long period of time (Gordon, 1988, Corver and Ownby, 1995).

$$CO_2 + H_2O \rightleftharpoons H_2CO_3$$

Factors influencing Soil Acidity Development

Soil acidity development is influenced by edaphic, climatic and biological factors.

Climate

Movement of excess water through soil rapidly in high rainfall areas is an important factor responsible for acid soil development. Leaching is predominant in coarse textured soils, causing loss of bases. Iron and aluminium derivatives are relatively insoluble. These are seldom leached out and contribute to surface acidity, particularly in laterite soils. Since, the effect of rainfall on acid soil development is very slow, it may take hundreds of years for new parent material to become acidic under high rainfall. When the weathering process is drastic, the subsoil and in many cases the whole profile may become acidic (Misra, 2004). The regions with annual rainfall of 1350 mm have acid soils with pH values 5 or even less. In hilly regions, where the loss of water through evaporation is less due to low temperature, the conditions for the development of acid soils are favourable, though the rainfall is not so high (Pal, 2009).

Removal of bases

Base unsaturated soils are acidic. Removal of basic cations take place through two processes:

i) Leaching loss of basic cations,

ii) Crop removal of basic cations

Soil acidity develops by leaching of basic elements such as Ca^{2+}, Mg^{2+}, K^+, and Na^+ from the soil profile, leaving the acidic elements: H^+, $Al^{3+}$and Mn^{2+}. The soils become more acid when excess H^+ and Al^{3+} ions replace basic cations such as Ca^{2+}, Mg^{2+}, K^+ and $Na^+$from the surface of the clays and soil humus. The basic cations are often leached below the root zone, leaving H^+ and Al^{3+} ions behind, because they are more strongly attached to the negative charges on the soil surface. (Tisdale *et al.*, 1990).

The persistent solvent action of carbonic acid (H_2CO_3) on mineral constituents in the soil is responsible for removing many base forming cations (Ca^{2+}, Mg^{2+}) by dissolution and leaching.

The extent of leaching loss of Ca varies between 84 and 224 kg ha^{-1} and that of Mg between 24 and 83 kg ha^{-1}.

The growing plants also contribute to soil acidification, as a major nutrient uptake process is to exchange H^+ ions at the root surface for needed basic cations (Schulte *et al.*, 2005). For every positive charge taken in on a cation, a root can maintain charge balance either by taking up a negative charge as an anion or by excluding a positive charge as a different cation. When they take up far more of certain cation (e.g.K^+,NH_4^+andCa^{2+}) than they do for anions (NO_3,SO_4). Plants usually exude H^+ ions into the soil solution to maintain charge balance.

In the first two of the following examples, plant uptake results in the addition of H^+ ions to the soil solution (Brady and Weil, 2012).

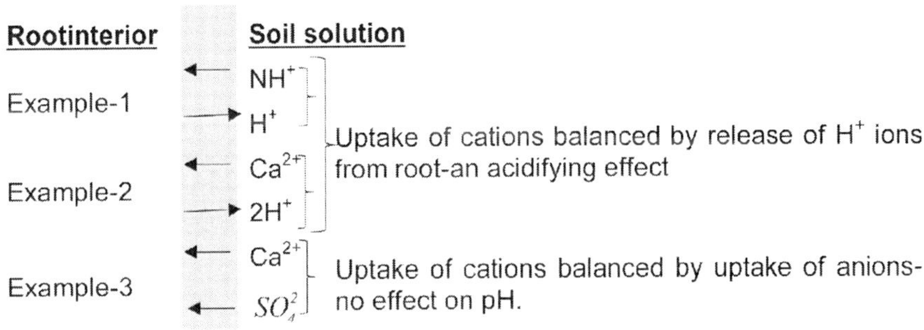

ii) Crop removal of bases

Cultivated crop removes variable amount of base forming elements such as Ca and Mg from the soil. Legume crops have greater acidifying effect on soil, because these crops take up more cations (Ca and Mg) than anions compared to non-legume crops (Table 1, Pattanayak, 2013).

Table 1: Nutrient uptake (kg ha^{-1}) by crops

Crop	Economic yield (kg ha^{-1})	Biomass (dry)(kg ha^{-1})	Nutrient uptake (kg ha^{-1}) N	P	K	Ca	Mg
Cereal							
Rice	5936	14046	121.0	23.0	142.0	47.1	30.3
Maize (cob)	13500	8330	92.3	17.3	108.2	44.1	20.1
Ragi	3880	8510	169.0	28.1	163.7	40.1	19.4
Pulse							
Greengram	1160	7080	82.3	16.3	93.1	115.2	34.6
Blackgram	970	6809	78.8	15.3	88.9	108.2	33.4
Arhar*	1420	5560	143.9	14.6	80.9	167.5	43.2
Oilseed							
Groundnut (Kharif)	1450	9770	257.9	20.6	110.6	85.9	15.8
Groundnut (Rabi)	2800	5310	143.6	12.7	99.1	80.1	18.1
Sunflower	2430	3627	137.0	7.8	83.0	58.3	18.9
Mustard (Residual)	660	2170	39.6	4.8	23.6	14.4	5.4
Vegetable							
Okra	4630	1150	127.6	14.6	80.0	41.9	18.6
Cowpea	3970	8000	271.4	22.0	59.3	38.1	12.3
Potato	21000	5890	152.4	11.0	179.8	39.0	10.1
Brinjal	17110	5280	130.3	15.3	75.0	40.3	12.7
Fodder							
Berseem	34400	5590	258.0	11.2	71.5	78.9	30.1

Vegetation

Vegetations influence the earth surface through uptake, transformation and redistribution of materials in the atmosphere, pedosphere and lithosphere (Schlesinger, 1997). Almost all plants leave an imprint (e.g., input of organic matter, respired CO_2 which enhance the weathering of rocks etc.), variations in size, growth rate, life span, allocation, tissue composition and many other attributes that affect cycling patterns and properties of soil differently (Jackson *et al.,* 2000). Location wise comparison of areas historically dominated by trees and grasses indicates that, soil pH and base saturation tend to be lower under forests with similar parent material and climate (Ugolini *et al.*, 1988).

Soil acidification is the outcome of the conversion of grasslands to forests based on the basic mechanisms of soil acidification, following the afforestation of grass land ecosystems, covering (i) organic acid inputs, (ii) soil respiration and (iii) sequestration/ complex formation and redistribution of cations. In very low temperature regions, acid soils develop easily. The foliage of coniferous trees, which generally grow in these areas, is devoid of alkali elements. Upon degradation of leaf litters, organic acids are released gradually and create acidity in soil.

Vegetation of coastal zones and marshy places facilitate development of acidity. The places where excessive vegetation is accompanied by heavy rainfall and water pounding, soils are very acidic in reaction. In these situations, organic materials do not get enough oxygen to decompose completely. These under–decomposed organic materials keeps on piling up and soil becomes very acidic,but at the sametime very rich in organic matter (Gordon, 1988).

Topography

The topography has an influence on soil formation, most likely through hydrological processes. The soil reaction (pH), $CaCO_3$, Ca and Mg as well as base saturation increases down the slope (Brubaker *et al.*, 1993). The O-horizon i.e., organic layer is more exposed to topographic control than the B-horizon. Slopy places with good drainage conditions are often acidic in nature. On hill slopes,the development of acid soils is easier. In river basin, acid soils generally do not develop. The plains with good drainage may also develop acidly in due course of time (Seibert *et al.,* 2007).

Acid rain

Industrial activities including combustion of coal and oil in electric power generation, the combustion of fuel in vehicles, the smelting of sulphur containing metal ores emit enormous quantities of nitrogen and sulphur containing gases into the atmosphere.The other sources of the gases include forest fires and burning of crop residues. Such environmental problem is transboundary in nature. The gases (mainly $SO_2$and oxides of nitrogen NO, NO_2, N_2O) react with water and other substances in the atmosphereto form HNO_3 and H_2SO_4. These two strong acids are then returned to the earth in rain (as well as in snow, fog, and dry deposition). This rain or precipitation is called Acid rain.

The normal rain water that is in equilibrium with atmospheric CO_2 has a pH of about 5.5. The pH of acid rain commonly is between 4.0 and 4.5, but may be as low as 2.0.

Major chemical processes for acid precipitation

$N_2 + O_2 \rightarrow 2NO$; $4NO + 2O_2 \rightarrow 4NO_2$; $2NO_2 + H_2O = HNO_2 + HNO_3$

NO + sunlight + OH (air) $\rightarrow HNO_3$

Volatile organic compounds + sunlight + H_2O (air) $\rightarrow H_2O_2$

SO_2 + Oxidants (from wet surfaces) $\rightarrow H_2SO_4$

S (in fuel) + $O_2 \rightarrow SO_2$; $SO_2 + HOH = H_2SO_3$;

SO_2 + OH + O_2 (in air) $\rightarrow H_2SO_4$

SO_2 + oxidants (from wet surfaces) $\rightarrow H_2SO_4$

The incoming strong acids mobilize aluminium in soil minerals and aluminium displaces Ca^{2+} and other non-acid cations from the exchange complex. The presence of strong acid anions (SO_4^{2-}, NO_3^-) facilitates the leaching of displaced Ca^{2+} ions. Soon Al^{3+} and H^+ ions, rather than Ca^{2+} ions, become dominant on the exchange complex, as well as in the soil solution and drainage water.

Forms of Soil Acidity

There are three common forms of soil acidity: (i) active acidity, due to H^+ ions in soil solution; (ii) salt replaceable (exchangeable) acidity, involving the aluminium and hydrogen that are easily exchangeable by other cations in a simple buffered salt solution, such as KCl; and (iii) residual acidity, which can be neutralized by limestone or other alkaline materials, but cannot be detected by the salt replaceable technique. These add up to form the total acidity of a soil. In addition, a less common, but some times very important form, namely **potential acidity**, can arise upon the oxidation of sulphur compounds in certain acid sulphate soils.

Active acidity

Soils are acid because of positively charged H^+ ions in the soil solution and on the surface of clay and organic matter particles that make up the soil.

Hydrogen ion (H^+) activity in the soil solution is referred to as active acidity. The term pH measures the active acidity. A pH value below 7.0 is alkaline, or non-acid. This pool is very small compared to the acidity in the exchangeable and residual pools.

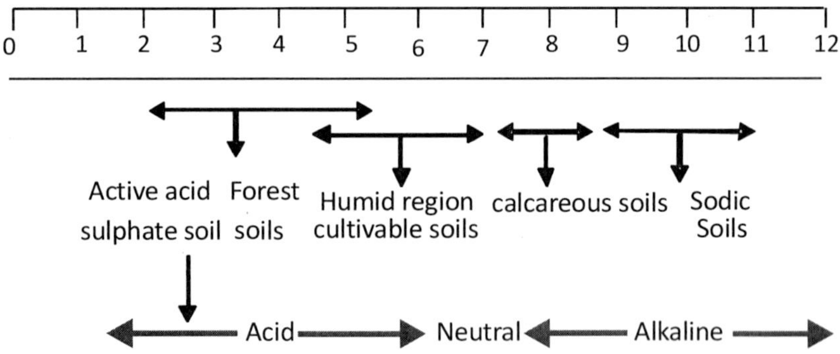

Fig. Ranges of pH typical for various types of soils

Only about 2 kg of $CaCO_3$ may be needed to neutralize the active acidity in the upper 15 cm of a hectare of an average mineral soil at pH 4.0 and 20 per cent moisture. Even so, the active acidity is extremely important, since, it determines the solubility of many substances and provides the soil solution environment, to which plant roots and microbes are exposed. Very acid soils contain aluminium ions in solution, which can add to the active acidity as they hydrolyze.

Exchangeable (salt replaceable) acidity

This is primarily associated with exchangeable aluminium and hydrogen ions, that are present in large quantities in very acid soils held by the negatively charged soil particles of clay and organic matter. These ions can be released into the soil solution by cation exchange with an unbuffered salt, such as KCl.

$$\boxed{\text{Clay}}\begin{matrix} H^+ \\ Al^{3+} \\ H^+ \end{matrix} + 4\ KCL \rightleftharpoons \boxed{\text{Clay}} \qquad + AlCl_3 + HCl$$

Soil solid   Soil solution   Soil solution   Soil solid

Once released into soil solution the Al^{3+}hydrolyzes to form additional H^+ (Tisdale *et al.*, 1990). The chemical equivalent of salt replaceable acidity in strongly acid soils is commonly thousand times that of active acidity in the soil solution. Even in moderately acid soils, the limestone needed to neutralize exchange acidity is more than 100 times of that needed to neutralize the active acidity. At a given pH value, exchangeable acidity is generally highest for smectites, intermediate for vermiculites and lowest for Kaolinite.

Residual acidity

Together, exchangeable (salt replaceable) and active acidity account for only a fraction of the total acidity. The remaining part i.e. residual acidity is generally associated with hydrogen and aluminium ions (including aluminium hydroxyl ions), that are bound in non-exchangeable forms by organic matter and clays. As the pH increases, the bound hydrogen dissociates and the bound Al^{3+} ions are released and precipitate as amorphous $Al(OH)_3^0$. These changes free the negative cation exchange sites and increase the CEC. The reaction with a liming material [e.g. $Ca(OH)_2$] shows how the bound hydrogen and aluminium can be released.

$$\boxed{Clay}\begin{matrix} Al^{3+} \\ H^+ \end{matrix} + Ca(OH)_2 \rightleftharpoons \boxed{Clay}\begin{matrix} Ca^{2+} \\ Ca^{2+} \end{matrix} + Al(OH)_3 + H_2O$$

Soil solid   Soil solution

The residual acidity is commonly much greater than either the active or salt-replaceable acidity. It may be 1000 times greater than the soil solution or active acidity in sandy soil and 50,000 or even 100,000 times greater in clayey soil high in organic matter. The amount of ground lime stone recommended to at least partly neutralize residual acidity in upper 15 cm of soil is commonly 5 to 10 metric tonnes per hectare.

Total acidity

For most soils (not potential acid sulphate soils), the total acidity that must be overcome to raise the soil pH to a desired value can be defined as:

Total acidity = Active acidity + salt – (replaceable) acidity + residual acidity.

Total acidity

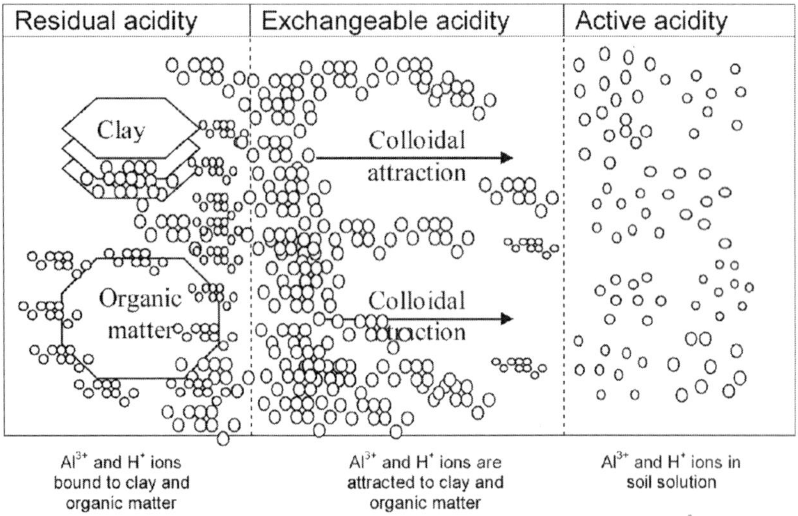

Table 2: Forms of acidity in soils under different soil order

Location	Total Acidity	pH dependent acidity	Total exchange acidity	Acidity due to to H^+	Acidity due to Al^{3+}
Alfisols		[cmol (P^+) kg^{-1}]			
Odisha[b]	10.3	8.70	1.60	0.45	1.15
West Bengal[a]	3.2	3.10	0.10	0.03	0.07
Himachal Pradesh [d]	67.5	63.80	1.80	-	-
Meghalaya[f]	13.4	12.08	1.32	0.50	0.82
West Bengal-2[g]	12.6	9.0	3.60	1.10	2.50
Inceptisols					
Odisha[b]	2.29	2.19	0.10	0.08	0.02
Odisha[h]	4.35	4.14	0.21	0.12	0.09
West Bengal[a]	5.98	5.50	0.48	0.19	0.29
Mizoram[c]	12.50	11.80	1.41	0.23	1.80
Nagaland[e]	7.29	6.72	0.56	0.05	0.51
Entisols					
Odisha[b]	10.76	10.65	0.11	0.11	-
West Bengal[a]	5.15	4.93	0.22	0.11	0.11
Mizoram[c]	12.24	11.63	1.64	0.32	1.32
Nagaland[e]	22.28	19.58	2.69	0.26	2.43
Ultisols					
Odisha[b]	5.30	4.70	0.60	0.30	0.30
Mizoram[c]	13.19	11.67	1.52	0.29	1.23
Nagaland[e]	27.20	20.85	6.34	0.60	5.74

(a) Chand and Mandal, (2000), (b) Misra *et al.* (1989), (c) Misra and Saithantuaanga (2000), (d) Subehia *et al.* (2005), (e) Dolui *et al.* (2010), (f) Venkatesh *et al.* (2002), (g) Dolui and Maity (2004), (h) Pattanayak (1992).

Soil pH and Exchangeable cations

Any particular soil would give somewhat different distribution pattern for exchangeable cations depending upon its pH.

A mineral soil (Fig. 1 B) with mixed mineralogy and a moderate organic matter level exhibits a moderate decrease in effective CEC, as pH is lowered, suggesting that pH-dependent charges and permanent charges each account for about half of the maximum CEC.At pH values above 5.5, the concentrations of exchangeable Al^{3+} and H^+ ions are too low and the effective CEC is essentially 100 per cent saturated with exchangeable non-acid cations (Ca^{2+}, Mg^{2+}, K^+ and Na^+ - base cations).

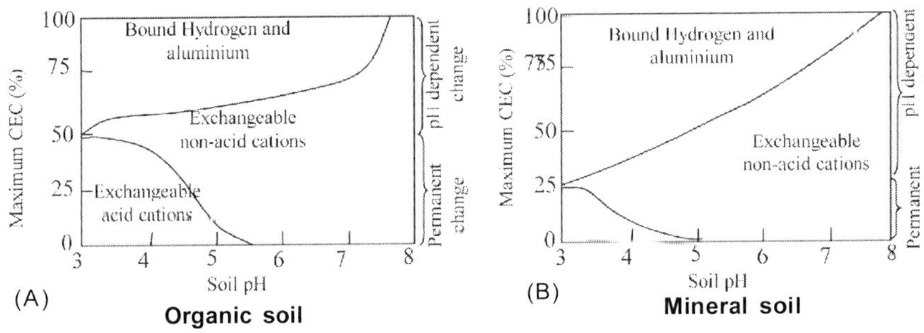

Fig. 1:General relationship between pH and cations held in exchangeable form

As pH drops (pH 7.0 to above 5.5), the effective CEC is reduced because H^+ ions and $Al(OH)_x^{y+}$ ions ($Al(OH)^{2+}$, $Al\ (OH)^+_2$etc) are tightly bound to some of the pH-dependent charge sites. As pH is further reduced from 5.5to 4.0, Al^{3+} ions, along with some H^+ ions, occupy an increasing portion of the remaining exchange sites. Exchangeable H^+ ions occupy a major portion of the exchangecomplex only at pH levels < 4.0.

The CEC of an organic soil is (Fig. 1A) dominated by pH dependent (variable) charges with only a small amount of permanent charge. Therefore, as pH is lowered, the effective CEC of the organic soil declines more drastically than the effective CEC of the mineral soil. At low pH levels, exchangeable H^+ ions are more prominent and Al^{3+} less prominent on organic soil than on mineral soil.

Acid Sulphate Soil

The acid sulphate soils are naturally occurring soils, sediments or organic substrates (e.g. peat), that are mostly under water logged conditions. These soils contain iron sulphide minerals (mainly pyrite-FeS_2) or, their oxidation products. In an undisturbed state below water table, acid sulphate soils are harmless. However, if these soils are drained for agriculture, forestry, or other development, air enters (O_2) the soil pores and both the sulphur (S^o, S^-, or S^{2-}) and the iron (Fe^{2-}) are oxidized changing **potential acid sulphate** soils into **active sulphate soils**. Ultimately, such soils earn their name by producing large and impressive quantities of sulphuric acid (H_2SO_4), resulting in soil pH value below 3.5 and in some cases 2.0 (Borek, 1994).

Characterization of acid sulphate soils

The acid sulphate soils are characterized by pH less than 4.0, containing oxidizing iron sulphides, with variable texture, often containing jarosite (a yellow colour mottle produced as byproduct of oxidation process) (Fitzpatrick *et al.*,1996).

Genesis and chemistry of acid sulphate soils

These soils have been formed within the past 10,000 years, after the rise of sea level adjacent to sea shore. When the sea level rose and inundated the land, sulphate (SO_4^{2-}) in sea water was mixed with land sediments containing iron oxides and organic matter. Under such anaerobic conditions, lithotrophic bacteria like *Thiobacillus ferrooxidans* formed iron sulphides (Pyrite). For pyrite formation, it requires supply of sulphur (sea water), anaerobic (O_2 free) condition, supply of energy for bacteria (usually rotting organic matter, e.g. mangrove forest plant leaves), system to remove reaction products for continuing forward reaction (tidal flushing system), source of iron (most often from terrestrial sediments) and a temperature greater than 10 °C (Potter, 1976). The principal reactions involved are:

$$\underset{\text{Pyrite}}{Fe^{(ii)}S_2^{-1}} + 3\frac{1}{2}O_2 + H_2O \qquad \underset{\text{Ferrous sulphate}}{Fe^{(ii)}S^{(vi)}O_4} + \underset{\text{Sulphuric acid}}{H_2S^{(vi)}O_4}$$

$$Fe^{(ii)}S^{(vi)}O_4 + \frac{1}{4}O_2 + 1\frac{1}{2}H_2O \rightleftharpoons Fe^{(iii)}OOH + H_2SO_4$$

Ferrous sulphate — Iron oxyhydro xides — Sulphuric acid

$$S^o \quad + \quad 1\frac{1}{2}O_2 + H_2O \rightarrow H_2SO_4$$

Elemental sulphur — Sulphuric acid

Oxidation of sulphur in pyrite and oxidation and hydrolysis of the $FeSO_4$, produce acidity. The sulphur is oxidized from a^{-1} to a^{+6} valence and the iron is oxidized from +2 to +3. In above reactions, two moles of acidity (one mole of H_2SO_4) are ultimately produced for each mole S that reacts, whether from FeS or elemental S. The oxidation reactions occur by chemical processes, but are generally facilitated by microorganisms (like bacterium *Thiobacillus ferroxidans* and Archaeon *Ferroplasma acidpilum*) ,that make them proceed thousand time faster, especially when warm, moist conditions favour microbial activity.

Borek (1994) presented the reaction processes in potential acid sulphate soils as :

$$[2FeS_2 + 9\,O_2 + 4\,H_2O \text{ ® } 8H^+ + 4SO_4 + 2Fe\,(OH)_3 \text{ (solid)}]$$

Sulphur derived from sea water under water logged anaerobic condition reduced by anaerobic micro-organisms.

$$\{SO_4^{2-} \xrightarrow[\text{Desulf tomaculum}]{\text{Desulfovibro}} \rightarrow S^2 \xrightarrow[Fe^{2+}]{Siltation} \quad FeS + FeS_2\}$$

Silt build up in the inundated soils and when such soils are exposed to air following retreat drainage of sea water, O_2 enters and the reactions occur are :

$$FeS \frac{O_2}{H_2O} \text{ FeS} + H_2SO_4 \text{ (near neutral reaction)}$$

At pH 3.0

$$FeS_2 + O_2 + H_2O \rightarrow H^+ + SO_4^{2-} + S^o + Fe\,(OH)_3 \text{ (solid)}$$

Thiobacillus ferro-oxidans

$$\text{At pH } 3.0\text{-}5.0: S^o \xrightarrow[H_2O]{O_2} SO_4^{2-}$$

Thiobacillus
thio-oxidans

Colour and Properties of Acid Sulphate Soils

The acid sulphate soils are often mid-dark grey to greenish grey in colour. These are soft and buttery with consistency of clay. Pale yellow mottles are often seen in the sub-soil layer. Upon drainage, these soils turn red. Acid sulphate soils possess strong muddy odour due to presence of ferrous and hydrogen sulphide. The unique character, which very often identifies an active acid sulphate soil, is the presence of typical yellow mottles, which resembles the mineral, jarosite.

Available N and P contents are usually low, while K and S are quite high. Due to seasonal changes, the salinity level is variable. Capillary action in soil maintain high level of salt concentration in surface layer.

The properties of acid sulphate soils studied at different locations have been presented in Table-3 (Beena and Manorama, 2013), Table 4, (Pal, 1991).

Table 3: Acidity characterization of acid sulphate soils of Kerala (Beena and Manorama, 2013)

Series	pH			Acidity [cmol(P^+) kg^{-1}]				
	H_2O	KCl	$CaCl_2$	Exch H^+	Exch.Al^{3+}	Exch	Hydrolytic	Potential
Thakazhi	3.24	2.67	2.95	2.25	4.55	6.8	93.7	100.5
Ambalapuzha	3.78	3.14	3.46	1.37	1.83	3.2	37.4	40.6
Purakad	4.03	3.32	3.70	1.25	3.55	4.8	72.5	77.3
Thotapally	4.58	3.48	3.99	1.37	2.83	4.2	51.4	55.6
Thuravur	3.79	3.17	3.43	5.75	2.25	8.0	45.8	53.8
Kallara	3.38	2.80	3.21	5.02	1.93	7.0	89.4	96.4

Table 4: Soil properties of acid sulphate soils of Sundarban areas of West Bengal, India (Pal, 1991).

Parameters	Value
pH	3.3-5.9
$ECe_{(dSm^{-1})}$	0.95-6.25
Organic carbon (g kg^{-1})	4.0-10.6
CEC [cmol(P^+) kg^{-1}]	18.5-27.5
Total acidity[cmol(P^+) kg^{-1}]	2.3-15.4
Hydrolytic acidity [cmol(P^+) kg^{-1}]	1.4-4.0
Texture	Clayey

Lime Potential

The measurement of soil pH can give variable results, since it is affected by suspension effect, soil-water ratio and electrolyte levels. Schofield and Taylor (1955) proposed the use of 0.01 M $CaCl_2$/KCl solutions for obtaining stable readings in pH measurement. However, instead of using single ion activity measurements, Schofield and Taylor suggested use of ion activity ratios for determination of soil acidity and base saturation. If the soil exchange complex is saturated with both H^+ and Ca^{2+} ions, at equilibrium as per Schofield's ratio law

$$\frac{(H^+)}{\left(\sqrt{Ca^{2+}}\right)} = \text{constant.}$$

By taking-log, the above ratio changes to

$$-\log \frac{(H^+)}{\left(\sqrt{Ca^{2+}}\right)} = \text{constant.}$$

$$\text{Or } -\log (H^+) - \left[-\log\left(\sqrt{Ca^{2+}}\right)\right] = \text{constant}$$

$$\text{Or pH } - \frac{1}{2} \text{ pCa} = \text{constant}$$

This equation is called, lime potential. It characterizes the composition of the exchange complex with respect to its saturation by H^+ and Ca^{2+} ions. Many methods for determination of lime requirement take into consideration the lime potential.

Lime requirement of Acid soils

The lime requirement of a soil is defined as the amount of liming material that must be added to raise the pH to some prescribed value, which is usually in the range of pH 6.0 to 7.0. For this purpose, potential acidity must be given due consideration along with the buffering action of given soil. The term, lime requirement is the amount of lime required to alleviate restrictions to crop yield. The removal of exchangeable aluminium, or its reduction to non-toxic level is the main consideration in liming process.

There appears to be no standard pH to decide as when to lime an acid soil, it varies from soil to soil and with crop species. Because Al^{3+} is often predominant phytotoxic, soluble and exchangeable cation in acid soils, affecting through its adverse influence on root growth, the nutrient and water uptake by plants (Miranda and Rowell, 1987), as index of its concentration is often used as an indication of whether lime is required or not, especially in tropical soils.

The standard incubation method includes addition of known amounts of $CaCO_3$ to measure the quantity of soil and incubating the mixtures for several months to allow the reaction to go to completion. The pH values of the soils are then measured and a buffer curve is drawn relating soil pH at the end of the incubation period to decide on the amount of $CaCO_3$to be added.

Earlier attempt made by Woodruff (1948) was for determination of lime requirement for routine analysis. This method was based on the principle, that acid in soil lowers the pH of a buffer solution and the pH change depends on the amount of acid added through soil. The problem is encountered in this method, when the test soil contains large amounts of Al. The method of Shoemaker, Mc lean and Pratt (the SMP method) was developed using a modified buffer (Shoemaker *et al.*, 1961). The decrease in pH in the buffer on reaction with soil was calibrated with incubated $CaCO_3$ standard method. The lime requirements were then found to raise the soil pH to 6.0, 6.4 or 6.8. These lime requirements may only be suitable for soils of similar acidic properties and CEC values (Adams, 1981). The buffer solution used was a dilute mixture of triethanol amine (TEA), p-nitrophenol, potassium chromate and calcium acetate.

A new buffer (TSM) was developed by Santanna *et al.* (2010) to reproduce the behaviour of SMP buffer in the estimate of potential acidity in Brazilian acid soils. The buffer in this method contains four weak bases TEA, MES [2-(N-morpholino) ethane sulfonic acid hydrate], calcium acetate and imidazole) and calcium chloride to control ionic strength. The new buffer is a reliable predictor of potential acidity and can replace the SMP buffer.

Coastal Acid Soils

The coastal zone in India represent transition from terrestrial to marine influence and vice versa. India has a 8129 km long coastline. Its peninsular region is bounded by the Arabian Sea on the west, the Bay of Bengal on the east and the Indian Ocean on the south. The area covering 50 kms from sea shore constitute coastal zone (Yadav 2001) and comprises of 10.78 M ha (Velayutham, *et al.*, 1998) constituting 5.5 per cent of the total land in the country spreading over 9 coastal states and 4 union territories.

The coastal soils in India are of varied types and are degraded due to salt accumulation, drainage congestion, droughtness, soil acidity etc. (Bandyopadhyay, *et al.* 2001). The salt affected soils with acidic characteristies are confined to soundarban delta of West Bengal and a part of Mahanadi delta in Odisha receiving rainfall ranging from 1400-1600 mm per annum. The salt affected soils of Malabar Coast, Kerala, Sundarban delta of West Bengal and some parts of Andaman and Nicobar Islands receiving annual rainfall ranging from 2000-3000 mm are highly acidie (acid sulphate soils). The characterstics of coastal acid saline soils of eastern India is presented in Table-5

The coastal soils of Eastern India (WB and Odisha) are very deep, fine loamy to fine texture red, imperfectly poorly drained, acidic to slightly alkaline in reaction with moderate to high salinity hazard (Sarkar *et al.,* 2001). Bandyopadhyay and Rao (2001) suggested for integrated nutrient for management coastal salt affected soils. Liming saline acidic soils integrated with biofertilizers application and chemical fertilizers based on soil test resulted in significant yield increase and nutrient uptake by rice, potato, green gram and sunflower crops grown as principal crops in coastal district Puri in Odisha. Table-6 (Pattanayak, 2011).

Table 5: Coastal soil charactics of Eastern India

Sl.No	Soil parameters	Physiographic Zone			
		Sundarban	Coastal plain of Bengal basin	Utkal plain Odisha	Mahanadi delta delta Odisha
1.	Texture	Fine loamy to clayey	Sandy	Fine loamy	Fine loamy
2.	Reaction (pH)	5.3 -8.1	5.8 - 8.2	4.2 - 7.6	4.5 - 6.3
3.	ECe (d Sm^{-1})	8.2 - 26.7	4.1-19.1	4.3 - 24.9	5.1-27.1
4.	Organic carbon (g kg^{-1})	1.4 -6.7	0.9 - 7.6	0.7 - 6.1	0.7 - 3.0
5.	CEC cmol (p+) Kg^{-1} soil	14.2 - 28.9	8.1-15.9	13.1-30.1	14.1-31.4
6.	Base saturation (%)	61 - 90	62 - 98	66 - 84	68 - 88
7.	Exch. cations cmol (p+) Kg^{-1}	Ca^{2+} Mg^{2+} Na^{+} K^{+}	Ca^{2+} Mg^{2+} Na^{+} K^{+}	Ca^{2+} Mg^{2+} Na^{+} K^{+}	Ca^{2+} Mg^{2+} Na^{+} K^{+}
8.	Classification	*Typic Fluraquents* *Typic Ustifluvents*	*Typic Usti Psaments* *Aquic Usti Psaments*	*Typic Tropaquents* *Aquic Tropaquents*	*Vertic tropaquepts* *Typic Ustropepts*

Table 6: Influence of scil amelioration, biofertilizers application and applicatior of soil test based fertilizers on yield and nutrient uptake by major crops grown insaline acid soils in Puri district in Odisha.

Sl.No	Crop	Yield (q ha^{-1})				Nutrient uptake (kg ha^{-1})											
						N				P				K			
		STD	STD +L (AM)	STD+ L+ BF	CD (p= 0.05)	STD	STD +L	STD+ L+ BF	CD (p= 0.05)	STD	STD+ L (AM)	STD+ L+BF	CD (p= 0.05)	STD	STD+ L (AM)	STD+ L+ BF	CD (p= 0.05)
1.	Rice (grain)	43.0	47.0	50.8	0.9	110.0	120.0	138.0	8.8	11.0	13.0	14.0	0.8	140.0	163.0	188.0	12.1
2.	Potato (Tuber)	148.0	159.0	177.0	16.0	126.0	139.0	150.0	8.4	8.7	10.0	11.9	1.2	152.0	169.0	185.0	14.0
3.	Green gram (Seed)	7.1	8.0	9.1	0.28	45.0	50.0	64.0	2.7	5.7	6.9	8.1	0.7	25.0	37.0	48.1	2.1
4.	Sunflower (Seed)	16.9	18.4	19.5	1.3	-	-	-	-	-	-	-	-	-	-	-	-
5.	Tomato (Fruit)	289.0	309.0	327.0	12.0	-	-	-	-	-	-	-	-	-	-	-	-

STD: Soil Test Dose, L: Lime as PMS, AM: Amendment, BF: Biofertilizers
Soil Texture: Sandy loam, pH: 5.11-5.15, Lime requirement: 5.5 t PMS ha^{-1}, EC 0.4-0.8 dSm^{-1} (kharif), 4.8-8.4 dSm^{-1} (summer)
Organic carbon: 3.4-4.7 kg^{-1} soil. Available N-230-248 kg ha^{-1}, P-14.1-17.4 kg ha^{-1}, K-110-1'7 kg ha^{-1} .$CaCl_2$ extractable –S: 4.8 -8.0 kg ha^{-1}, Boron 0.18 mg kg^{-1} soil. Exch. Ca 2.1 and Exch. Mg 1.2 cmol (p+) Kg^{-1} soil.
Sources of lime: Paper Mill Sludge (@0.2LR mixed with FYM): BF: Biofertilizer Azotobacter Azospirillum and PSB 1:1:1 ratio, 3 kg each ha^{-1}.Rice (BF): Bluegreen algae (@15 kg ha^{-1})

References

Adams, A. (1981) Alleviating chemical toxicities, liming acid soils. In Modifying the root environment to reduce crop stress (eds. G.F. Arkin and H.M. Taylor). *American SocietyofAgricultural Engineers Monograph,* 4:269.

Ano, A.O.andUbochi, C.I. (2007) Neutralization of soil acidity by animal manures: mechanism of reaction. *African Journal of Biotechnology*, 6:364-368.

Asghar, M. and Kanehiro, Y. (1980) Effect of sugarcane trash and Pineapple residue on soil pH, redox-potential, extractable Al, Fe, and Mn. *Tropical Agriculture (Trinidad)* 57:245-258.

Bandyopadhyay, B.K., Sen, H.S., Maji, B. and Yadav, J.S.P. (2001) Saline and alkali soils of India and their management. *ISCAR Monograph Series1* pp 1-72.

Bandyopadhyay, B.K. and Rao, D. L. N. (2001) Integrated Plant Nutrient Management in Saline soils. *Journal of the Indian Society CoastalAgricultural Research 19* (1&2):35-58

Beena,V.I. and Manorama, T.K.C.(2013) Acidity characterization in acid sulphate soils of Kerala. 8th International Symposium on Plant Soil Interactions, UAS, Bangalore, October 18-20, pp.390-391.

Borek, S.L. (1994) Effect of humidity on pyrite oxidation. In Environmental Geo-chemistry of Sulfide Oxidation; Alpers, C.N., Blowers, D.N., ed.; American Chemical Society: Washington, D.C., 31-44.

Brady, N.C. and Weil, R.C. (2012) In the Nature and Properties of soils 14th Edition. Published by Dorling Kindersley (India). Pvt. Ltd. Pearson Education in South Asia.

Brubaker, S.C., Jones, A.J., Lewis, D.T. and Frank, K. (1993) Soil properties associated with landscape position. *Soil Science Society of America Journal* 57(1): 235-239.

Carver, B.F. and Ownby, J.D. (1995). Acid soil tolerance in wheat. *Advances in Agronomy* 54: 117-173.

Chand, J.P. and Mandal, B. (2000) Nature of acidity in soils of West Bengal. *Journal of the Indian society of Soil Science* 48:20-26.

Dolui, A.K., Kar, S. and Ghosal, M. (2010). Characterization and lime requirement estimation of Acid soils of Nagaland. *Journal of the Indian Society of Soil Science* 58(2): 147-153.

Dolui, A.K.and Maity, B. (2004). Nature of soil acidityin relation to different forms of Fe and Al of some *Alfisols* of West Bengal. *Journal of the Indian Society of Soil Science* 52(4): 415-420.

Fitzpatrick, R.W., Fritsch, E. and Self, P.G. (1996) Interpretation of soil features produced by ancient and modern process in degraded land scapes; V. Development of Saline Sulfide features in non-tidal seepage areas. *Geoderma* 69:1-29.

Gordon, G.V. (1988). Causes and effects of soil acidity Oklahoma Cooperative Extension Facts 2239, Division of Agricultural Sciences and National Resources. Oklahoma StateUniversity (OSU)

Hue, N.V., Craddock, G.R. and Adams, F. (1986) Effect of organic acids on aluminium toxicity in sub-soils. *Soil Science Society of America Journal* 50:28-34.

Jackson, R.B., Schenk, H.J., Jobbagg, E.G., Canadell, J., Colello, G.D. and Dickinson, R.F. (2000). Below ground consequences of vegetation change and their treatment in models. *Ecologyical Applications* 10: 470-483.

Lindsay, W.L. and Stephenson, H.F. (1959). Nature of the reactions of monocalcium phosphate monohydrate in soils: 1. The solution that reacts with the soil. *Soil Science Society of America Proceedings*23:12-18.

Miranda, L.N. and Rowell, D.L. (1987) The effects of lime and phosphorus on the function of wheat roots in acid top soils and sub-soils. *Plant and Soils*104(2): 253-262.

Misra, U.K. (2004) Acid soils and its management. *Journal of the Indian Society of Soil Science*. 52(4): 332-343.

Misra, U.K.; Satapathy, S.and Panda, N. (1989) Characterization of some acid soils of Orissa.1 Nature of Soil Acidity. *Journal of the Indian Society Soil Science*37:22.

Misra, U.K.and Saithantuaanga, H. (2000) Characterization of acid soils of Mizoram. *Journal of the Indian Society of Soil Science* 48:437-446.

Pattanayak, S.K. (2011) Annual Report of All India Network Project on Biodiversity in Biofertilizers, Directorate of Research, O.U.A.T., Bhubaneswar.

Pal, A.K. (1991) Acid sulophate soils of Sundarbans. In Acid Soils of India, ICAR, New Delhi, pp.264-265.

Pal, Indrani (2009) Rainfall Trends in India and their impact on Soil Erosion and Land Management. Ph.D. thesis submitted to Department of Engineering at the University of Cambridge, U.K.

Pattanayak, S.K. (1992). Effect of long term fertilizers and manure application to a medium land acid laterite on nutrient transformation and changes in some soil properties. Ph.D. thesis submitted to IARI, New Delhi.

Pattanayak, S.K.(2013) Annual and QRT Report, All India Network Project on Biodiversity in Biofertilizers, Directorate of Research, OUAT, Bhubaneswar.

Potter, T. (1976) The problems of fish culture associated with acid sulphate soils and methods for their improvement. In Report on the ASEAN Seminar/Workshop on Shrimp Culture, 15-23 No.1976, Uoilo, Philippines, Manila, ASEAN National Coordinating Agency of the Philippines.

Santanna, M.A., Kaminski, J., dos Santos, D.R., dos Anjos, J., Cella, C. and Calegari, D.S.P. (2010). A new buffer that imitates the SMP solution for determining potential acidity of Brazillian soils. Symposium 3.1.1 Further development of soil evaluation methods. In Proc. 19th World Congress of Soil Science, Soil Solutions for a Changing World, 1-6 August, 2010, Brisbane, Australia, pp.13-16.

Sarkar, D., Sahoo, A.K., Sah, K.D. and Gajbhiye, K.S. (2001) Coastal Soils of Eastern India, Their characteristics, potentials and limitations towards Alternate Land Use. *Journal of the Indian Society Coastal Agricultural Research* 19(1&2): 80-83.

Schlesinger, W.H. (1997) Biogeochemistry, an Analysis of Global Change. Academic Press, San Diego.

Schutte, E.F., Walsh, L.M., Kelling, K.A. (2005). Management of Wisconsin soils Cooperative Extension Publication A3588 University of Wisconsin-Extension.

Seibert, J. Stendhl, J. and Sorensen, R. (2007). Topographical influences on soil properties in boreal forests. *Geoderma* 141:139-148.

Shoemaker, H.E.,Mc Lean, E.O. and Pratt, P.F.(1961) Buffer methods for determining lime requirement of soils with appreciable amounts of extractable aluminium. *Soil Science Society of America Proceedings* 25: 274-277.

Subehia, S.K., Verma, S. and Sharma, S.P. (2005) Effect of long term use of chemical fertilizers with or without organics on forms of acidity, P adsorption and crop yields in an acid soil. *Journal of the Indian Society of Soil Science* 53(3): 308-314.

Ugolini, F.C., Dahlgren, R.A., Shoji, S.and Ito, T. (1988). An example of andolization and podzolization as revealed by soil solution studies, Southern Hakkoda, Northern Japan. *Soil Science* 145:111-125.

Velayutham, M., Sarkar, D., Reddy, R.S., Natarjan, A., Shira Prasad, C.R., Challa, O., Harindranath, C.S., Shyamapura, R.L., Sharma, J.P. and Bhattacharaya, T. (1998) Soil resources and their potentials in coastal areas of India. *In* Frontiers of Research and its Application in coastal Agriculture, held at GAU, Navsari, Gujarat, India, 16-20 the September 1998

Venkatesh, M.S., Majumdar, B., Kailash, K. and Patiram (2002). Effect of P., FYM and lime and yield, P uptake, by maize and forms of soil acidity in typicHapludalf of Meghalaya. *Journal of the Indian Society of Soil Science* 50(3): 254-258.

Yadav, J.S.P. (2001) Resource Management for Higher productivity and Ecological security in the coastal Ecosystem of India. *Journal of the Indian Society for Coastal Agricultural Research* 19 (1&2):10-29.

4

Genesis and Classification of Acid Soils of India

A.K. Sahoo and Dipak Sarkar

Acid soils are found in India in about 93 million hectares (Table 1) and they occur in the Himalayan region, the great Indo-Gangetic and Brahmaputra plains, peninsular plateau and uplands and the coastal plains including island eco-systems under varying environmental conditions of landscape, geology, climate and vegetation. Conditions which are conducive for the formation of acid soils, such as high rainfall coupled with high temperature encouraging intense weathering and loss of alkali and alkaline earth metals with a simultaneous accumulation of iron and aluminium hydroxy compounds or the formation of free acidic substances in the humid forest foot hill regions, exist in some region or other in almost all the states in India. A soil with pH less than 7.0 is deemed as an acid soil (Soil Survey Staff,1951). However, for characterization and classification, acid soils may be defined as those having pH less than 5.5 in 1:1 water extract (Soil Survey Staff, 1999).

Genesis of Acid Soils

There are five major factors viz. climate, vegetation, parent material, topography and human interference responsible for formation of acid soils.

Table 1: State wise distribution of Acid soils in India (area in 000 ha)

Sl.No.	District	Strongly acidic (pH <4.5)	Moderately acidic (pH 4.5-5.5)	Slightly acidic (pH 5.5-6.5)	Total	TGA	% of TGA
1.	Andhra Pradesh	0.0	0.0	2827.5	2827.5	27504.5	10.3
2.	Arunachal Pradesh	4775.9	1742.7	268.8	6787.4	8374.3	81.1
3.	Assam	23.5	2331.2	2332.7	4687.5	7843.8	59.8
4.	Bihar	0.0	36.7	2324.9	2361.6	9416.3	25.1
5.	Chhattishgarh	156.4	5930.1	4386.6	10473.0	13480.5	77.7
6.	Goa	3.6	113.7	191.1	308.3	370.2	83.3
7.	Himachal Pradesh	0.0	157.0	1620.6	1777.6	5567.3	31.9
8.	Jammu & Kashmir	0.0	93.3	1480.1	1573.4	22223.6	7.1
9.	Jharkhand	0.0	999.6	5772.1	6771.7	7971.4	84.9
10.	Karnataka	0.0	61.4	3254.7	3316.1	19179.1	17.3
11.	Kerala	138.0	2789.6	753.2	3680.7	3886.3	94.7
12.	Madhya Pradesh	0.0	1124.7	10601.8	11726.5	30864.1	38.0
13.	Maharastra	0.0	240.0	4332.6	4572.6	30771.3	14.9
14.	Manipur	426.9	1437.2	325.1	2189.2	2232.7	98.1
15.	Meghalaya	0.0	1186.3	1054.4	2240.7	2242.9	99.9
16.	Mizoram	0.0	1267.6	777.3	2044.9	2108.1	97.0
17.	Nagaland	118.9	1483.3	55.7	1657.9	1657.9	100.0
18.	Odisha	0.0	261.6	8409.7	8671.3	15570.7	55.7
19.	Sikkim	278.9	323.4	2.8	605.0	709.6	85.3
20.	Tamil Nadu	264.0	347.3	4294.5	4905.8	13005.8	37.7
21.	Tripura	56.6	749.0	237.2	1042.8	1048.6	99.5
22.	Uttar Pradesh	0.0	0.0	337.5	337.5	24104.6	1.4
23.	Uttarakhand	0.0	1183.6	2300.6	3484.2	5336.5	65.3
24.	West Bengal	0.0	555.6	4199.7	4755.3	8875.2	53.6
25.	Others*	0.0	0.0	0.0	0.0	64381	0.0
26.	Total	6242.6	24414.6	62141.2	92798.4	328726.3	28.2
27.	Area (%)	1.9	7.4	18.9	28.2	100.0	

Others* : Delhi, Gujarat, Haryana, Punjab, Rajasthan, A&N Island, Chandigarh, D&N Haveli, Daman & Diu, Lakshadweep and Pondicherry.
Source: Maji *et.al*. 2012

Climate: Acid soils are generally found in the areas of high rainfall. In these regions, evaporation of water from the soil surface is less than precipitation. Excess amount of water moves downward in the soil and meets with the ground water table. This downward moving water carries a large amount of soluble salts from the soil surface. A reasonable amount of these salts is a must for maintaining neutral soil reaction. In absence of these salts, surface soil is rendered acidic. Some compounds of iron and aluminium are resistant to leaching and, therefore, left on the surface. These compounds are acidic in nature and impart acidity to the surface soil (eg. laterite soils).In India, it is believed that the regions with acid soils do not receive more than 750 mm annual rainfall. The regions with annual rainfall of 1350 mm have acid soils with pH value 5 or even lower. In hilly regions, where the loss of water through evaporation is very low due to low temperature, the conditions for the development of acid soils are very favorable although the rainfall is not so heavy.

Vegetation: In the regions of very low temperature, acid soils can develop easily. The foliage of coniferous trees, which grows in these regions are devoid of alkali elements. When the leaf litter of coniferous trees on ground is degraded, organic acids are released, which gradually make the soil, acidic. Vegetation of coastal zones and marshy places facilitates development of acidity. The places, where excessive vegetation is accompanied by heavy rainfall and water pounding, soils are very acidic in reaction. In these places, plant residues are constantly submerged in water and they do not get enough air to decompose completely. This under-decomposed organic material keeps on piling up and soil becomes very acidic but at the same time, these are rich in organic matter.

Parent material: Some young soils exhibit acidic nature even in areas of medium rainfall and vegetation. These soils reflect the nature of rocks from which they have developed. Some rocks (parent material) are acidic in nature (e.g. igneous rocks). After weathering of these rocks, acidic constituents dominate the composition of soil. The rocks of soft and simple composition are especially suitable for development of acidic soils as they allow quick percolation of water through them and have fewer amounts of alkaline elements and salts.

Topography: Sloppy terrain with good drainage conditions often develop acidic environment. On hill slopes, development of acid soils is easier. Acid soils do not develop generally in river basins. The plains with good drainage may also develop slight acidity in due course of time.

Human interference: Excessive use of water or keeping the field submerged for a long time accompanied by improper drainage may lead to the development of acidity in soil. Continuous use of acid forming fertilizers (e.g. ammonium sulphate, ammonium chloride, etc) for years may also lead to considerable reduction in soil pH in cultivated areas.

Acid Soil regions

The pedons representing acid soils from different regions of the country are described and classified, which show considerable heterogeneity in the morphological, physical and chemical characteristics primarily due to variations in the factors of soil formation operating in these regions. Distribution of acid soils in different region in India along with their classification is presented in table 2 and fig.1.

Acid Soils of Himalayan region

Acid soils in the Himalayan region extend from Kashmir in the north-west to Sikkim, Arunachal Pradesh and other states in the north-east with a wide variation in the macro-and micro-climate, physiography and vegetation.Physiographically, the Himalayan region of acid soils consists of (i) greater and higher Himalayas with mountain and valleys at high altitudes consisting of metamorphic rocks (ii) middle and lower Himalayas comprising varied sedimentary and metamorphic rocks (iii) sub-Himalayan zone consisting of Siwalik sediments of tertiary period. The climate, in general, can be characterized as sub-tropical in low altitudes (<1000 m), warm to cool temperate (wet) between 1000 and 2000 m, cool temperate between 2000 and 3000 m, temperate (alpine) between 3000 and 4500m, and above 4500 m in the Himalayan meadow up to the snow line.There is considerable variation in vegetation due to differences in climate and altitude.

The soils in the eastern region of Himalayas are developed on gneiss and alluvium and are distributed in northeastern hills of Nagaland. Mizoram, Manipur, Meghalaya, Arunachal Pradesh and Tripura. They are well drained, loamy to clayey, and moderate to

Table 2: Classification of acid soils of India and their distribution

Broad soil group	Soil Great Group (2006)	pH	Distribution
Laterites	Plinthaqults, Plinthustults and Plinthudults,Plinthustalfs	4.8-7.0	Karnataka, Madhya Pradesh,Chhattishgarh Odisha, West Bengal, Kerala, Assam, Jharkhand
Lateritic soils	Paleudalfs,Hapludalfs,Hapludults, Kanhaplohumults,Kandihumults, Dystrustepts, Ustorthents, Haplustepts, Haplustalfs, Rhodustalfs, Paleustalfs	5.0-7.0	Kerala, Odisha, West Bengal, Assam, North-Eastern States, Jharkhand
Mixed red and yellow soils	Dystrustepts, Udifluvents, Hapludalfs Haplustepts, Haplustalfs, Paleustalfs	5.5-6.5	Karnataka,Madhya Pradesh, Chhattisgarh,erstwhile UP, Odisha, Chotanagpur and Santhal Praganas of Jharkhand
Ferruginous red soils	Paleudalfs,Udifluvents,Udorthents, Ulticintergrades, Epiaquepts, Haplustalfs, Paleustalfs, Rhodustalfs	5.0-6.8	Tamil Nadu,parts of Karnataka, Maharastra, Andhra Pradesh, Odisha, Jharkhand, Goa
Ferruginous red and gravely soils	Haplustalfs, Rhodustalfs, Haplustepts Ustorthents, Kandihumults, Udorthents, Dystrudepts, Paleudults	5.2-6.7	Assam, North Eastern States,WB, Odisha, erstwhile UP, Andaman &Nicobar Islands, Chhattisgarh
Brown forest soils of Western Himalayas	Hapludolls,Udifluvents,Udorthents, Dystrustepts, Dystrudepts	5.5-6.0	Upper region of western Himalayas, Jammu and Kashmir, Himachal Pradesh, Uttaranchal
Brown forest podzolic soils	Hapludolls,Udorthents, Udifluvents Eutrudepts, Dystrustepts, Dystrudepts,	5.1-5.4	Upper region of western Himalayas, Jammu and Kashmir, Himachal Pradesh
Brown forest soils of Central Himalayas	Hapludults,Udorthents,Udifluvents, Dystrustepts, Dystrudepts	5.2-5.7 Uttarakhand	Lower region of Central Himalayas, erstwhile UP,
Brown forest podzolic soils of Central Himalayas	Hapludalfs, Kandihumults, Kanhapludults, Kandiudalfs, Dystrudepts, Eutrudepts, Haplohumults	4.7-6.0	Eastern Himalayas, North-Eastern States of Himalayan region,

Contd.

Broad soil group	Soil Great Group (2006)	pH	Distribution
Foot hill soils	Argiaquolls, Epiaquolls and Endoaquolls,	4.5-6.5	Jammu and Kashmir, West Bengal, erstwhile UP
Peat soils	Histosols	3.5-4.5	Kerala, Assam
Alluvial soils	Ustifluvents, Udifluvents, Dystrudepts,	4.2-5.8	Assam, West Bengal, Bihar
Coastal alluvial soils	Ustifluvents, Fluvaquents, Sulfaquents, Sulfaquepts, Epiaquents, Endoaquents, Udifluvents	3.5-5.8	West Bengal (Sunderbans), Kerala, Some pockets of Tamil Nadu
Degraded saline and acid saline soils	Fluvaquents, Sulfaquents, Halaquepts	5.2-6.6	West Bengal, Odisha, Parts of Tamil Nadu and Kerala.

Sources : Murthy *et.al.*, 1976; Staff NBSS & LUP, 2002; Panda *et.al.*,2009; Soil Survey Staff USDA, 2006.

strongly acid with low cation exchange capacity and low to moderate base saturation. These soils are classified as Typic Kanhapludults, Typic Paleustults, Typic Paleudults, Typic Haplustepts, Typic Kandiudults, Umbric Dystrochrepts, Typic Kandihumults, Typic Hapludolls, Humic dystrudepts, Typic Hapludolls, Typic Haplumbrepts, typic Udorthents, Typic Argiudolls, Humic Pachic Dystrudepts. They are cultivated for millets, potato, maize, rice and plantation crops like tea, coffee, rubber, and horticultural crops like pineapple and orange

Studies in the Himalayan region have been made by various workers such as Raychaudhuri *et.al.*, (1963), Yadav (1963), Dhir (1967), Biswas (1968), Nayak *et.al.*,(2002), Sahoo *et.al.*,(2010) besides several others. Representative acid soil profiles from different regions of the Himalayas are described and classified below.

Soils of Eastern Himalayas

Profile Location : Village : Phyangla, District : East Sikkim, Sikkim

Physiographic Position: Ridge and Summit, 15-30% slope

Rainfall : 3500 mm

Land use: Forest

Classification: Fine loamy ,mixed, thermic *HumicDystrudepts*

Morphometic Characteristics

Horizon	Depth (cm)	Geology &Parent material	Drainage	Colour	Texture (USDA)	Coarse fragments (>2 mm) %
A	0-17	Gneiss	Moderately well drained	10 YR 3/2 M	cl	5
Bw1	17-38			10YR 3/3 M	cl	10
Bw2	38-66			7.5 YR 4/4 M	sicl	10
Bw3	66-88			10 YR 3/6M	sicl	10
Bw4	88-120			10 YR 3/4M	sicl	30
Bw5	120-150			10 YR 3/4M	cl	50

Physico -chemical Characteristics

Depth(cm)	pH		OC(%)	Clay (%)	CECcmol (p+)kg^{-1}	B.S.(%)
	KCl (1:1)	Water (1:2.5)				
0-17	-	4.1	2.5	30.6	15.1	49
17-38	-	4.2	2.5	33.7	15.6	48
38-66	-	4.7	0.9	35.0	10.0	45
66-88	-	4.8	0.8	36.2	10.2	44
88-120	-	4.9	0.8	39.0	10.5	45
120-150	-	5.1	0.7	40.0	12.2	46

Source: Das *et.al.*, 2003

Soils of eastern Himalayas

Profile Location: Village: Brang, District : South Sikkim, Sikkim

Physiographic Position: Hills side slope, 30-50% slope

Rainfall: 3500 mm

Land Use: Mostly under cultivation of maize

Classification: Loamy-skeletal,mixed, thermic *Typic Hapludolls*

Morphometic Characteristics

Horizon	Depth (cm)	Geology & Parent material	Drainage	Colour	Texture (USDA)	Coarse fragments (>2 mm) %
Ap	0-20	Gneiss	Excessively drained	10 YR 3/2 M	sil	15
Bw1	20-36			10YR 3/3 M	sil	25
Bw2	36-64			10 YR 5/6 M	l	35
C11	64-87			5 YR 6/3M	l	35
C12	87-100			5 YR 6/3M	sl	40

Physico -chemical Characteristics

Depth(cm)	pH		OC(%)	Clay (%)	CECcmol (p+)kg^{-1}	B.S.(%)
	KCl (1:1)	Water (1:2.5)				
0-20	-	4.3	2.6	14.5	10.6	52
20-36	-	4.5	1.8	16.5	9.1	56
36-64	-	4.8	1.5	14.5	7.0	57
64-87	-	5.0	0.9	12.5	5.6	61
87-100	-	5.1	0.8	8.5	4.1	61

Source: Das *et.al.*, 2003

Soils of eastern Himalayas

Profile Location: Barbung, P.S. : Kurseong, District: Darjiling, West Bengal

Physiographic Position: Hills, very steep slopes, 30-50% slope

Rainfall: 3000 mm

Land Use: Natural and planted forest

Classification: Loamy- skeletal,mixed. Thermic *Typic Haplumbrepts*

Morphometic Characteristics

Horizon	Depth (cm)	Geology &Parent material	Drainage	Colour	Texture (USDA)	Coarse fragments (>2 mm) %
A1	0-12	Gneiss	Well drained	10 YR 3/2M	l	10
AB	12-50			10 YR 3/2M	l	10
B2	50-70			10 YR 3/3M	l	10
B3	70 86			10 YR 4/4M	cl	20
C	86-185			10 YR 5/6M	l	20

Physico -chemical Characteristics

Depth(cm)	pH		OC(%)	Clay (%)	CECcmol (p+)kg^{-1}	B.S.(%)
	KCl (1:1)	Water (1:2.5)				
0-12	-	4.5	3.22	22.1	44	44
12-50	-	4.6	1.31	20.1	49	49
50-70	-	4.6	1.94	18.1	49	49
70-86	-	4.6	1.28	32.1	52	52
86-185	-	4.7	0.56	26.1	56	56

Source: Nayak *et.al.*, 2001

Soil of eastern Himalayas

Profile Location: Village: Gobardhana, P.S.: Ramnagar, District: Paschim Champaran, Bihar

Physiographic Position: Hills and side slopes, 3-8% slope

Rainfall: 1448 mm.

Land Use: Thin forest,shisham, babul

Classification: Coarse-loamy, mixed, hyperthermic *Typic Udorthents*

Morphometic Characteristics

Horizon	Depth (cm)	Geology &Parent material	Drainage	Colour	Texture (USDA)	Coarse fragments (>2 mm) %
A	0-18	Alluvium	Well drained	10 YR 3/1 M	sl	-
AC	18-48			10YR 6/4 M	sl	-
C1	48-99			10 YR7/3 M	ls	-
2C2	99-125			10 YR 6/3M	l	-
2C3	125-155			10 YR 4/3M	sl	-

Physico -chemical Characteristics

Depth(cm)	pH		OC(%)	Clay (%)	CECcmol (p+)kg^{-1}	B.S.(%)
	KCl (1:1)	Water (1:2.5)				
0-18	-	6.0	0.37	19.6	8.6	66
18-48	-	6.1	0.24	15.2	6.4	67
48-99	-	6.1	0.15	11.8	4.8	56
99-125	-	6.0	0.10	20.6	7.6	52
125-155	-	6.2	0.08	19.0	8.0	68

Source: Sahoo *et.al.*, 2002

Soils of western Himalayas

Profile Location: Near Aharbal R.H.,Anantnag, District : Jammu & Kashmir

Physiographic Position: Strongly sloping –moderately steep lands, 10-25% slope

Rainfall: 3000 mm.

Land use: Forest

Classification : Coarse silty, mixed, mesic *Typic Argiudolls*

Morphometic Characteristics

Horizon	Depth (cm)	Geology &Parent material	Drainage	Colour	Texture (USDA)	Coarse fragments (>2 mm) %
A11	0-12	Granite	Well drained	10YR 3/3M	l	5.0
A12	12-24			10YR 3/3M	sil	9.0
AB	24-46			10YR 3/3M	sil	12.5
Bt1	46-67			10 YR 4/4M	sil	14.0
Bt2	67-100			10 YR 4/4M	sil	42.5

Physico -chemical Characteristics

Depth(cm)	pH		OC(%)	Clay (%)	CECcmol (p+)kg^{-1}	B.S.(%)
	KCl (1:1)	Water (1:2.5)				
0-12	-	6.42	3.92	7.5	5.98	71
12-24	-	6.30	1.89	5.3	2.17	96
24-46	-	6.14	1.11	9.8	6.25	90
46-67	-	5.70	1.01	13.3	8.97	79
67-100	-	5.45	0.99	19.5	12.24	80

Source: Maji *et.al.*, 2012

Soils of north east Himalayas

Profile Location: Village : Tarin, Tahsil: Zero (Sadar),District :Lower Subansiri, Arunachal Pradesh

Physiographic Position: Lower hill slopes, Elevation: 1550 m

Rainfall: 2500 mm

Land use: Moderately dense pine forest

Classification: Fine-loamy, mixed, thermic *Humic Pachic Dystrudepts*

Morphometic Characteristics

Horizon	Depth (cm)	Geology &Parent material	Drainage	Colour	Texture (USDA)	Coarse fragments (>2 mm) %
A	0-17	Sandstone	Well	10YR 3/2M	cl	-
Bw1	17-35			10YR 3/3 M	cl	-
Bw2	35-60			10YR 3/3 M	cl	15
Bw3	60-83			10YR 4/4 M	cl	20
BC	83-125			10YR 5/4 M	cl	-
C	125-165			10YR 5/3 M	sl	-

Physico -chemical Characteristics

Depth(cm)	pH		OC(%)	Clay (%)	CECcmol (p+)kg^{-1}	B.S.(%)
	KCl (1:1)	Water (1:2.5)				
0-17	4.1	5.1	5.47	34.0	23.9	25
17-35	4.0	4.8	2.74	35.5	17.6	24
35-60	4.0	4.8	1.33	33.0	13.0	10
60-83	3.9	5.2	0.57	31.3	11.6	10
83-125	3.9	5.1	0.41	29.5	11.1	10
125-165	3.4	5.0	0.25	19.5	8.7	13

Source: Nayak *et al*., 2002

Soils of Purvanchal

Profile Location :Village : Langol Hill, Tahsil: Imphal District : Imphal, Manipur

Physiographic Position: Lower hill slopes (15 – 25%)

Rainfall: **1800 mm**

Land use: Grass land

Classification: Fine, mixed, thermic *Typic Kandiudults*

Morphometic Characteristics

Horizon	Depth (cm)	Geology &Parent material	Drainage	Colour	Texture (USDA)	Coarse fragments (>2 mm) %
A	0-15	Sandstone	Well	5YR 4/6M	sic	2
Bt1	15-35			5YR 4/6M	sic	3
Bt2	35-63			5YR 5/6M	c	3
Bt3	63-92			5YR 5/6M	c	4
Bt4	92-122			5YR 5/6M	c	5
Bt5	122-150			5YR 5/6M	c	5

Physico -chemical Characteristics

Depth(cm)	pH		OC(%)	Clay (%)	CECcmol (p+)kg^{-1}	B.S.(%)
	KCl (1:1)	Water (1:2.5)				
0-15	3.9	4.7	2.4	49.2	12.7	28
15-35	4.1	5.0	2.0	52.4	14.7	11
35-63	4.0	5.0	1.2	58.3	14.1	13
63-92	4.0	4.5	0.9	57.4	13.9	11
92-122	3.9	4.5	0.7	55.3	14.7	12
122-150	4.1	4.7	0.4	52.8	13.0	9

Source: Sahoo *et al*., 2010

Soils of Purvanchal

Profile Location: Village: Shibbari, Tahsil : Manu, District : North Tripura, Tripura

Physiographic Position : Medium relief parallel ridge

Rainfall: 2427mm

Land Use: Partly cultivated for single crop and mostly under thin forest

Classification: Fine, kaolinitic, hyperthermic *Typic Kandihumults*

Morphometic Characteristics

Horizon	Depth (cm)	Geology &Parent material	Drainage	Colour	Texture (USDA)	Coarse fragments (>2 mm) %
A1	0-18	Shales,siltstone	Well drained	10 YR 3/3 M	cl	-
Bt1	18-33			7/5YR 4/4 M	c	-
Bt2	33-74			5 YR 4/6 M	c	-
Bt3	74-95			5YR 4/6 M	c	-
Bt4	95-120			5YR 5/6 M	c	-
Bt5	120-145			5YR 5/6 M	c	-
Bt6	145-170			7.5 YR 5/6 M	c	-

Physico -chemical Characteristics

Depth(cm)	pH		OC(%)	Clay (%)	CECcmol (p+)kg^{-1}	B.S.(%)
	KCl (1:1)	Water (1:2.5)				
0-18	4.0	4.8	1.4	32.3	23	69
18-33	4.0	4.9	1.3	45.5	21	57
33-74	3.8	4.5	1.1	42.3	23	31
74-95	3.4	4.2	0.9	48.3	16	20
95-120	3.7	4.5	0.5	46.3	17	26
120-145	3.6	4.2	0.7	46.3	18	26
145-170	3.7	4.5	0.6	41.3	22	29

Source: Bhattacharyya *et.al.*, 2004

Soils of Purvanchal

Profile Location: Village: Chhailengta, Tahsil: Chhailengta, District: North Tripura, Tripura

Physiographic Position: Undulating plains with low mounds and narrow valleys

Rainfall: 2427mm

Land Use: Mostly cultivated viz.paddy

Classification: Fine-loamy,mixed, hyperthermic *Oxyaquic Dystrudepts*

Morphometic Characteristics

Horizon	Depth (cm)	Geology &Parent material	Drainage	Colour	Texture (USDA)	Coarse fragments (>2 mm) %
A1	20-30	Sandstone	Moderately	7/5YR3.5/5M	l	-
3Bw1	30-66	& alluvium	Well drained	10 YR 4/3M	sic	-
3Bw2	66-85			10YR 4/4 M	sicl	-
3Bw3	85-110			10YR 5/4 M	sic	-
3Bw4	110-160			10YR 5/4 M	sicl	-

Physico -chemical Characteristic

Depth(cm)	pH		OC(%)	Clay (%)	CECcmol (p+)kg^{-1}	B.S.(%)
	KCl (1:1)	Water (1:2.5)				
0-20	3.7	4.2	0.64	18.9	4.3	49
20-30	3.8	4.4	0.92	26.5	6.8	32
30-66	3.7	4.4	0.84	42.8	14.0	25
66-85	3.8	4.5	0.66	38.5	13.0	25
85-110	3.9	4.6	0.64	41.5	13.8	25
110-160	3.9	4.5	0.45	39.5	13.3	24

Source: Bhattacharyya *et.al.,*2004

Soils of Purvanchal

Profile Location: Village: Limakang, **District:** Senapate, Manipur

Physiographic Position: Strongly position hill, 15-30% slope

Rainfall :1353 mm

Land Use: Thin forest

Classification: Fine, mixed, thermic *Umbric Dystrochrepts*

Morphometic Characteristics

Horizon	Depth (cm)	Geology &Parent material	Drainage	Colour	Texture (USDA)	Coarse fragments (>2 mm) %
A1	0-17	Tertiary old sedimentary, shale	Excessively drained	10 YR 3/2 M	sicl	-
A2	17-34			10YR 4/3 M	sic	-
B21	34-65			10 YR 4/4 M	sic	-
B22	65-95			10 YR 5/6 M	sic	-
B3	95-125			10 YR 5/6 M	sicl	-

Physico -chemical Characteristics

Depth(cm)	pH		OC(%)	Clay (%)	CECcmol (p+)kg^{-1}	B.S.(%)
	KCl (1:1)	Water (1:2.5)				
0-17	3.9	5.0	3.9	40.0	17.30	14
17-34	3.9	4.9	3.0	42.5	15.70	8
34-65	3.9	4.8	11.4	45.0	11.40	10
65-95	3.9	4.8	0.92	46.0	8.70	11
95-125	3.8	4.8	0.56	28.5	5.80	13

Source: Sen *et.al.*, 2006

Soils of Purvanchal

Profile Location : Village: Suongpeh, District : Churachandpur, Manipur

Physiographic Position: Moderately steep sloping hill slopes, 8-15% slope

Rainfall: 2652 mm

Land use: Moderately dense forest

Classification: Fine, mixed, thermic *Typic Hapludults*

Morphometic Characteristics

Horizon	Depth (cm)	Geology &Parent material	Drainage	Colour	Texture (USDA)	Coarse fragments (>2 mm) %
A1	0-17	Tertiary old sedimentary, shales	Excessively drained	10 YR 5/3 M	sicl	7
Bt1	17-35			7.5YR 5/4 M	sic	9
Bt2	35-50			7.5 YR 5/6 M	c	10
Bt3	50-70			7.5 YR 5/6 M	c	15
C	70-110			7.5YR 5/6 M	sicl	25

Physico -chemical Characteristics

Depth(cm)	pH		OC(%)	Clay (%)	CECcmol (p+)kg^{-1}	B.S.(%)
	KCl (1:1)	Water (1:2.5)				
0-17	3.8	4.9	2.50	32.2	13.80	34
17-35	3.7	4.9	0.80	42.3	11.00	24
35-50	3.8	4.8	0.40	55.8	13.00	22
50-70	3.9	5.0	0.40	49.6	11.20	22
70-110	3.9	5.0	0.40	39.7	9.60	13

Source: Sen *et.al.*, 2006

Soils of hot-moist subhumid Purvanchal ecosystem

Profile Location : Village : Phulomi, District : Kohima

Physiographic Position : Undulating hills

Rainfall : 1120.3 mm.

Land use: Forest

Classification : Fine, mixed, hyperthermic *Typic Paleudults*

Morphometic Characteristics

Horizon	Depth (cm)	Geology & Parent material	Drainage	Colour	Texture (USDA)	Coarse fragments (>2 mm)%
A1	0-10	Shale and sandstone	Somewhat excessively drained	10YR 3/3	sicl	5
A2	10-21			10YR 4/4	sicl	10
Bw1	21-32			10YR 4/6	cl	-
Bw2	32-57			10YR 5/6	sic	10
2Bt3	57-100			7.5 YR 5/6	sic	-
2Bt4	100-150			10 YR 5/6	c	15

Physico -chemical Characteristics

Depth(cm)	pH		OC(%)	Clay (%)	CECcmol (p+)kg^{-1}	B.S.(%)
	KCl (1:1)	Water (1:2.5)				
0-10		4.6	2.50	33.5	10.37	16.9
10-21		4.4	1.30	35.5	9.80	40.8
21-32		4.5	1.13	38.5	9.07	8.6
32-57		4.3	0.98	40.0	9.50	8.4
57-100		4.7	0.33	49.7	8.60	33.9
100-150		4.7	0.37	45.0	9.50	29.4

Source: Maji *et.al.*,2004

Soils of warm perhumid Purvanchal ecosystem

Profile Location : Village : Thuipuzumi, District : Phek

Physiographic Position : Steeply sloping hill slopes

Rainfall : 1686.2 mm.

Land use: Forest

Classification : Fine, mixed, thermic *Humic Dystrudepts*

Morphometic Characteristics

Horizon	Depth (cm)	Geology & Parent material	Drainage	Colour	Texture (USDA)	Coarse fragments (>2 mm) %
A1	0-15	Shale and sandstone	Somewhat excessively drained	10YR 5/2	sicl	20
Bw1	15-31			10YR 5/4	sic	20
Bw2	31-60			10YR 4/4	c	40
R	60+			Rock layer		

Physico -chemical Characteristics

Depth(cm)	pH		OC(%)	Clay (%)	CECcmol (p+)kg^{-1}	B.S.(%)
	KCl (1:1)	Water (1:2.5)				
0-15	-	5.1	3.88	37.0	22.56	21.1
15-31	-	5.0	1.03	46.0	24.20	37.0
31-60	-	5.1	0.80	50.0	12.06	47.1
60+	Bed rock					

Source: Maji *et.al.*, 2004

Soils of Meghalaya plateau

Profile Location :Village :Umkrem, District : West Khasi Hills, Meghalaya

Physiographic Position: Hill slopes, 8-15% slope

Rainfall :1145 mm

Land Use: Open scrub

Classification: Fine, mixed, thermic *Typic Kandiudults*

Morphometic Characteristics

Horizon	Depth (cm)	Geology & Parent material	Drainage	Colour	Texture (USDA)	Coarse fragments (>2 mm) %
A1	0-15	Gneiss	Excessively drained	10 YR 4/3 M	cl	5
BW	15-29			10YR 4/4 M	cl	5
Bt1	29-52			5 YR 4/6 M	c	-
Bt2	52-100			5 YR 4/6 M	c	-
Bt3	100-135			2.5YR 4/6 M	c	15
Bt4	135-175			2.5YR 4/6 M	c	30

Physico -chemical Characteristics

Depth(cm)	pH		OC(%)	Clay (%)	CECcmol (p+)kg^{-1}	B.S.(%)
	KCl (1:1)	Water (1:2.5)				
0-15	-	4.8	2.46	39.1	8.7	40
15-29	-	5.1	1.36	39.1	8.0	29
29-52	-	5.3	0.56	49.3	7.2	29
52-100	-	5.5	0.30	54.9	6.8	29
100-135	-	5.4	0.18	55.6	6.8	29
135-175	-	5.3	0.12	50.5	6.8	28

Source: Singh *et.al.*, 2005

Soils of Meghalaya plateau

Profile Location : Village : Myngkre farm, District : Jaintia hills, Meghalaya

Physiographic Position: Hill slope, 8-15% slope

Rainfall: 5539 mm

Land Use: Mixed forest

Classification: Fine, mixed, hyperthemic *Humic Dystrudepts*

Morphometic Characteristics

Horizon	Depth (cm)	Geology & Parent material	Drainage	Colour	Texture (USDA)	Coarse fragments (>2 mm) %
A1	0-19	Gneiss	Excessively drained	10 YR 3/3 M	cl	60
BW	19-35			7.5YR 4/6 M	sc	60
Bt1	35-75			7.5 YR 4/6 M	c	20
Bt2	75-100			7.5 YR 5/6 M	c	20
Bt3	100-125			7.5YR 5/6 M	cl	10

Physico -chemical Characteristics

Depth(cm)	pH		OC(%)	Clay (%)	CECcmol (p+)kg^{-1}	B.S.(%)
	KCl (1:1)	Water (1:2.5)				
0-19	-	5.0	1.96	35.0	8.15	24
19-35	-	4.9	1.34	37.0	7.52	15
35-75	-	5.0	0.51	40.0	3.20	20
75-100	-	4.5	0.24	40.5	4.70	25
100-125	-	5.3	0.19	38.0	4.70	29

Source: Singh *et.al.*,2005

Acid Soils of Peninsular India

Acid soils are distributed in almost all the states except the western states, where the rainfall is low. Peninsular India, the geology of which is morphologically polygenetic and complex, extends from the southern margin of the great plains up to the coastal areas and plains. The climate is of sub-tropical and tropical monsoon type. The rainfall ranges from 600 to 4000 mm with mean annual temperature ranging between 22 and 30°C. The acid soils of peninsular India are represented by broad groups of laterites and red and yellow soils. Studies on laterite and red soils have been made by various workers (Tamhane and Karale 1967; Raychaudhuri and Govinda Rajan, 1971; Gowaikar 1973; Manickam and Krishnamoorthy 1973; Deshpande 1973, Sehgal *et.al.*, 1998).

Red and Lateritic soils occurring in the peninsular India are distributed in the states of Kerala, Tamil Nadu, Karnataka, Andhra Pradesh, Madhya Pradesh, Chhatisgarh, Maharashtra, Goa,

Odisha, West Bengal and Pondicherry. These are developed on granite, gneiss, ferruginous sandstone and schists of Achaean Period. At places, they are developed on basalt and shales and in alluvium and marine deposits. Soils of the hot Semi-arid region in Eastern Ghats and Deccan Plateau (Agro-Eco Region AER 7) are distributed in different districts of Telangana in Andhra Pradesh, and Eastern Ghats. They occur on moderate to gently sloping lands .These soils are developed from granite-gneiss and calcic gneiss. The soils, in general, are coarse-textured, slightly acid, with low CEC and moderate to high base saturation. They are classified as Udic Rhodustalfs, Typic Haplustepts, Typic Haplustalfs and Kandic Paleustalfs. These soils are dominantly cultivated for rainfed crops such as sorghum, pearl millet, pigeon-pea and groundnut, and under irrigated conditions, for sugarcane and cotton.

Red soils of Hot Semi-arid Eco-region in Eastern Ghat(AER 8) occur on moderate to gently sloping or nearly level lands and are distributed in Eastern Ghats, Tamil Nadu uplands, Karnataka plateau in association with black or grey coloured soils. They are well drained, comparatively coarse textured, slightly acid to neutral, with moderate to high base saturation. They are classified as Udic Rhodustalfs,Typic Kandiustalfs, Kandic Paleustalfs and Typic Haplustepts.

The acid soils of Hot Sub-humid Eco-region in Eastern Plateau and Eastern Ghats (AER 12) are observed in Eastern (Chotanagpur) plateau and Eastern Ghats of Orissa, Jharkhand and West Bengal. These soils occur on gently to very gently sloping landforms and are developed from ferruginous sandstone. They are fine loamy, acid, and have low to moderate cation exchange capacity and moderate base saturation. They are classified as Typic Rhodustalfs, Ultic Paleustalfs, Ultic Haplustalfs, Typic Haplustepts, Lithic Ustorthents and Typic Haplustalfs. The soils are generally under forest vegetation and, at places cultivated to upland rice, maize and finger millet.

Representative soils of Peninsular India are described below:

Soils of Eastern Ghat

Profile Location: Village :Singaraju,Tahsil :R. Udayagiri, District : Gajapati , Odisha

Physiographic Position : Granite and Granite gneiss landscape-hills, 3-8 % slope

Rainfall:1600mm.

Land use: Forest; sal,kendu, simul

Classification : Fine loamy, mixed, iso-hyperthermic *Ultic Haplustalfs*

Morphometic Characteristics

Horizon	Depth (cm)	Geology & Parent material	Drainage	Colour	Texture (USDA)	Coarse fragments (>2 mm) %
A	0-13	Granite gneiss	Well drained	7.5 YR 5/2D	sl	
Bt1	13-28			7.5 YR4/4M	scl	-
Bt2	28-60			7.5 YR4/4M	scl	-
Bt3	60-90			5 YR 4/6M	scl	4
Bt4	90-127			5 YR 4/4M	scl	7

Physico -chemical Characteristics

Depth(cm)	pH		OC(%)	Clay (%)	CECcmol (p+)kg^{-1}	B.S.(%)
	KCl (1:1)	Water (1:2.5)				
0-13	-	5.3	0.36	18.7	9.3	59
13-28	-	5.6	0.28	26.5	10.4	58
28-60	-	5.8	0.17	30.4	11.3	64
60-90	-	5.8	0.20	31.7	10.6	62
90-127	-	5.8	0.12	27.1	9.4	65

Source: Sarkar *et.al.*, 2005

Soils of Garhjats Hills

ProfileLocation : Village :Bhojpur,Tahsil : Kuchinda, District :Sambalpur, Odisha

Physiographic Position : Granite gneiss landscape hillocks, 1-3% slope

Rainfall :1566 mm.

Land Use: Pasture

Classification: Fine loamy, mixed, hyperthermic *Typic Haplustepts.*

Morphometic Characteristics

Horizon	Depth (cm)	Geology & Parent material	Drainage	Colour	Texture (USDA)	Coarse fragments (>2 mm) %
A	0-14	Granite gneiss	Moderately well drained	10 YR 5/6M	scl	-
Bw1	14-37			10 YR 5/6M	scl	-
Bw2	37-64			10 YR 6/8M	scl	-
Bw3	64-97			10 YR 7/8M,	scl	-
Bw4	97-150			10 YR 7/4M,	scl	-

Physico -chemical Characteristics

Depth (cm)	pH		OC(%)	Clay (%)	CECcmol (p+)kg^{-1}	B.S.(%)
	KCl (1:1)	Water (1:2.5)				
0-14	-	4.9	0.54	28.1	6.9	51
14-37	-	5.6	0.50	29.4	7.0	60
37-64	-	6.1	0.38	29.7	6.9	78
64-97	-	5.6	0.27	30.2	7.4	70
97-150	-	5.2	0.19	29.5	7.6	78

Source: Sarkar *et.al.*, 2005

Soils of Dandakaranaya

Profile Location :Village :Kochniamba,Tahsil : Umarkot,District: Nowrangpur, Odisha

Physiographic Position: Plain, 1-3% slope

Rainfall:1350 mm.

Land Use: Forest

Classification: Fine-loamy, mixed, hyperthermic *Typic Haplustalfs*

Morphometic Characteristics

Horizon	Depth (cm)	Geology & Parent material	Drainage	Colour	Texture (USDA)	Coarse fragments (>2 mm) %
A	0-15	Quartzite	Well drained	5 YR 4/6M	ls	-
Bt1	15-38			7.5 YR 4/6M	scl	-
Bt2	38-67			7.5 YR 5/6M	scl	6
Bt3	67-122			10 YR 5/3M	c	-
Bt4	122-150			10 YR 5/2M	c	-

Physico -chemical Characteristics

Depth(cm)	pH		OC(%)	Clay (%)	CECcmol (p+)kg^{-1}	B.S.(%)
	KCl (1:1)	Water (1:2.5)				
0-15	-	4.9	0.21	10.9	4.4	52
15-38	-	5.3	0.19	29.4	8.9	56
38-67	-	5.8	0.15	33.2	13.6	61
67-122	-	6.0	0.14	49.3	19.4	66
122-150	-	6.0	0.11	47.6	19.0	68

Source: Sarkar *et.al.*, 2005

Soils of Eastern Dandakarnya Plateau

Profile Location : Village : Urhje, Tahsil : Mohala District : Rajnandhgaon, Chhattishgarh

Physiographic Position : Undulating to rolling plateau, 3-8% slope

Rainfall :1400 mm

Land Use: Paddy, Wheat,Gram

Classification: Fine, kaolinitic, isohyperthermic *Typic Haplustalfs*

Morphometic Characteristics

Horizon	Depth (cm)	Geology & Parent material	Drainage	Colour	Texture (USDA)	Coarse fragments (>2 mm) %
A1	0-8	Granite-Gneiss	Excessively drained	10 YR 5/3 10 YR 3/3	scl	-
A12	8-20			7.5YR 3/4M	c	-
Bt1	20-37			7.5YR 4/4	c	-
Bt2	37-55			7.5YR 4/4	c	-
Cr	55-100 +	....................Weathered parent material....................				

Physico -chemical Characteristics

Depth(cm)	pH KCl (1:1)	pH Water (1:2.5)	OC(%)	Clay (%)	CECcmol (p+)kg^{-1}	B.S.(%)
0-8	-	5.5	1.23	35.2	10.5	74
8-20	-	5.4	0.96	38.5	13.6	78
20-37	-	5.2	0.78	43.2	14.2	73
37-55	-	5.4	0.62	46.5	14.7	76
55-100 +	....	Weathered parent material	....			

Source: Tamgadge *et.al.*, 2002

Soils of Eastern Dandakarnya Plateau (Sedimentary)

Location : Village : Sanjhari, Dhamtari, District : Raipur, Madhya Pradesh

Physiography Position : Undulating plateau (slightly dissected)

Rainfall : 1200 mm

Land use: Forest,

Classification: Fine-loamy, kaolinitic,iso hyperthermic *Typic Haplustepts*

Morphometic Characteristics

Horizon	Depth (cm)	Geology & Parent material	Drainage	Colour	Texture (USDA)	Coarse fragments (>2 mm) %
A	0-12	Sedimentary	Well drained	5 YR 5/3D 5 YR 3/3 M	scl	-
Bw1	12-35			5 YR 4/4D 5 YR 3/4M	scl	-
Bw2	35-63			2.5 YR 4/4D 2.5 YR 3/4M	scl	-
Cr.	63-70+	....	Weathered parent material	....		

Physico -chemical Characteristics

Depth(cm)	pH KCl (1:1)	pH Water (1:2.5)	OC(%)	Clay (%)	CECcmol (p+)kg^{-1}	B.S.(%)
0-12		5.0	1.48	30.4	7.50	85
12-35		5.2	1.21	30.6	8.50	88
35-63		5.2	1.01	30.6	8.50	88
63-70+	....	Weathered parent material	....			

Source: Tamgadge *et.al.*, 2002

Soils of Eastern Dandakarnya Plateau (Laterite)

Location : Village : Arjal, Tahsil : Mohala, District : Rajnandgaon, Madhya Pradesh

Physiographic Position: Undulating plateau , 3-8% slope

Rainfall : 1400 mm

Land Use: Forest

Classification: Fine-loamy, kaolinitic, isohyperthermic *Typic Haplustalfs*

Morphometic Characteristics

Horizon	Depth (cm)	Geology & Parent material	Drainage	Colour	Texture (USDA)	Coarse fragments (>2 mm) %
A	0-15	Laterite	Well drained,	10 YR 4/4D10	scl	-
Bw	15-38			7.5 YR 3/4 M	scl	-
B21	38-65			7.5 YR 3/4 M	sc	-
22	65-96			7.5 YR 4/4 M		
Cr.	96+	....................Weathered parent material....................				

Physico -chemical Characteristics

Depth(cm)	pH		OC(%)	Clay (%)	CECcmol (p+)kg^{-1}	B.S.(%)
	KCl (1:1)	Water (1:2.5)				
0-15		5.4	0.96	30.6	11.6	64
15-38		5.4	0.80	35.3	13.2	71
38-65		5.5	0.76	38.9	14.1	67
65-96		5.6	0.64	39.5	15.6	71
96+	——Weathered parent material——					

Source: Tamgadge *et.al.*, 2002

Soils of Telangana Plateau

Profile Location: Village: Bhutpur, Taluk: Medak, District: Medak, Andhra Pradesh

Physiography Position: Gently and moderately sloping uplands, 5-15% slope

Rainfall :790-850 mm

Land use: Cultivated to sorghum (jower), pigeon pea (tur), paddy and forest.

Classification : Fine , mixed, semiactive,isohyperthermic *Rhodic Paleustalfs*

Morphometic Characteristics

Horizon	Depth (cm)	Geology & Parent material	Drainage	Colour	Texture (USDA)	Coarse fragments (>2 mm) %
A	0-16	Peninsular	Well drained	7.5 YR 4/4 M	ls	25
Bt1	16-40	gneiss		2.5YR 3/4 M	sc	30
Bt2	40-73			2.5YR 3/4	sc	30
Cr	73-83			Weathered granite gneiss		

Physico -chemical Characteristics

Depth(cm)	pH		OC(%)	Clay (%)	CECcmol (p+)kg^{-1}	B.S.(%)
	KCl (1:1)	Water (1:2.5)				
0-16		4.0	4.8	0.50	9.6	3.1
16-40		3.8	4.9	0.38	38.0	9.7
40-73		4.2	5.3	0.27	42.1	11.2

Source: Reddy *et.al.*, 2005

Soils of Talangana Plateau

Profile Location: Village : Sidhapur, Taluk : Sangareddy, District : Medak, Andhra Pradesh

Physiography Position : Gently sloping uplands, 3-8 % slope

Rainfall :712-940 mm

Land use: Large area under eucalyptus plantation

Classification : Loamy-skeletal,mixed, isohyperthermic *Typic Ustorthents.*

Morphometic Characteristics

Horizon	Depth (cm)	Geology & Parent material	Drainage	Colour	Texture (USDA)	Coarse fragments (>2 mm) %
A	0-11	Peninsular gneiss	Somewhat excessively drained	7.5 YR 4/6 M	s	25
A21	11-32			7.5YR 4/6 M	ls	30
A22	32-71			5YR 4/6 M	ls	30
C	71-80			Weathered granite		

Physico -chemical Characteristics

Depth(cm)	pH		OC(%)	Clay (%)	CECcmol (p+)kg^{-1}	B.S.(%)
	KCl (1:1)	Water (1:2.5)				
0-11	4.4	5.3	0.15	5.5	1.4	50
11-32	4.3	5.2	0.12	11.1	3.6	56
32-71	4.3	5.3	0.09	10.9	3.7	46

Source: Reddy *et.al*, 2005

Soils of Mahanadi Basin

Profile Location :Village :Titilagarh, Tahsil :Titilagarh,District: Balangir, Odisha

Physiographic Position: Foot hills, 3-8% slope

Rainfall :1382 mm.

Land use & vegetation :Forest

Classification :Fine-loamy, mixed, hyperthermic *Ultic Haplustalfs*

Morphometic Characteristics

Horizon	Depth (cm)	Geology & Parent material	Drainage	Colour	Texture (USDA)	Coarse fragments (>2 mm) %
A	0-16	Granite gneiss	Well drained	5YR 3/4M	scl	-
Bt1	16-34			5YR 4/4M	scl	-
Bt2	34-66			5YR 4/4M	scl	-
Bt3	66-114			5YR 4/4M	scl	-
Bt4	114-150			5YR 3/4M	scl	-

Physico -chemical Characteristics

Depth(cm)	pH KCl (1:1)	pH Water (1:2.5)	OC(%)	Clay (%)	CECcmol (p+)kg^{-1}	B.S.(%)
0-16	-	4.3	0.08	23.3	10.5	48
16-34	-	4.6	0.07	29.7	14.2	51
34-66	-	5.1	0.23	33.0	16.2	57
66-114	-	5.2	0.35	32.8	16.2	57
114-150	-	5.2	0.42	32.3	16.8	58

Source: Sarkar *et.al.,* 2005

Soil of Chottanagpur Plateau (Dharwar Landscape)

Profile Location: Village: Rebda, P.S.: Barachati, District : Gaya, Bihar

Physiographic Position : Dharwar landscape, 8-15% slope

Rainfall :1131mm.

Land use: Moderately dense forest

Classification: Loamy-skeletal,mixed, hyperthermic *Lithic Ustorthents*

Morphometic Characteristics

Horizon	Depth (cm)	Geology & Parent material	Drainage	Colour	Texture (USDA)	Coarse fragments (>2 mm) %
A	0-15	Quartzite and schist	Somewhat excessively drained	5YR 4/4 M	scl	38
C	15-33			5YR 4/4 M	sl	72
R	33+	——————-Hard rock——————				

Physico -chemical Characteristics

Depth(cm)	pH KCl (1:1)	pH Water (1:2.5)	OC(%)	Clay (%)	CECcmol (p+)kg^{-1}	B.S.(%)
0-15	-	5.4	0.30	30.3	9.5	60
15-33	-	6.1	0.12	19.1	6.0	68
33+	Hard rock					

Source: Sahoo *et al*., 2002

Soils of Chottanagpur Plateau (Granite gneiss landscape)

Profile Location: Village: Hirhi, Block: District :Lohardaga, Jharkhand

Physiographic Position : Gently sloping upland (Tanr), of plateau region

Rainfall : 1300 mm.

Land use: Millets

Classification : Fine, mixed, hyperthermic *Typic Paleustalfs*

Morphometic Characteristics

Horizon	Depth (cm)	Geology & Parent material	Drainage	Colour	Texture (USDA)	Coarse fragments (>2 mm) %
Ap	0-13	Granite gneiss	Well drained	5YR 6/6 D 5YR 4/4 M	sl	-
Bt1	13-31			5YR 4/4 M	cl	-
Bt2	31-52			5YR 4/6 M	cl	-
Bt3	52-82			2.5YR 3/6 M	c	-
Bt4	82-100			2.5YR 3/6 M	cl	3
Bt5	100-129			2.5YR 4/6 M	cl	10
Bt6	129-160			2.5YR 4/6 M	cl	12

Physico -chemical Characteristics

Depth(cm)	pH		OC(%)	Clay (%)	CECcmol (p+)kg^{-1}	B.S.(%)
	KCl (1:1)	Water (1:2.5)				
0-13	3.7	4.3	0.44	10.8	4.5	53
13-31	3.9	4.9	0.42	30.6	8.9	63
31-52	4.3	5.3	0.40	39.7	11.8	71
52-82	4.7	5.7	0.36	41.5	12.6	76
82-100	4.9	5.9	0.25	39.9	12.2	77
100-129	5.0	6.3	0.22	38.3	11.6	78
129-160	5.2	6.2	0.18	36.7	11.4	78

Source: Sahoo *et.al.*, 2010

Soils of Eastern Baghelkhand Plateau

Location: Village: Bogra, Tahsil: Batoli, District: Ambikapur, Chhattishgarh

Physiographic Position : Isolated hills with pediments, 15-30% slope

Rainfall :1400 mm

Land Use: Forest

Classification: Clayey-skeletal, mixed, hyperthermic *Typic Ustorthents*

Morphometic Characteristics

Horizon	Depth (cm)	Geology & Parent material	Drainage	Colour	Texture (USDA)	Coarse fragments (>2 mm) %
A	0-9	Granite-Gneiss	Excessively drained	10 YR 5/4 M	c	-
AC	9-50			2.5YR 4/6 M	c	-
Cr	50+		Weathered parent material			

Physico -chemical Characteristics

Depth(cm)	pH		OC(%)	Clay (%)	CECcmol (p+)kg^{-1}	B.S.(%)
	KCl (1:1)	Water (1:2.5)				
0-9	-	4.9	0.42	50.4	12.0	78
9-50	-	5.1	0.42	53.2	13.6	91
50+ Weathered parent material						

Source: Tamgadge *et.al.,* 2002

Soils of North Deccan Plateau

Profile Location : Village : Pariwara, Tahsil : Nainpur, District : Mandla, Madhya Pradesh

Physiography Position : Satpura Range (Granite), Hills with pediments, 3-8% slope

Rainfall :1400 mm

Land Use: Forest

Classification :Fine-loamy, kaolinitic, hyperthermic *Typic Ustochrepts.*

Morphometic Characteristics

Horizon	Depth (cm)	Geology & Parent material	Drainage	Colour	Texture (USDA)	Coarse fragments (>2 mm) %
A	0-17	Quartzite gneiss	Well drained	5YR 4/6M	sc	-
B1	17-43			2.5Y 4/6M	sc	-
B2	43-72			2.5YR 3/6 M	scl	-
B22	72-110			2.5YR 3/6 M	scl	-
Cr	110+			Weathered parent material		

Physico -chemical Characteristic

Depth(cm)	pH		OC(%)	Clay (%)	CECcmol (p+)kg^{-1}	B.S.(%)
	KCl (1:1)	Water (1:2.5)				
0-17	-	4.9	0.93	38.6	12.8	88
17-43	-	5.4	0.90	40.6	12.8	88
43-72	-	5.4	0.65	41.6	14.2	94
72-110	-	5.7	0.48	38.6	12.8	88
110+			Weathered parent material			

Source: Tamgadge *et.al.*, 1999

Soils of Deccan Plateau

Profile Location: Village : Bakka, District : Betul, Madhya Pradesh

Physiography Position: Satpura Range (Sedimentary), 3-8% slope

Rainfall :1122 mm

Land Use: Forest

Classification :Fine-loamy, kaolinitic, isohyperthermic *Typic Haplustalfs*

Morphometic Characteristics

Horizon	Depth (cm)	Geology & Parent material	Drainage	Colour	Texture (USDA)	Coarse fragments (>2 mm) %
A	0-8	Sandstone	Moderately well drained	10 YR 5/3 D 10YR 4/3M	sl	-
Bw	8-27			10YR 3/2 M	scl	-
Bt1	27-48			10YR 3/2 M	sc	-
Bt2	48-76			10YR 5/3 M	sc	-
Bt3	76-101			10YR 4/3M	sc	-
C	101-125+			10YR 4/3M	sc	-

Physico -chemical Characteristics

Depth(cm)	pH		OC(%)	Clay (%)	CECcmol (p+)kg^{-1}	B.S.(%)
	KCl (1:1)	Water (1:2.5)				
0-8	-	5.5	0.97	18.1	10.8	88
8-27	-	5.4	0.66	37.1	25.4	93
27-48	-	5.5	0.62	37.6	28.8	90
48-76	-	5.8	0.46	38.4	29.0	89
76-101	-	5.9	0.39	43.1	33.8	89
101-125+	-	6.0	0.31	45.1	33.8	83

Source: Tamgadge *et.al.*, 1999

Soils of Central Highland, Vindhyan Range (Sedimentary)

Profile Location : Village : Amraunya, Tahsil : Bijawar, District: Chhattarpur, Madhya Pradesh

Physiography Position : Plain land with isolated, 3-8% slope

Rainfall :1050 mm

Land Use: Forest

Classification : Fine, kaolinitic,ohyperthermic *Typic Haplustalfs*

Morphometic Characteristics

Horizon	Depth (cm)	Geology & Parent material	Drainage	Colour	Texture (USDA)	Coarse fragments (>2 mm) %
Ap	0-12	Sandstone	Well drained	7.5 YR 6/6D	scl	-
Bw	12-36			7.5 YR 3/2 M	c	-
Bt1	36-65			5 YR 4/4 M	c	-
Bt2	65-90			5YR 4/3M	sc	-
Cr	90+	Weathered parent material				

Physico -chemical Characteristic

Depth(cm)	pH		OC(%)	Clay (%)	CECcmol (p+)kg^{-1}	B.S.(%)
	KCl (1:1)	Water (1:2.5)				
0-12		5.3	0.68	38.3	13.5	84
12-36		5.4	0.63	44.5	18.2	85
36-65		5.6	0.51	47.2	19.6	89
65-90		5.7	0.44	49.8	22.6	86
90+	Weathered parent material					

Source: Tamgadge *et.al.*, 1999

Soils of Eastern Baghelkhand Plateau (Quartize & Schist)

Profile Location: Village: Chansura, Tahsil: Bandhagarh, District: Shahdol, Madhya Pradesh

Physiography Position : Undulating land 360m above MSL,3-8% slope

Rainfall : 1100 mm

Land Use: Forest

Classification : Fine, kaolinitic, hyperthermic *Typic Rhodustalfs*

Morphometic Characteristics

Horizon	Depth (cm)	Geology & Parent material	Drainage	Colour	Texture (USDA)	Coarse fragments (>2 mm) %
A	0-12	Sandstone	Well drained	7.5 YR 7/6 D7.5 YR 4/4 M	c	-
Bw	12-31			5 YR 5/6 M	c	-
Bt1	31-46			5 YR 5/8 M	c	-
Bt2	46-66			2.5 YR 3/6 M	c	-
Bt3	66-96			2.5 YR 3/6 M	c	-
Bt4	96-140+			2.5 YR 3/6 M	c	-

Physico -chemical Characteristics

Depth(cm)	pH		OC(%)	Clay (%)	CECcmol (p+)kg^{-1}	B.S.(%)
	KCl (1:1)	Water (1:2.5)				
0-12	-	5.6	1.44	53.8	28.0	88
12-31	-	5.3	1.28	56.8	29.6	87
31-46	-	5.3	0.78	60.8	30.0	92
46-66	-	5.2	0.62	63.8	34.0	90
66-96	-	5.2	0.62	63.8	34.0	90
96-140+	-	5.2	0.62	63.8	34.0	90

Source: Tamgadge *et.al.*, 1999

Soils of Central Sahyadri

Profile Location: Village: Gavane, Tahsil: Sattari, District : North Goa, Goa

Physiographic Position :undulating lands, 15-30% slope

Rainfall : 3000 mm.

Land Use: Coconut, Banana

Classification: Clayey-skeletal,mixed, *Typic Haplustepts*

Morphometic Characteristics

Horizon	Depth (cm)	Geology & Parent material	Drainage	Colour	Texture (USDA)	Coarse fragments (>2 mm) %
Ap	0-12	Modeartely deep	Somewhat excessively drained	7.5 YR 3/3 M	c	35
Bw1	12-34			5YR 3/4 M	c	45
Bw2	34-55			5 YR3/4 M	c	45
Bw3	55-85			10 YR 3/4M	c	35
Cr	85+			-	-	-

Physico -chemical Characteristics

Depth(cm)	pH		OC(%)	Clay (%)	CECcmol (p+)kg^{-1}	B.S.(%)
	KCl (1:1)	Water (1:2.5)				
0-12	4.5	5.3	2.10	40.0	14.0	50
12-34	4.4	5.4	1.99	42.5	14.6	49
34-55	4.5	5.6	1.80	47.5	14.7	50
55-85	4.7	5.8	1.76	46.4	14.3	61
85+						

Source: Harindranath *et.al.*, 1999

Soils of Western Hills

Profile Location : Village : Thevalapuram, Taluk : Kottarakkara, District : Kollam, Kerala

Physiography Position : Mounds,uplands and hills, 3-33 % slope

Rainfall :1200 mm

Land Use: Principal use is for growing rubber

Classification : Clayey-skeletal,kaolinitic, isohyperthermic *Ustic Kanhaplohumults*

Morphometic Characteristics

Horizon	Depth (cm)	Geology & Parent material	Drainage	Colour	Texture (USDA)	Coarse fragments (>2 mm) %
Ap	0-16	Khondalite	Well drained	5 YR 3/4M	scl	60
AB	16-38			5YR 4/4M	scl	60
Bt1	38-72			5YR 4/6 M	c	60
Bt2	72-103			2.5 YR 4/6M	c	60
Bt3	103-130			2.5 YR 4/6M	c	60
R	130	Hard rock				

Physico -chemical Characteristics

Depth(cm)	pH		OC(%)	Clay (%)	CECcmol (p+)kg^{-1}	B.S.(%)
	KCl (1:1)	Water (1:2.5)				
0-16		4.2	4.9	1.72	24.9	3.3
16-38		4.3	4.6	1.28	29.3	2.9
38-72		4.2	4.7	0.79	48.1	3.1
72-103		4.1	4.9	0.46	48.9	2.7
103-130		4.1	4.6	0.49	50.5	2.8

Source: NBSS & LUP Staff, 2006

Soils of Western Hills

Profile Location : Village : Kanjirapally, Taluk : Kanjirapally, District : Kottayam, Kerala

Physiography Position :Uplands, Mounds and hills, 3-33 % slope

Rainfall :3500 mm

Land Use : Principal use is for growing rubber

Classification : Clayey-skeletal,kaolinitic, isohyperthermic *Ustic Kandihumults.*

Morphometic Characteristics

Horizon	Depth (cm)	Geology & Parent material	Drainage	Colour	Texture (USDA)	Coarse fragments (>2 mm) %
Ap	0-13	Charnockite	Well drained	5 YR 3/3	gsc	30
Bt1	13-32			5YR 3/4	gsc	35
Bt2	32-56			2.5YR 3/6	gsc	40
Bt3	56-83			2.5 YR 3/6	gsc	40
Bt4	83-112			2.5 YR 4/6	gsc	35
Bt5	112-150			2.5 YR 4/6	gsc	40

Physico -chemical Characteristics

Depth(cm)	pH		OC(%)	Clay (%)	CECcmol (p+)kg^{-1}	B.S.(%)
	KCl (1:1)	Water (1:2.5)				
0-13	4.3	4.4	2.24	33.0	1.9	17
13-32	4.3	4.4	1.63	39.5	1.2	8
32-56	4.3	4.5	1.17	38.0	0.9	7
56-83	4.4	4.5	0.95	40.5	0.3	6
83-112	4.4	4.4	0.83	36.0	0.4	8
112-150	4.4	4.4	0.06	34.5	0.6	13

Source: NBSS & LUP Staff, 2006

Acid Soils of Eastern Plains

In the eastern plains, acid soils are mostly confined to Assam Valley although they occur to some extent in eastern Bihar,West Bengal and Tripura in the lower Gangetic plains.The Assam Valley is built mostly by the aggradation work of the Brahmaputra and its tributaries. The general level of the valley ranges from 130 m in the east to 30 m in the west. Geologically, the valley is built by the deposition of alluvium upon the sag formed during the rise of the Himalayas (Singh, 1971). Due to peculiar topography and land elevation accompanied by high precipitation, most of the soils of Assam are acidic (Panda, 2003). The climate is humid sub-tropical, rainfall varying from 200 to 350 cm. The high rainfall results in increasing organic matter, Fe and Al oxides, depth of clay enrichment and $CaCO_3$ accumulation,decrease in soil pH and silicon-sesquioxide molar ratio (Chakravarty, 1977). The increase in sesquioxide controls the behavior of ions in solution and increases the acidity of soils. Under subtropical climate, rapid decomposition of organic matter also contributes towards soil acidity development.

The vegetation in the valley is pre-dominantly dense mixed semi-evergreen, evergreen and wet deciduous types. The red soils of Hot Sub-humid (Moist) to Humid (with Inclusion of Per-humid) Eco-region (AER 15) occur on level to very gently sloping landforms in Assam, Tripura and West Bengal plains. They are classified as Ustic Paleustalfs, Typic Udipsamments, Aeric Endoaquepts and Aquic Udifluvents. These soils are cultivated to rice and jute, and plantation crops like tea, and horticultural crops like pineapple, citrus and banana. These soils are suitable for tea, citrus and banana, and moderately suitable for maize.

Representative acid soil profiles of Eastern plain are presented below:

Soil of Brahmaputra Valley

Profile Location :Village :Chengalimari, District : Sonitpur, Assam

Physiographic Position: Gently sloping recent flood plain, 1-3% slope

Rainfall :1750 mm

Land Use: Rice, Potato

Classification: Mixed, hyperthermic *Typic Udipsamments*

Morphometic Characteristics

Horizon	Depth (cm)	Geology & Parent material	Drainage	Colour	Texture (USDA)	Coarse fragments (>2 mm)
0-15		Alluvium	Well drained	10 YR 2/2M	sl	-
C1	15-33			10YR 2/2 M	sl	-
C2	33-45			10 YR 3/3M	ls	-
C3	45-120			10YR 7/3 M	s	-

Physico -chemical Characteristics

Depth(cm)	pH		OC(%)	Clay (%)	CECcmol (p+)kg^{-1}	B.S.(%)
	KCl (1:1)	Water (1:2.5)				
0-15	-	4.6	3.76	10.0	9.2	39.0
15-33	-	4.8	2.37	9.0	7.59	40.5
33-45	-	5.3	0.06	8.0	2.99	66.9
45-120	-	5.9	0.30	5.0	1.38	77.1

Source: Vadivelu *et.al.* 2004

Soil of Brahmaputra Valley

Profile Location: Village :*Dharamtul, District : Morigaon, Assam*

Physiographic Position: Nearly level uplands, 0-1% slope

Rainfall :1860 mm

Land Use: Paddy

Classification: Fine, mixed, hyperthermic *Aeric Endoaquepts*

Morphometic Characteristics

Horizon	Depth (cm)	Geology & Parent material	Drainage	Colour	Texture (USDA)	Coarse fragments (>2 mm) %
Ap	0-10	Alluvium	Imperfect to poorly drained	10 YR 4/1 M	c	-
Bw1	10-28			10YR 4/2 M	c	-
Bw2	28-40			10 YR 4/2 M	c	-
Bw3	40-80			10YR 5/3 M	c	-
Bw4	80-160			10YR 5/6 M	c	-

Physico -chemical Characteristics

Depth(cm)	pH		OC(%)	Clay (%)	CECcmol (p+)kg^{-1}	B.S.(%)
	KCl (1:1)	Water (1:2.5)				
0-10	-	4.5	1.6	73.0	18.6	38
10-28	-	4.4	1.0	68.5	14.0	33
28-40	-	4.5	0.6	60.0	8.2	44
40-80	-	4.0	0.2	61.0	7.8	28
80-160	-	4.7	0.1	62.5	8.0	45

Source: Vadivelu *et.al.* 2004

Soils of Indo-Gangetic Alluvial Plain

Profile Location: Village: Daraboyjot , P.S. : Phansidewa, District: Darjiling, West Bengal

Physiographic Position: Lower piedmont plain, 0-1% slope

Rainfall :2193 mm

Land use: Paddy, Bamboo and Grass

Classification: Coarse –loamy,mixed,hyperthermic *Aquic Udifluvents.*

Morphometic Characteristics

Horizon	Depth (cm)	Geology & Parent material	Drainage	Colour	Texture (USDA)	Coarse fragments (>2 mm) %
Ap	0-25	Alluvium	Imperfectly drained	10YR4/2M	sl	-
C1	25-55			10YR4/1M	ls	-
C2	55-89			10YR4/1M	sl	-
C3	89-113			10YR 4/1M	sl	-
2C4	113-135		———Sand layer———			

Physico -chemical Characteristics

Depth(cm)	pH		OC(%)	Clay (%)	CECcmol (p+)kg^{-1}	B.S.(%)
	KCl (1:1)	Water (1:2.5)				
AP	-	4.5	0.33	10.9	2.4	47
C1	-	4.8	0.26	8.9	4.0	50
C2	-	5.0	0.24	10.9	4.0	51
C3	-	5.2	0.30	12.9	5.0	54
2C4	-	5.9	0.11	6.8	4.8	62

(Nayak *et.al.*, 2001)

Acid Soils of Coastal Plain and Island Ecosystem

The soils of Hot Sub-humid to Semiarid Ecosystem in Eastern coastal Plain (AER 18) are distributed in Eastern coastal plains of Andhra Pradesh, Odisha, Pondicherry, Tamil Nadu and West Bengal and are developed on sandstone. These soils are classified as Rhodic Kandiustalfs, Acric Endoaqualfs, Typic Haplustalfs and are cultivated mainly to rice and coconut. In low lying areas, fish and prawn culture is also a practice. These soils are moderately suitable for coconut and rice. Productivity is low to medium with the limitations of low PAWC, moisture stress, nutrient imbalance and surface crust formation.

The acid soils (red and lateritic) of Hot Humid-per-humid Eco-region in Coastal Plain (AER 19) are distributed in coastal lands of Gujarat, Karnataka, Kerala, Maharashtra, Tamil Nadu and Union Territory Goa, Daman and Diu. These soils are developed on laterite and are well drained, moderately deep to deep acid soils with moderate base saturation .These are classified as Ustic Haplohumults, Ustic Palehumults, Ustic Kandihumults, Typic Haplustepts, Lithic Ustothents and Typic Haplorthox. These soils have high potentials for the cultivation of export-oriented plantation crops like pepper, rubber, mango, coffee, tea, cocoa, pineapple, cashewnut and arecanut and rice. These soils are prone to severe soil erosion with problems of depletion of nutrients, due to intense leaching, resulting in nutrient imbalance and iron and aluminium toxicity due to low soil pH.

The acid soils (red) of Hot Humid to Perhumid Island Ecoregion in Islands of Andaman and Nicobar (AER 20) are developed on sandstone, shales and marine deposits. These are poor to moderately drained, and at places well drained, loamy, slightly acid, with low cation exchange capacity, and are classified as Tropofluvents and Typic Haplustalfs. These soils are cultivated to coconut, arecanut, tapioca, pepper and oil palm.They are suitable for coconut and moderately suitable for oil palm and tapioca. Soil erosion and inundation of coastal areas are the major constraints.

Acid sulphate soils of coastal plain occur mostly in Kerala, in the west coast, deltaic areas of Sunderbans in West Bengal and some pockets of Tamil Nadu. In soils, under tidal influence, the reaction of sulfuric acid with iron bearing minerals yield basic iron sulphate called as jarosite [$KFe_3 (SO_4)_2 (OH)_6$] which is conspicuous in most acid sulphate soils. Physiographically, the region having acid soils in the coasts and Sunderbans is characterized by the sandy beach, coastal sand-dunes or mud flats, alluvial tracts along rivers, deltas, lagoons and estuary. These areas are subject to periodical tidal action. The climate in coastal areas is humid tropical with annual precipitation of more than 3000mm. Sunderban area is sub-tropical humid with rainfall of 1500 to 2000mm.The natural vegetation consists of casuarina and mangrove swamps. The coastal soils are the members of Fluvents, Sulfaquepts and their intergrades.

Some representative pedons are described below:

Soils of Utkal Plain

Profile Location :Krishnachandrapur,Tahsil : Baripada, District: Mayurbhanj, Odisha

Physiographic Position :Plain, 3-8% slope

Rainfall :1600 mm.

Land Use: Paddy

Classification: Fine, mixed, iso-hyperthermic *Acric Endoaqualfs.*

Morphometic Characteristics

Horizon	Depth (cm)	Geology & Parent material	Drainage	Colour	Texture (USDA)	Coarse fragments (>2 mm) %
Ap	0-14	Granite	Poorly	10 YR 6/3M	l	-
Btg1	14-41	gneiss	drained	10 YR 5/1M	c	-
Btg2	41-67			10 YR 7/2M	c	-
Btg3	67-96			10 YR 7/2M	c	-
Btg4	96-127			10 YR 7/2M	cl	5
Btg5	127-150			10 YR 6/1M	cl	15

Physico -chemical Characteristics

Depth(cm)	pH		OC(%)	Clay (%)	CECcmol (p+)kg^{-1}	B.S.(%)
	KCl (1:1)	Water (1:2.5)				
0-14	-	4.5	0.78	25.0	10.6	48
14-41	-	4.9	0.20	40.4	14.1	51
41-67	-	5.1	0.19	47.6	17.7	53
67-96	-	5.5	0.15	48.8	19.4	58
96-127	-	5.6	0.12	39.3	19.9	67
127-150	-	5.8	0.11	37.0	19.9	62

(Sarkar *et al.*, 2005)

Soils of Coastal Plain

Saline acid sulphate soils

Profile Location: Village: Arikulam, District : Kozhikode State : Kerala

Physiographic Position: Flat low land (0-1% slope)

Rainfall : 3600 mm.

Land Use: Marshy land

Classification : Fine, mixed, iso-hyperthermic *Typic Sulfaquents*

Morphometic Characteristics

Horizon	Depth (cm)	Geology & Parent material	Drainage	Colour	Texture (USDA)	Coarse fragments (>2 mm) %
A1	0-9	Alluvium (Laterite out wash)	Poor	10YR 4/4M	c	-
C1	9-20			10YR 5/4M	c	-
C2	20-36			10YR 4/4M	c	-
C3	36-70			10 YR 5/1M	c	-
C4	70+			10 YR 2/2M	c	-

Physico -chemical Characteristics

Depth(cm)	pH		ECe (ds/m)	OC (%)	Clay (%)	CECcmol (p+)kg^{-1}	B.S. (%)
	KCl (1:1)	Water (1:2.5)					
0-9	-	4.4	43.6	2.32	50.0	18.7	68
9-20	-	4.8	12.6	2.09	40.2	23.4	55
20-36	-	4.8	11.4	1.40	47.1	24.4	49
36-70	-	4.5	8.4	1.51	65.0	26.1	69
70+	-	3.4	38.6	23.54	79.5	69.2	53

(Bhargava & Abrol, 1984)

Soils of Coastal Plain

Saline acid sulphate soils

Profile Location: Village: Ambalpusha, District : Alappuzha, State : Kerala

Physiographic Position: Depressed low lying land

Rainfall:2600 mm.

Land Use: Rice

Classification: Fine loamy, mixed, isohyperthermic, *Typic Sulfaquepts*

Morphometic Characteristics

Horizon	Depth (cm)	Geology & Parent material	Drainage	Colour	Texture (USDA)	Coarse fragments (>2 mm) %
Ap	0-14	Alluvium (Recent out wash)	Poor	10YR 4/4M	cl	-
Bwg1	14-32			10YR 5/1M	cl	-
Cg1	32-46			7.5YR 6/0M	sil	-
Cg2	46-100			7.5 YR 7/0M	sl	-

Physico -chemical Characteristics

Depth(cm)	pH		OC(%)	Clay (%)	CECcmol (p+)kg^{-1}	B.S.(%)
	KCl (1:1)	Water (1:2.5)				
0-14	4.1	2.9	1.80	28.0	19.2	60
14-32	4.2	2.0	0.87	29.0	20.8	63
32-46	4.3	2.1	0.52	19.4	19.6	72
46-100	4.5	1.5	0.29	17.6	10.6	70

(Bhargava & Abrol, 1984)

Soils of Coastal Plain of Sundarbans

Profile Location : Village : Simultala, Block: Basanti , District : South 24 Parganas

Physiographic Position : Very gently sloping land

Rainfall : 1500 mm.

Land use: Paddy

Classification: Fine, mixed,acidic, hyperthermic *Sulfic Endoaquepts*

Morphometic Characteristics

Horizon	Depth (cm)	Geology & Parent material	Drainage	Colour	Texture (USDA)	Coarse fragments (>2 mm) %
Ap	0-25	Alluvium	Poor	5Y 5/1	sic	-
B11	25-60			5Y 5/1 2.5 YR 3/4	sic	-
B12	60-130			2.5Y 3/0 2.5 Y 8/6	sic	-
C	130-180			2.5Y 4/0	sic	-

Physico -chemical Characteristics

Depth(cm)	pH		OC(%)	Clay (%)	CECcmol (p+)kg^{-1}	B.S.(%)
	KCl (1:1)	Water (1:2.5)				
0-25	4.2	4.9	0.86	-	29.0	-
25-60	3.2	3.7	0.71	-	30.4	-
60-130	3.1	3.5	0.93	-	22.3	-
130-180	5.0	5.4	0.32	-	23.4	-

Bandyopadhyay and Maji, 1995

Soils of Coastal Plain of Sundarbans

Profile Location : Village : Canning, Block: Canning , District : South 24 Parganas

Physiographic Position : Very gently sloping land

Rainfall : 1500 mm.

Land use: Paddy

Classification : Fine, mixed, hyperthermic *Typic Endoaquepts*

Horizon	Depth (cm)	Geology & Parent material	Drainage	Colour	Texture (USDA)	Coarse fragments (>2 mm) %
Ap	0-30	Alluvium	Poor	2.5 Y 6/22.5 Y 7/6	sic	-
B1	30-85			2.5 Y 6/22.5 Y 7/6	sic	-
B21	85-140			2.5Y 6/4 5 YR 5/6	sic	-
B22	140-210			5YR 5/210 YR 5/6	sic	-

Physico -chemical Characteristics

Depth(cm)	pH		OC(%)	Clay (%)	CECcmol (p+)kg^{-1}	B.S.(%)
	KCl (1:1)	Water (1:2.5)				
0-30	6.0	6.5	0.54	-	17.2	-
30-85	6.6	7.0	0.21	-	18.5	-
85-140	6.3	6.8	0.27	-	17.7	-
140-210	5.5	5.9	0.80	-	22.4	-

Bandyopadhyay and Maji, 1995

Soils of Konkan Coast

Profile Location :Village :Harchil,Tahsil : Ratnagiri, District : Ratnagiri, Maharastra

Physiographic Position : Undulation lands with narrow valleys.

Rainfall :2617 mm

Land Use: Nachni&Karonda

Classification :Loamy, mixed, iso-hyperthermic *Lithic Ustorthents*

Morphometic Characteristics

Horizon	Depth (cm)	Geology & Parent material	Drainage	Colour	Texture (USDA)	Coarse fragments (>2 mm) %
Ap	0-18	Laterite	Well drained	10 YR 4/4 10 YR 3/4	cl	10.0
Cr.	18-35 +	Weathered laterite				

Physico -chemical Characteristics

Depth(cm)	pH		OC(%)	Clay (%)	CECcmol (p+)kg^{-1}	B.S.(%)
	KCl (1:1)	Water (1:2.5)				
0-18	4.7	5.2	2.99	34.4	16.2	91.2
18-35 +	Weathered laterite					

(Challa *et.al.*, 1999)

Soils of Konkan Coast

Location :Village :Janvali, Tahsil : Kankauli, District : Sindhudurg, Maharastra

Physiography : Undulating lands with narrow valleys, 8-15% slope

Rainfall :4099 mm

Land Use: **Paddy,** Nachni /forest (Bamboo, Teak)

Classification: Fine loamy, mixed, iso-hyperthermic, *Typic Haplustepts*

Morphometic Characteristics

Horizon	Depth (cm)	Geology & Parent material	Drainage	Colour	Texture (USDA)	Coarse fragments (>2 mm) %
A	0-15	lateritic	Well drained	5 YR 3/4M	cl	10
B1	15-32			2.5YR 3/4M	cl	20
B2	32-50			2.5YR 3/6M	scl	20
C	50+	Pebbles, cobbles and quartzite				

Physico -chemical Characteristics

Depth(cm)	pH		OC(%)	Clay (%)	CECcmol (p+)kg^{-1}	B.S.(%)
	KCl (1:1)	Water (1:2.5)				
0-15	4.9	5.3	1.99	38.4	15.0	39.0
15-32	4.9	5.3	1.21	37.4	14.6	59.2
32-50	5.0	5.3	0.76	39.0	13.4	71.9
50+	Pebbles,cobbles and quartzite					

(Challa *et.al.*, 1999)

Soils of Konkan Coast

Profile Location : Village : Merces, Tahsil : Tiswadi, District : North Goa, Goa

Physiographic Position : Undulating lands , 3-8 % slope

Rainfall : 3200 mm

Land Use: Coconut, Ficus spp, mango

Classification: Fine- loamy,mixed, *Typic Ustropepts*

Morphometic Characteristics

Horizon	Depth (cm)	Geology & Parent material	Drainage	Colour	Texture (USDA)	Coarse fragments (>2 mm) %
Ap	0-15		Well drained	7.5YR 4/4D 7.5 YR 3/4 M	sl	0.0
2A2	15-40			7.5YR4/4D 7.5YR 3/4M	s	0.0
2Bw1	40-74			10 YR4/4 M	scl	0.0
2Bw2	74-108			10 YR 5/4M	scl	0.0
2BC	108-145			10YR 5/3 M	sl	0.0

Physico -chemical Characteristics

Depth(cm)	pH		OC(%)	Clay (%)	CECcmol (p+)kg^{-1}	B.S.(%)
	KCl (1:1)	Water (1:2.5)				
0-15	4.1	4.5	0.77	16.3	8.1	71
15-40	4.6	4.8	0.28	7.5	3.9	27
40-74	4.2	5.0	0.27	26.2	2.8	44
74-108	4.1	5.2	0.26	27.2	4.2	55
108-145	4.2	5.9	0.20	19.0	6.9	84

(Harindranath *et.al.*, 1999)

Soil Acidity as Diagnostic Criterion in Soil Taxonomy

In the 7th Approximation soil classification system, base saturation percentage is used as a definitive criterion in separating Alfisols (Gray-Brown Podjolics) from Ultisol (Red- Yellow Podjolics) and Umbric epipedons (acidic dark coloured surface horizons) from Mollic epipedons (other dark colour surface horizons).

Percentage base saturation based on CEC determined as sum of basic cations plus $BaCl_2$- TEA acidity has been used as the criterion of Ultisol order and that based on NH_4OAc-CEC has been used as the criterion for classification at the great group level of Inceptisols and Oxisols (USDA, 2006) (Table 3). A positive correlation between soil pH and base saturation percentage was reported in the classification system of Indian Soils.

The soil classification in the subgroup level of some soil orders is also based on the base saturation percentage. The name of the suborders with the prefix Umbric, Ultic, Dystric, etc. is mostly based on base saturation percentage in the soils.

The relation between soil pH and soil classification in subgroup level with the prefix Sulfic is mostly based on soil pH values viz. Sulfic Cryaquepts, Sulfic Endoaquepts, Sulfic Tropaquepts etc.

Management in relation to physical properties

Problems of crop production arising from undesirable physical properties of acid soils of India include (Gupta and Nagarajarao, 1982) viz. (i) high percolation loss of water and nutrients in light textured acid red and laterite soils, (ii) crust formation in red soils having more than 20% clay adversely affecting seedling emergence, (iii) quick drying of surface layer in shallow soils with *morrum* at shallow depth, (iv) hardening of soils and presence of compact

Table 3. Percentage base saturation as diagnostic criterion in soil taxonomy

Soil classification		Diagnostic criterion: Base saturation (%)	Depth
Order	Great Group		
Inceptisols	Dystrocryepts	<50	In some part of 25 to 75 cm
	Dystrustepts	<60	In all horizon within 25 to 75 cm
	Dystroxerepts	<60	In all horizon within 25 to 75 cm
	Eutrudepts	60 or more	In some part of 25 to 75 cm
	Eutrogelepts	60 or more	In some part of 25 to 75 cm
Oxisols	Eutraquox	35 or more	In all horizon within 125 cm
	Eutroperox	35 or more	In all horizon within 125 cm
	Eutrustox	35 or more	In all horizon within 125 cm
	Eutrotorrox	35 or more	In all horizon within 125 cm
	Eutrudox	35 or more	In all horizon within 125 cm
Ultisols	All Great Groups	<35	180 cm below surface or to 125 cm below upper boundary of argillic horizon or immediately above lithic contact or paralithic contact.

layers and clay pans at shallow depth as in Alfisols and Entisols having bulk density more than 1.7 g crn^{-3} and thus offering high mechanical impedance to root growth, (v) unsuitability of structureles puddled soils for non-rice *Rabi* crops (vi) erodibility in red soil areas having compact subsoil, (vii) drainage problems in acid sulphate marshy soils developed in lagoons and also in alluvial soil of deltaic and coastal regions, (viii) the requirement of high frequent irrigation in light textured porous acid laterite and old alluvial soils and (ix) crop water stress due to low water retentivity and poor root growth in strongly acid soils

Compaction of medium and light textured soils by heavy rollers or loaded tractors or puddling of soils with finer texture brings down the percolation loss of water and nutrients and thereby increase the rice yield in acid soils (Mandal, 1984). In crusting soils, application of straw mulch on seed lines increases the emergence of cotton, ragi, soybean, cowpea and horsegram due to increase in moisture retention and reduction of soil strength and hence reduces rain drop impact force. (Gupta and Nagarayarao, 1982).

For rice culture, Lal (1985) suggested minimum tillage for soils with high activity clays and low permeability, puddling for moderately permeable soils with medium activity clays and soil compaction for highly permeable soils with low clay activity.

Rain water management is an important step in overcoming drought in red and laterite soil areas. Under unirrigated conditions, profile (lateritic sandy loam acidic soil, Kharagpur) moisture use by gram, mustard and safflower crops was higher due to conventional than strip-row (minimum) tillage (Mittra *et al.*, 1992). However, the trend was reverse in respect of yield and water use efficiency.

Management in relation to chemical properties

Use of soil amendment for correction of soil acidity: Liming is practiced to neutralize the low magnitude active acidity due to hydrogen and aluminum ions in the soil solution and part of the exchange acidity. The effect of lime on change in pH and CEC of low pH acid Alfisol in Odisha is shown in table 4.

Table 4: Effect of lime on change in soil chemical properties of acid Alfisol, Odisha

Location	Initial pH	Initial CECcmol(p)kg^{-1}	Lime*	
			pH	CECcmol(p)kg^{-1}
Jashipur	4.6	7.8	6.3	8.2
Semkunta	5.0	9.0	6.2	9.2
Bhubaneswar	4.4	2.8	6.0	3.4

Source: Sahoo and Patnaik (1990)
*lime @0.5 LR.

The effect of liming and fertilization on crop yields (q ha^{-1}) in some acid soils of India is shown in Table 5.

Table 5: Effect of liming and fertilization on crop yields (q ha^{-1}) in some low pH soils of India

Acid soil region/State	Crop	FP*	FP+Lime	100% NPK	100%+Lime NPK	Per cent response to liming
Jorhat, Assam (pH 4.9-5.2)	Rapeseed	7.3	8.6 (17.8)**	10.5 (43.8)	12.9 (76.7)	23
Ranchi, Jharkhand (pH 4.5-5.2)	Pea	28.6	32.5 (13.8)	42.6 (49.3)	51.2 (79.4)	27
Vellayani, Kerala (pH 4.8-5.2)	Blackgram	3.5	44 (25.6)	4.0 (14.8)	5.6 (58.2)	38
Bhubaneswar,) Odisha (pH 4.0-4.5)	Groundnut	8.6	12.5 (45.3)	14.3 (66.3)	17.9 (108.1)	23
Kalyani, West Bengal (pH 4.5-5.0)	Mustard	4.8	6.5 (35.4)	7.0 (45.8)	9.1 (89.6)	65

Source: Sharma and Sarkar (2005)
*FP – Farmers practice
**Figures in paranthesis are percent increases over FP

- *Controlling Fe-toxicity in rice.* Rice grown in medium and low land situations in acid lateritic soils suffer from iron toxicity owing to lateral flow of soluble ferrous iron from the adjacent uplands. Tall *Indica rice* varieties have some degree of tolerance to high concentration of Fe^{++}. The problem is further aggravated when nutrient imbalances are created owing to deficiencies of K,P,S and Mn. Applications of lime sludge at 0.25 LR dose every alternate year and of 100-60-60 NPK (+S) were quite effective is raising the yield to normal level (Anonymous, 1985). Application of potash @120-160kg K_2O/ha was also effective in reducing the toxicity both in susceptible and tolerant varieties. Application of high dose or K fertilizer perhaps help in creating an oxidizing zone around rice rhizosphere grown under Fe toxic condition, prevents reduction of Fe3+ - Fe2+ and restricts Fe uptake by rice, thereby resulting in an improved yield (Mitra and Sahoo, 1992).

- *Response to sulphur and micronutrients.* Application of sulphur and micronutrients proves beneficial in raising productivity of acid soil region. Groundnut is grown extensively in *rabi* season in light to medium textured soils are not only deficient in calcium but also sulphur because of their low organic matter content. Application of 30 kg S as gypsum raised the yield, shelling percentage and oil content of kernel both in unlimed an limed lateritic soil.

Management of Acid Sulphate Soils

Acid sulfate soils are among the major problem soils, that require appropriate methods of amelioration. Since, the soils are situated in lowland and suitable for rice culture, amelioration measures for wetland rice include:

Leaching and Drainage: The accumulation of soluble salts and aluminum, when acid sulfate soils are submerged can sometimes reach toxic levels. Removal of these substances is therefore necessary and can be accomplished by leaching and drainage. In practice, however, leaching is not normally possible due to the impermeability of the soil and the low elevation of the land. The acid sulphate soils are saline and hence removal of soluble salts by suitable drainage is also needed for successful rice production.

Submergence: Although, the sulphate concentration in the acid sulphate soils is high, chances of sulphate toxicity is minimum, since there is no marked soil reduction due to flooding and submergence even in the presence of sustainable amount of organic matter. This is more so, because of the presence of large amount of Fe, which would be instrumental in activating the sulphide.

Liming: Several researchers have reported liming to be the simplest way of increasing the pH of the soil. The combination of leaching and liming however had the greatest effect in ameliorating such soils.

Peat and Marshy Acid Sulphate Soils

Acid peaty and marshy soils occur in Assam, Kerala, coastal Odisha, south east coast of Tamil Nadu and tarai region of U.P. and Bihar.

While smectite is the dominant clay mineral in peaty soils, kaotinite associated with smectite dominate the clay fraction of acid sulphate soils of Kerala. The acid sulphate soils are characterized by deep-black colour, heavy texture and high acidity. These soils are extremely acidic, pH of the soil, when dry, going below 4.0. These soils are put to paddy cultivation by providing mud-embankments round the fields (Subramoney, 1971).

Management of Acid Sulphate Soils of Sunderbans

An extensive area of potential or active acid sulphate soils with strong acidity (pH 3.5) has been identified in the coastal saline tract of West Bengal, where crop productivity is limited owing to P deficiency and Al toxicity associated with strong acidity. Information on low cost management practices to improve the productivity of such soils are very meagre. Results of field experiment on such soil (pH 3.5) with rice 'MW- 10' (end of November to mid June) as first crop followed by rice 'IR 30' (June to December) as the second crop showed that Mussoorie rock phosphate applied 20 days before seed sowing and mixed thoroughly with soil by ploughing proved superior to single superphosphate at all levels of P (40, 80 and 120 kg P/ha) both in direct and residual effect (Laskar and De, 1989). Field experiment conducted on a soil (pH 3.3) in coastal area (Nirdeshkali) of West Bengal showed that grain yield of rice 'MW 10' was highest when lime (9.11 tonnes $CaCO_3$/ha) corresponding to 10 times of exchangeable Al was applied (Laskar, 1990). The results, further indicated that the yield was maximum with lime equivalent equal to or less than the amount theoretically needed to neutralise Fe and Al. Exchangeable Al serves as a good estimator of lime required for economic productivity of rice.

Conclusions

The acid soils occur in high rainfall zones comprising of hilly/ mountains and coastal regions. The highly leached soils are generally poor in fertility and water holding capacity. Highly acidic soils are restricted to the Himalayan ecosystem, red and lateritic regions of India comprising both the southern and eastern plateau region and some pockets in the greater plains of the country. The eastern Himalayan ecosystem comprising of North-Eastern region and a part of sub-Himalayan West Bengal and Jharkhand is having more extensive area under acid soils as compared to those of

Western Himalayan region covering Himachal Pradesh, Uttarakhand and Jammu & Kashmir. The Peninsular India and some areas under coastal ecosystem show formation of acid soils particularly in coastal Tamil Nadu, Kerala, Karnataka and Goa.

The acid soils with pH values less than 4.5 covers only 1.9 percent of the total geographical area (TGA) of the country while moderately acid soils having pH in the range of 4.5 – 5.5 covers 7.4 percent of the TGA. However, area with pH value less than 5.5 are problematic and are critically degraded soils and exhibit deficiencies of phosphorus, calcium, magnesium and molybdenum; toxicities of iron, aluminium and manganese; poor microbial activity leading to low nitrogen and sulphur availability thereby threatening the sustainability of the production system.

The genesis of acid soils has been attributed to several pedogenic and climatological factors. Variation in geological formation and nature of parent materials, topographical features, soil hydrological conditions, rainfall and temperature are the main causative factors for formation of the acid soils.

The soils are classified into the orders viz. Ultisol, Alfisol, Mollisol, Inceptisol and Entisol. Among the soil Greatgroups, Dystrudepts, Hapludolls, Haplumbrepts, Udorthents, Kandiudults, Kandihumults, Hapludults occur in the Eastern and North Eastern Himalayas, while Argiudolls are prominent in the Western Himalayas. Haplustalfs, Haplustepts and Ustorthents are common in Peninsular India. Acid soils of Eastern plain comprise Endoaquepts, Udifluvents and Ustipsamments. However, coastal plain and Island ecosystem is conspicuous with the occurrence of Sulfaquepts, Sulfaquents, Endoaqualfs, Ustorthents and Haplustepts.

Lastly, knowledge about genesis and classification of acid soils is of immense importance in order to develop appropriate soil management practices and ameliorative measures for raising the productivity of such soils suffering due to the problem of soil acidity and other associated constraints.

References

Bandyopadhyay, A,K. (1989).Acid sulphate soils of the Sundarbans (West Bengal), *Journal of the Indian Society of Soil Science*. 37:416

Bandyopadhyay, A.K. (1987a). Acid sulphate soils of the Sundarbans (West Bengal), *Journal of the Indian Society of Soil Science*. 35:55

Bandyopadhyay, A.K. (1987b). Acid sulphate soils of the Sundarbans (West Bengal), *Journal of the Indian Society of Soil Science*. 57:547

Bandyopadhyay, B.K. and Maji, B. (1995). Nature of acid soils of Sunderban delta and suitability for classifying them as acid sulphate or potential acid sulphate soils, *Journal of the Indian Society of Soil Science*. 43:251-255

Bhargava, G.P. and Abrol,I.P. (1984). Morphology and characteristics of some acid soils in Kerala State. *Journal of the Indian Society of Soil Science*. 32:137-145.

Bhattacharyya, T., Sarkar, Dipak, Dubey, P.N., Ray, S.K., Gangopadhyay, S.K., Baruah, U. and Sehgal, J. (2004). Soil Series of Tripura, NBSS Publ. No.111, *NBSS & LUP, Nagpur*, pp.115.

Challa, O., Gajbhiye, K.S. and Velayutham, M. (1999). Soil Series of Maharashtra, NBSS Publ. No.79, *NBSS & LUP, Nagpur*, pp.428.

Chakravarty, D.N. (1977). Ph.D. Thesis, Punjab Agricultural University, Ludhiana.

Das, T.H., Sarkar, Dipak and Gajbhiye, K.S. (2003). Soil Series of Sikkim, NBSS Publ. No.105, *NBSS & LUP, Nagpur*, pp.86.

Deshpande, T.L.(1973). Bull. No.11, *Indian Society of Soil Science,IARI,New Delhi.*

Gowailkar, A.S. (1973). *Journal of the Indian Society of Soil Science*. 21:343

Gupta, R.P. and Nagarajarao, Y. (1982). *Review of Soil Research in India, Part 1., pp.60, 12th* InternationalCongress of Soil Science., New Delhi.

Harindranath, C.S., Nair, K.M., Srinivasan, S., Krishnan, P. and Gajbhiye, K.S. (2002). Soil Series of Goa, NBSS Publ. No.92, *NBSS & LUP, Nagpur*, pp.112.

Laskar, B.K., Debanath, N.C., De, G.K. and Basak, R.K.(1991). Acid sulphate soils of the Sundarbans (West Bengal). *Journal of the Indian Society of Soil Science*. 40:390.

Maji, A.K., Baruah, U., Dubey, P.N., Verma, T.P., Butte, P.S., Shilu, Kika and Angami, V. (2004). Soil Series of Nagaland, NBSS Publ. No.109, NBSS & LUP, Nagpur, pp. 126.

Maji, A.K., Reddy, Obi G.P. and Sarkar, Dipak (2012). Acid Soils of India- Their extent and spatial variability, NBSS Publ. No.145, NBSS & LUP (ICAR), Nagpur, pp. 138.

Manickam, T.S. and Krishnamoorthy, K.K. (1973). Bull. No.11, *Indian Society of Soil Science*, New Delhi.

Mittra, G.N. and Sahoo, S.K. (1992). Effect of potassium application in alleviating iron toxicity and nutrient uptake by rice in lateritic soils. Proceedings of workshop on micronutrient, 22-23 January, 1992, pp.269-272.

Murthy, R.S., Shankaranarayana, H.S. and Hirekerur, L.R. (1976). Distribution, genesis and classification of acid soils of India, In: Acid Soils of India, Bull. No.11, *Journal of the Indian Society of Soil Science., New Delhi*, pp.1-17.

Nayak, D.C., Sarkar, Dipak and Velayutham, M.(2001). Soil Series of West Bengal, NBSS Publ. No.89, *NBSS & LUP, Nagpur*, pp.260.

Nayak, D.C., Gangopadhyay, S.K., Sarkar, Dipak and Sen, T.K. (2002). Characteristics and classification of some acid soils of Lower Subansiri district of Arunachal Pradesh, *Agropedology*, 12:112-211.

Panda, N. (2003). Acid soils of Eastern India In:Characterization and Sustainable Management of Acid Soils of Eastern India, Winter School 2003 (ICAR), Department of Soil Science & Agricultural Chemistry, OUAT, Bhubaneswar.

Panda, N., Sarkar, A.K. and Chamuah, G.C. (2009). Soil acidity, In: Fundamentals of Soil Science (2nd Edition), *Indian Society of Soil Science*, pp.317-328.

Raychaudhuri, S.P. and Govinda Rajan, S.V. (1971). *Review of Soil Research in India, Indian Society of Soil Science*, p. 107.

Reddy, R.S., Budilal, S.L., Ramesh Kumar, S.C., Naidu, L.G.K. and Krishnan, P. (2005). Soil Series of Medak district, Andhra Pradesh, NBSS Publ.127, *NBSS & LUP, Nagpur*, pp.102.

Sahoo, A.K., Sarkar, Dipak and Gajbhiye, K.S. (2002). Soil Series of Bihar, NBSS Publ. No.89, *NBSS & LUP, Nagpur*, pp.289.

Sahoo, A.K., Sarkar, Dipak, Singh, S.K. and Butte, P.S. (2010). Soil Resource Mapping of Lohardaga district, Jharkhand for Land Use Planning, Paper presented in National Seminer on "Issues in Land Resource Management: Land Degradation, Climate change and Land Use Diversification, NBSS &LUP, Nagpur.

Sahoo, A.K., Sarkar, Dipak, Baruah, U. and Butte, P.S. (2010). Characterisation, classification and evaluation of soils of Langol Hill, Manipur for rational land use planning. *Journal of the Indian Society of Soil Science*. 58:355-362.

Sarkar, Dipak, Sah, K.D., Sahoo, A.K. and Gajbhiye, K.S.(2005). Soil Series of Orissa, NBSS Publ. No.119, *NBSS & LUP, Nagpur*, pp.254.

Sarkar, A.K. (2005). Managing natural resources for increasing agricultural production in Eastern India, Journal of the Indian Society of Soil Science, 53:435-447.

Sehgal, J., Challa, O., Thampi, C.J., Maji, A.K. and Naga Bhushana, S.R. (1998). Red and lateritic soils of India, *In* Red and Lateritic Soils of World (Vol.2), ed. J.L. Sehgal, W.E.Blum and K.S.Gajbhiye, pp.1-18.

Sen, T.K., Baruah, U., Sarkar, Dipak, Maji, A.K. and Patil, V.P. (2006). Soil Series of Manipur, NBSS Publ. No.134, *NBSS & LUP, Nagpur*, pp.53.

Sharma, P.D. and Sarkar, A.K. (2005). Managing acid soils for enhancing productivity, National Resource Management Division (ICAR), Krishi Anusandhan Bhavan-II, Pusa Campus, New Delhi, Tech. Bull. pp.23.

Singh, R.L.(1971). *India, A Regional Geography*, National Geographic Society of India, Varanasi.

Singh, R.S., Baruah, Utpal; Sarkar, Dipak; Butte, P.S. and Gajbhiye, K.S. (2005). Soil Series of Meghalaya, NBSS Publ. No.121, *NBSS & LUP, Nagpur*, pp.86.

Soil Survey Staff USDA (2006). *Keys to Soil Taxonomy*, Xth edition, pp.332.

Staff, NBSS & LUP (2002). Soils of India, NBSS Publ. No.94, *NBSS & LUP, Nagpur*, pp.130.

Staff, NBSS & LUP (2006). Soils Series of Kerala, NBSS Publ. No.136, *NBSS & LUP, Nagpur*, pp.136.

Subramoney, N. (1971). *In Soil and Water Research in India in Retrospect and Prospetc*, pp. 123, ICAR, New Delhi.

Tamgadge, D.B., Gajbhiye, K.S., Velayutham, M. and Kaushal, G.S. (1999). Soil Series of Madhya Pradesh, NBSS Publ. No.78, *NBSS & LUP, Nagpur*, pp.523.

Tamgadge, D.B., Gajbhiye, K.S. and Pandey, G.P. (2002). Soil Series of Chhattisgarh, NBSS Publ. No.85, *NBSS & LUP, Nagpur*, pp.183.

Tamhane, R.V. and Karale, R.L. (1967). *Journal of the Indian Society of Soil Science*. 15:269.

Soil Survey Staff (1951) Soil Survey Manual.Agriculture Handbook 18.U.S.Department of Agriculture,Washington DC.503pp.

Soil Survey Staff(1999)Soil Taxonomy.Agriculture Handbook 436.2nd edition.U.S.Department of Agriculture,Washington DC. 869pp.

Vadivelu, S., Sen, T.K., Bhaskar, B.P., Baruah, U., Sarkar, Dipak; Maji, A.K. and Gajbhiye, K.S. (2004). Soil Series of Assam, NBSS Publ. No.101, NBSS & LUP, Nagpur, pp.229.

5

Field Studies on Acid Soils of India

A.K. Sarkar, S.P. Sharma, T.C. Baruah, Patiram, B. Mandal, S.K. Pattanayak, V.K.Venugopal and S.C. Talashilkar

Many farmer participatory field experiments were designed and conducted in different states of India having large area under acid soils, with locally available soil ameliorants to study their effect on soil health and crop performance. This work was done during 2000 & 2005 under an ICAR network project on "Soil characterization and resource management of acid soil regions for increasing productivity". The details of the field locations are presented in table 1.

These experiments were conducted in cultivator's fields. The treatments used in these experiments comprised of the following: Farmer Practice (FP), Farmers Practice + Lime, 100% NPK and 100% NPK + Lime (2000-2003). In another set of farmers field experiments (2004 – 2005), the treatments were : Farmer Practice, 100% NPK, 50% NPK + Lime and 50% NPK + Lime + FYM. Each field experiment was replicated in 10 farmer's fields. Altogether 871 such experients were conducted during the five years in different states (Table- 1). Crops were raised to maturity following the recommended package and practices and grain yields were recorded. Soils, before and after harvest of crop and plant samples were collected for laboratory analysis.

Two pronged approach of liming acid soils to raise soil pH and adding fertilizers to ensure adequate supply of nutrients to crops

Table 1: Details of field experiment conducted under ICAR Network Project on Acid Soils

Network Center	Location of field trials	Soil pH	Crops	Liming materials	No. of trials conducted
Bhubaneshwar(Odisha)	Gajamarh, Parbati & Sankarpur	4.0 to 4.5	Groundnut, Pigeonpea	Paper mill sludge	135
Dapoli(Maharastra)	Dapoli	5.0 to 5.3	Groundnut/ Cowpea	Lime stone	85
Jorhat (Assam)	Biswanath, Dogaon pabhoi, Kumalia of Sonitpur district	4.9 to 5.2	Rapeseed/ Green gram	Lime sludge	62
Kalyani (West Bengal)	Midnapur (Jhargram) & Purulia	4.5 to 5.0	Mustard/ Wheat	Basic slag	124
Meghalaya (ICAR Complex NEH Region, Umiam)	Mawpam and Umraoi – maidan villages of Rivhoi district	4.5 to 5.0	Mustard/ Groundnut/ Maize	Lime	79
Palampur (H.P.)	Bandla, Saperoo Jhikli bheth village of kangra district	5.0 to 5.5	Maize, Wheat, Soybean Gobhi Sarson	Lime	210
Ranchi (Jharkhand)	Dumka district (Jama block), Pithoria block (Ranchi)	4.5 to 5.2	Maize + Pigeonpea / Groundnut +Pigeonpea, Blackgram, pea	Lime	121
Vellayani (Kerala)	Athiyannoor & Pathencode panchayats	4.8 to 5.2	Cowpea, Blackgram as intercrop in coconut based systems	Lime,Phospho-gypsum	55
Total					871

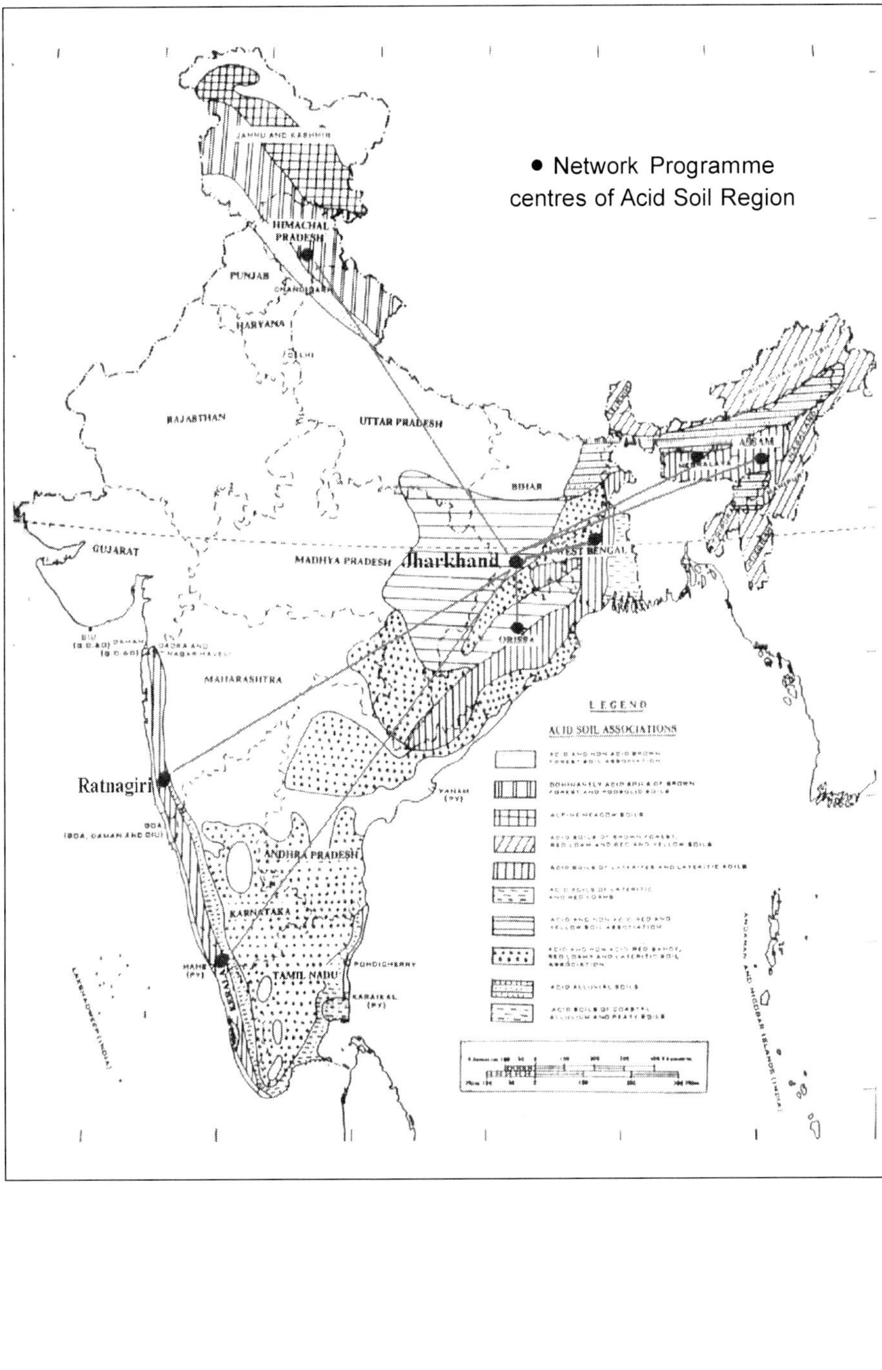
• Network Programme
centres of Acid Soil Region
HIMACHAL
PRADESH
PUNJAB
HARYANA
RAJASTHAN
UTTAR PRADESH
BIHAR
ASSAM
GUJARAT
MADHYA PRADESH
Jharkhand
WEST BENGAL
ORISSA
MAHARASHTRA
LEGEND
ACID SOIL ASSOCIATIONS
Ratnagiri
YANAM
(PY)
ANDHRA PRADESH
KARNATAKA
MAHE
(PY)
TAMIL NADU
PONDICHERRY
KARAIKAL
(PY)

were used. Liming was done to neutralize the low magnitude active acidity due to H^+ and Al^{3+} ions in the soil solution and part of the exchange acidity. The active acidity influences most the plant root and the microbes around the rhizosphere. Depending on the lime requirement (LR) of soils, 1/10th of LR dose was applied in farmers fields. Thus, the total quantity of lime applied to each crop ranged from 2 to 4 quintals/ ha depending on soil texture, soil pH, per cent soil organic carbon and clay content.

Liming materials

The availability of cheap and effective liming materials is crucial for the success of acid soil amelioration programme. The agricultural grade limestone powder and marketable lime are effective, but may not be economical and available in plenty at all the places. A number of industrial by-products like basic slag, lime sludge, phosphogypsum , press mud etc. are rich sources of calcium and could serve as cheap liming materials in some areas. The availability of basic slag is around 5-8 million tones per annum with different steel industries. Tata steel, Jamshedpur generates about 10 lakh tones of basic slag annually. Likewise, the paper mills in Assam, Nagaland, West Bengal, Odisha, Madhya Pradesh and Andhra Pradesh produce around 2 lakh tones of lime sludge per annum. Lime stone deposits in Arunachal Pradesh (140 million tones), Manipur (46 million tones), Assam (159 million tones), Meghalaya (1400 million tones) and Nagaland (357 million tones) are potential sources for producing agricultural grade lime at affordable price (Table 2).

Table 2: Availability of liming materials in India

Acid Soil Region / State	Liming material	Quantity available (million tones)	Cost/tone (Rs.)
Assam	Limestone	15.0	4,000
Himachal Pradesh	Marketable lime	-	4,000
Jharkhand	Limestone/Basic slag	1.0	1,000
Kerala	Lime shells	4.0	6,000
Maharastra	Marketable lime	0.2	2,000
Meghalaya	Limestone	14.0	1,000
Odisha	Paper mill sludge	0.2	500
West Bengal	Basic slag	0.3	1,000

Crop responses to liming and fertilization

Important crops of each region were grown without lime and with lime + fertilizers recommended for the crop. Lime application at the rate of 2-4 q/ha was done in the furrows in direct seeded crops along with fertilizer at the time of sowing . Results reveal that, Lime alone increased the yield of different crops from 13.83 to 49.44 per cent with a mean value of 27.87 per cent (Table- 3). Response of lime was fairly good in groundnut + pigeonpea cropping sequence in Odisha (49.44%) and Jharkhand (29.96%). The yield increase was marked in case of liming over farmer's practice on maize in Himachal Pradesh (15.98%) and Meghalaya (29.96%), on rapeseed (17.81%) and summer green gram (13.88%) in Assam, on wheat (16.22%) in Himachal Pradesh and on groundnut (17.61%) in Maharashtra.

The recommended application of fertilizer (100% NPK) further, increased the crop yields by 97 per cent over that in farmer practice. Response of crops to lime was higher when lime was applied with recommended dose of fertilizer (Table- 3). This good response to fertilizer application lent credence to commonly held view point that fertilizer use in crop production was inadequate in acid soil regions of the country. Liming with recommended fertilizer dose (100% NPK + Lime) increased the yield of crops from 49.26 to 189.22 per cent (mean 97.10 per cent) over the farmer's practice.

Table 3: Response of liming on Yield (qha^{-1}) of crops in acid soils regions

States	Crop	FP	FP + Lime	100% NPK + Lime
Assam	Rape seed	7.30	8.60 (17.81)	12.90 (76.71)
	Summer green gram	10.09	11.49 (13.88)	15.06 (49.26)
Himachal	Maize	23.46	27.21 (15.98)	38.30 (63.26)
	Wheat	17.26	20.06 (16.22)	32.77 (89.86)
Jharkhand	Maize+ Pigeonpea	39.30	51.37(30.71)	75.22 (91.40)
	[(Maize equivalent yield Pea (pod)]	28.56	32.51(13.83)	51.24 (79.41)
Maharastra	Groundnut	14.20	16.70 (17.61)	24.30 (71.13)
Meghalaya	Maize	10.58	13.75 (29.96)	30.60 (189.22)
	Mustard	0.62	1.48 (138.71)	4.91 (691.94)
Odisha	Groundnut + Pigeonpea (Groundnut equivalent yield)	18.00	26.90 (49.44)	37.60 (108.89)
West Bengal	Mustard	-	7.95(with 100 % NPK)	13.10 (64.78)
All states	Average of all crops	15.50	19.82 (27.87)	30.55 (97.10)

( ) % yield increase over farmer's practice (FP)

The response to the combined application of NPK and lime was more than the simple additive effect of fertilizer and lime, indicating synergy or complementary effect between fertilizer use and liming. The effect of liming and fertilization on crop yields in acid soils reveals the benefits one can derive by the adoption of such practice (Table- 3).

Effect of liming with fertilizer and FYM on crop yields

The crop yields with 50% recommended NPK + Lime @ 2-4 qha^{-1} were generally, equal or considerably higher to the yield with 100% recommended NPK. The yield of cereal crops like maize, wheat and maize equivalent yield in maize + pigeon pea intercropping system was slightly lower than yield of crop with 100% NPK fertilizer application. Pulses viz., summer green gram (18.2%) in Assam, pea green pod (32.3%) in Jharkhand, cowpea (23.3%) and black gram (26.6) in Kerala and groundnut (50.0%) in Meghalaya showed a considerable increase in yield with 50% recommended NPK + Lime over 100% NPK without lime application (Table- 4). Liming enhanced the nutrient use efficiency by crops and resulted in saving of chemical fertilizers to the extent of 50 per cent. Application of manure with 50% recommended NPK + Lime increased the yield of crops from 6 to 82 per cent over 100% NPK application (Table- 4). Conjunctive use of lime and FYM was beneficial and showed higher yields with better soil health.

Table 4: Effect of lime, fertilizer and FYM application on crop yield (qha^{-1}) in acid soils

States	Crop	100 % NPK	50 % NPK + Lime	50% NPK + Lime + FYM
Assam	Rapeseed	9.7	10.1 (4.1)	11.9 (22.68)
	Summer green gram	4.4	5.2 (18.2)	6.8 (54.5)
Himachal	Maize	34.0	33.1 (-2.7)	36.1 (6.2)
	Wheat	27.9	23.7 (-15.0)	29.8 (6.8)
Jharkhand	Maize + Pigeon pea (Maize equivalent yield	69.0	65.0 (-5.8)	82.5 (19.6)
	Pea (pod)	38.4	50.8 (32.3)	69.7 (81.5)
Kerala	Cowpea	8.6	10.6 (23.3)	11.3 (31.4)
	Blackgram	6.4	8.1 (26.6)	9.2 (43.7)
Meghalaya	Maize	30.5	30.3 (-0.7)	36.4 (19.3)
	Groundnut	14.2	21.3 (50.0)	25.4 (78.9)
Odisha	Groundnut	22.5	23.6 (4.9)	26.6 (18.2)
	Pigeon pea	12.0	12.2 (1.7)	14.6 (21.7)
West Bengal	Mustard	8.2	8.4 (2.4)	11.2 (36.6)
	Wheat	16.7	17.1 (2.4)	22.1 (32.3)
All states	Average of all crops	27.1	22.8 (-15.9)	28.1 (3.7)

( ) % yield increase over 100 % NPK.

Economic Benefits of Liming and Fertilization of Acid Soils

Liming and fertilization proved to be economical for different crops in acid regions (Table- 5). The benefit: cost ratio increased in the order: liming alone (farmer's practice + lime) < addition of fertilizers (100% NPK) and < conjunctive use of fertilizers and lime (100% NPK + Lime). Benefit: cost ratio varied in farmer practice, Farmer Practice + Lime, 100% NPK and 100% NPK + Lime, respectively, from 0.7 to 2.4, 0.6 to 3.2, 1.2 to 3.0 and 1.4 to 4.3 in different crops (Table- 5). Lime application with recommended (100%) NPK proved more economical with mean return Rs.2.54 per rupee investment in all crops grown in different states . The returns of Rs.1.41, 1.78 and 2.11, respectively, in Farmer Practice, Farmer Practice + Lime and 100% NPK on per rupee investment showed the beneficial effect of lime application with recommended dose of NPK.

The application of 50% NPK + Lime was more economical than 100% NPK without lime,which was revealed by relatively higher benefit: cost ratios in former than the latter (Table- 6). Benefit: cost ratio in crops grown with application of 50% NPK + lime was higher and varied from 1.58 (cowpea in Maharashtra) to 5.20

Table- 5: Benefit: Cost ratio of liming and fertilization of Crops in acid soils

Acid Soil Region / State	Crop	Farmer's practice (FP)	FP + Lime	100% NPK	100% NPK + Lime
Assam	Rapeseed	2.30	2.72	2.77	3.40
	Summer green gram	2.36	2.69	2.71	3.19
Himachal	Maize	1.15	0.97	1.99	1.73
	Wheat	0.73	1.02	1.46	1.79
Jharkhand	Maize + Pigeonpea (Maize equivalent yield)	1.52	1.85	1.97	2.37
	Pea (pod)	1.70	1.87	2.39	2.78
Kerala	Blackgram	1.47	1.63	1.36	1.84
Maharastra	Groundnut	1.25	1.40	1.61	1.76
Meghalaya	Maize	1.00	1.27	3.58	3.45
	Mustard	1.00	0.63	1.32	2.35
Odisha	Groundnut	0.76	1.03	1.17	1.37
	Pigeonpea	1.13	1.48	1.59	1.80
	Groundnut + Pigeonpea	1.64	2.27	2.56	2.93
West Bengal	Mustard	1.70	3.20	3.00	4.30
	Wheat	1.40	2.70	2.20	3.00
Mean of all crops		1.41	1.78	2.11	2.54

(groundnut in Meghalaya) as compared to 100% NPK. Average B: C ratio of crops in acidic soil region was 2.81 with 50% NPK + Lime (@ 2-4 qha^{-1}) while mean return was comparatively less (2.33) at recommended dose of fertilizer without lime.

Table 6: Benefit: Cost ratios of liming along with half the recommended dose of fertilizer for Crops in acid soils

Acid Soil Region / State	Crop	100 %NPK	50 % NPK + Lime
Assam	Rapeseed	2.20	2.28
	Summer green gram	1.39	1.46
Himachal Pradesh	Maize	1.19	1.65
	Wheat	1.46	1.62
Jharkhand	Maize + Pigeon pea (Inter cropping)	2.31	2.50
	Pea (pod)	2.16	2.93
Kerala	Cowpea	2.34	2.95
	Black gram	2.72	3.64
Maharastra	Cowpea	1.66	1.58
Meghalaya	Maize	3.97	4.54
	Groundnut	3.22	5.20
Odisha	Groundnut + Pigeon pea	2.80	2.82
West Bengal	Mustard	3.00	3.60
	Wheat	2.20	2.60
Mean of all crops		2.33	2.81

Liming in relation to soil properties and nutrient uptake by crops

Soil and plant samples from the experimental sites were analysed for assessment of changes in the soil properties and uptake of nutrients by crops.

Assam

In the field sites of two villages viz., Pabhoi and Dagaon, lime sludge was applied as per recommended treatment combinations and green gram crop was sown in 2nd week of April. In the trial plots, fertilizer was applied @ 15-35-10 kg ha^{-1} for N, P_2O_5 and K_2O. Surface soil (0-15 cm) samples collected after harvest of the crop were analyzed for important soil parameters (Table 7).

Table 7: Effect of liming, FYM and fertilizer on pH, organic carbon and nutrient status of soil after harvest of summer green gram.

Treatments	pH	O.C. (gkg^{-1})	Avail. N ($kg\ ha^{-1}$)	Avail. P_2O_5 ($kg\ ha^{-1}$)	Avail. K_2O ($kg\ ha^{-1}$)	Ca Cmol (p^+) kg^{-1}	Mg (p^+) kg^{-1}	Al Cmol (p^+) kg^{-1}
FP	4.9	9.8	375	24.3	113	0.89	0.68	0.66
100% NPK	4.9	9.3	415	29.5	125	0.92	0.70	0.67
50% NPK + Lime	5.1	10.3	393	27.2	118	1.11	0.73	0.63
50% NPK + Lime + FYM	5.1	11.3	411	27.8	119	1.12	0.73	0.61

Mean value of 19 trials.

From the results, it is apparent that pH of the soil was highest in both the plots where lime had been applied. Among the four treatments, org. C (%) was highest in FYM (50% NPK + Lime + FYM) treated plot and lowest in plot with recommended fertilizer (100% NPK) application. Available N content was high in all cases, with lowest value with farmer practice and highest with recommended NPK treated plots. Both, available P and K content of soil was in low to medium range, while among the four treatments, highest content of available P and K was recorded in 100% NPK treated plots.

Exch. Ca, Mg and Al content of soils show that the range of exchangeable Ca, Mg and Al varied from 0.89 to 1.12, 0.68 to 0.73 and 0.61 to 0.67 Cmol (p^+)kg^{-1}, respectively in different treatments. Results show that lime sludge application along with FYM improved the soil fertility.

After harvest of the crop, plant samples were analysed for total N, P, K, Ca and Mg (Table- 8). Results show increased uptake of nutrients in the treatment with 50% NPK + Lime + FYM followed by 50% NPK + Lime, 100% NPK and farmer practice in decreasing order.

Table 8: Effect of liming, FYM and fertilizer on nutrients uptake ($kg\ ha^{-1}$) by summer green gram seed

Treatments	Nutrients				
	N	P	K	Ca	Mg
FP	28.7	3.4	4.8	7.4	3.8
100% NPK	33.3	4.2	5.8	9.3	4.1
50% NPK + Lime	33.7	4.3	5.9	11.5	4.6
50% NPK + Lime + FYM	36.9	5.2	6.8	12.0	5.1

Himachal Pradesh

The effect of different treatments on the total uptake of N, P, K, Ca and Mg by Maize is given in table- 9.

Table 9: Effect of liming on nutrients uptake (grain + straw) by maize (kg ha^{-1}).

Treatments	Nutrients				
	N	P	K	Ca	Mg
FP	53.81	10.66	44.04	14.33	9.54
FP + lime	61.73	15.07	52.32	17.61	11.79
100% NPK	73.00	19.26	62.20	24.57	13.31
100% NPK + Lime	79.02	21.28	64.66	31.04	14.96

Mean uptake data of all experiments conducted in Bandla, Saperoo & Jhiklibheth villages reveal that lime application increased N uptake by maize significantly. The lowest N uptake (53.81 kg ha^{-1}) was recorded in the farmer's practice , while the highest N uptake (79.02 kg ha^{-1}) , in 100% NPK + Lime treated plot. Minimum uptake of P (10.66 kg ha^{-1}), K (44.04 kg ha^{-1}), Ca (14.33 kg ha^{-1}) and Mg (9.54 kg ha^{-1}) was recorded under the farmer's practice while the highest uptake of P, K, Ca and Mg was found with 100% NPK + Lime.

The treatments also influenced the uptake of N, P, K, Ca and Mg by Wheat (Table 10). Among the four treatments, lowest uptake of 32.22, 10.18, 27.41, 7.92 and 3.58 kg ha^{-1} of N, P, K, Ca and Mg, respectively was recorded under farmer practice. Crop uptake of N, P, K, Ca and Mg was highest with 100% NPK + Lime treatment, which were 63.96, 24.07, 56.60, 21.72 and 8.47 kg ha^{-1},respectively.

Table 10: Effect of liming on nutrients uptake (grain + straw) by wheat (kg ha^{-1}).

Treatments	Nutrients				
	N	P	K	Ca	Mg
FP	32.22	10.18	27.41	7.92	3.58
FP + Lime	40.66	13.26	34.41	10.17	5.34
100% NPK	58.67	20.77	50.21	16.92	7.61
1000% NPK + Lime	63.96	24.07	56.60	21.72	8.47

Mean uptake data of all expts. conducted in Bandla, Saperoo & Jhilkibheth villages.

In a second set of experiment with half dose of recommended fertilizer alongwith lime and FYM application, the total uptake (grain + straw) of N, P, K, Ca and Mg by maize and wheat indicated that lime increased the uptake of N, P, K, Ca and Mg in both crops.

Under farmer practice, removal of N, P, K, Ca and Mg was 55.80, 10.25, 44.72, 13.89 and 10.11 kg ha^{-1} by maize (Table- 11) and 33.38, 9.23, 21.26, 9.69 and 2.92 kg ha^{-1} by wheat crop (Table- 12) . The highest nutrient uptake by both the crops was recorded in the treatment receiving 50% NPK + Lime + FYM. N, P, K, Ca and Mg uptake with this treatment was 80.37, 16.63, 73.55, 19.51 and 14.78 kg ha^{-1} by maize and 59.81, 18.54, 44.01, 18.47 and 6.52 kg ha^{-1} by wheat crop, respectively.

Table 11: Effect of lime, FYM and fertilizer on nutrient uptake by maize (kg ha^{-1})

Treatments	Nutrients				
	N	P	K	Ca	Mg
FP	55.80	10.25	44.72	13.89	10.11
FP + Lime	63.41	12.30	55.25	17.24	11.83
50% NPK	68.69	13.75	60.84	15.96	12.03
50% NPK + Lime	74.84	15.60	67.86	18.32	14.01
50% NPK + Lime + FYM	80.37	16.63	73.55	19.51	14.78

Mean value of all expts. conducted in Bandla and Saperoo villages.

Table 12: Effect of liming, FYM and fertilizer on nutrient uptake by wheat

Treatments	Nutrients				
	N	P	K	Ca	Mg
FP	33.38	9.23	21.26	9.69	2.92
FP + Lime	40.90	12.18	28.41	12.82	4.63
50% NPK	45.50	14.85	34.80	14.24	5.27
50% NPK + Lime	51.80	17.60	36.27	16.43	5.77
50% NPK + Lime + FYM	59.81	18.54	44.01	18.47	6.52

Mean value of all expts. conducted in Bandla and Saperoo villages.

The soil properties viz., pH, org. C, available nutrients and exchangeable cations were monitored after harvest of maize crop (Table 13).

Table 13: Effect of liming on soil reaction (pH), organic carbon and available nutrients status in soil after harvest of maize crop.

Treatments	pH	O.C. (gkg^{-1})	Avail. N (kg ha^{-1})	Avail. P_2O_5 (kg ha^{-1})	Avail. K_2O (kg ha^{-1})	Ca Cmol (p^+) kg^{-1}	Mg Cmol (p^+) kg^{-1}
FP	5.44	11.83	269.88	18.06	143.50	3.81	1.27
FP + Lime	6.04	12.54	270.88	20.95	148.75	5.15	1.50
100% NPK	5.45	12.85	325.75	24.38	171.38	4.64	1.46
100% NPK + Lime	6.13	12.86	329.00	26.86	171.38	6.19	1.91

Mean value of all expts. conducted in Bandla, Saperoo & Jhiklibheth villages.

Use of lime with farmer's practice and 100% NPK use, significantly increased soil pH, showing the ameliorating effect of lime. Lime application did not show any marked change in the organic carbon status of the soils. Available N content of soil increased significantly with application of recommended NPK compared to the unfertilized plots. Available Phosphorus content of soil increased from 18.06 to 20.95 kg ha^{-1} and from 24.38 to 26.80 kg ha^{-1} due to lime application, in farmer's practice and recommended NPK application, respectively. Lime application did not have much effect on available K content of soil, however,it increased the exchangeable calcium and magnesium status of soil.

Jharkhand

The uptake of Ca, Mg and P in maize + pigeonpea cropping system increased with lime+NPK over Farmers practice from 31.70 to 119.40, 18.20 to 75.80 and 4.85 to 15.12 kg ha^{-1}, respectively. Lime application increased the uptake of nutrients over farmer practice and also over 100% NPK application, respectively from 31.70 to 67.70 and 65.40 to 119.40 kg ha^{-1}. Similar trend was also recorded in Mg and P uptake by maize + pigeonpea cropping system (Table 14).

Table 14: Effect of lime application on total nutrients uptake (kg ha^{-1}) by maize + Pigeonpea inter-cropping system.

Treatments	Maize + Pigeonpea		
	Ca	Mg	P
FP	31.70	18.20	4.85
FP + Lime	67.70	39.20	8.15
100% NPK	65.40	49.30	9.94
100% NPK + Lime	119.40	75.80	15.12

In a second set of experiment, nutrient uptake by crops was found lowest under farmer's practice viz., Ca (28.1 kg ha^{-1}), Mg (19.7 kg ha^{-1}), and P (6.76 kg ha^{-1}) by maize + pigeon pea cropping system (Table- 15). Highest uptake of Ca (64.7 kg ha^{-1}) and P (16.59 kg ha^{-1}) by the system was recorded in plots receiving 50% NPK + FYM + Lime, while Mg uptake was highest (71.9 kg ha^{-1}) in the treatment with 50% NPK + Lime.

Table- 15: Effect of lime application on total nutrient uptake (kg ha^{-1}) by maize + pigeonpea inter- cropping system

Treatments	Maize + Pigeonpea		
	Ca	Mg	P
FP	28.1	19.7	6.76
100% NPK	51.3	36.3	10.80
50% NPK +Lime	45.6	71.9	16.49
50% NPK+FYM+ Lime	64.7	52.8	16.59

Soil properties viz., pH, org. C, exchangeable cations and phosphorus were monitored after harvest of pigeonpea . Use of lime with 50% NPK alone and that with FYM, significantly increased the pH of soil showing ameliorating effect of lime in acid soil. Liming alone did not result in the build up to org. C in soil, while , but lime + FYM applied with half recommended dose of fertilizer resulted in a considerable build up of org. C (5.6 g kg^{-1}) in soil. Lime application increased exchangeable calcium and magnesium content of soil. Build up of Ca, Mg and phosphorus was more when lime was applied with FYM and half of the recommended dose of fertilizer (Table- 16).

Table 16: Exch. Ca, Mg and P content in soil after harvest of pigeonpea in maize + pigeonpea inter- cropping system.

Treatments	pH	OC ($g\,kg^{-1}$)	Ca{Cmol (p^+) kg^{-1}}	Mg{Cmol (p^+)kg^{-1}}	Avail. P ($kg\,ha^{-1}$)
FP	4.5	2.9	1.50	0.69	9.10
100% NPK	4.6	2.8	1.46	0.65	9.10
50% NPK +Lime	5.0	2.7	2.71	0.71	13.61
50% NPK+FYM+ Lime	5.0	5.6	2.82	0.67	15.60

Kerala

The soil samples from experimental plots were analysed for different soil parameters after harvest of black gram (Table- 17). Soil pH was high in treatment which received lime. Lime treatment resulted in lower values of exchangeable acidity compared to no-lime treatment. Higher N, P, K and exchangeable Ca and Mg in soil were recorded with lime treatment.

Table 17: Effect of liming on soil reaction (pH), organic carbon and nutrient status of soil after harvest of black gram.

Treatments	pH	Exch. Acidity {Cmol (p^+)kg^{-1}}	O.C. ($g\,kg^{-1}$)	N	P	K	Exch. Ca Cmol	Exch. Mg (p^+)kg^{-1}
				($kgha^{-1}$)				
FP	5.7	0.6	12.8	296	38	77	0.78	0.47
FP + Lime	6.2	0.4	9.7	349	34	99	1.62	0.85
100% NPK	6.2	0.4	9.4	380	33	127	2.45	0.65
100% NPK + Lime	5.7	0.6	11.9	338	38	89	0.95	0.43

Mean value of 15 trials.

Even though, the available nutrient status of the soil was influenced by application of lime, this was not reflected in nutrient concentration of N, P and K in plant (Table- 18). But, Ca and Mg concentration in plant sap was higher in limed plots as compared to un-limed plots. Highest Ca (1.08%) and Mg (0.40%) concentration in black gram was recorded in the treatment receiving 100% NPK + Lime.

Table 18: Effect of lime application on nutrient concentration (%) in black gram.

Treatments	Nutrient concentration (%)				
	N	P	K	Ca	Mg
FP	1.42	0.15	0.93	0.87	0.20
FP + Lime	1.68	0.15	0.96	1.02	0.30
100% NPK	1.48	0.16	0.84	1.08	0.40
100% NPK + Lime	1.49	0.16	1.05	0.78	0.20

Mean value of 15 trials.

In a second set of experiment on cowpea crop with combination of half the recommended fertilizer with lime and FYM, soil pH was (5.88 and 5.83, respectively) higher as compared to pH of 4.64 under farmers practice. The effect of lime on exchangeable acidity was similar to that on soil pH. There was significant influence on soil potassium, while soil phosphorus availability was not much influenced by lime application(Table- 19).Lowest concentration of all the nutrients was recorded under farmer's practice (Table- 20).

Table 19: Effect of liming, FYM and fertilizer on soil reaction (pH), organic carbon and available nutrients status of soil after harvest of cowpea.

Treatments	pH	Exch. Acidity {Cmol (p^+) Kg^{-1}}	O.C. (g kg^{-1})	K (kg ha^{-1})	P (kg ha^{-1})	Ca {Cmol (p^+) Kg^{-1}}	Mg {Cmol (p^+) Kg^{-1}}
FP	4.64	0.82	11.7	80.6	15.8	2.04	0.38
100% NPK	4.53	0.72	9.6	121.7	27.9	1.74	0.42
50% NPK +Lime	5.88	0.34	10.0	139.6	26.2	2.88	0.65
50% NPK+ FYM + Lime	5.83	0.51	8.2	134.3	32.3	2.61	0.73

Mean value of 15 trials.

Table 20: Effect of liming, FYM and fertilizer on nutrient concentration in cowpea.

Treatments	Nutrient concentration (%)				
	N	P	K	Ca	Mg
FP	1.47	0.47	1.67	0.61	0.17
100% NPK	1.56	0.54	1.56	0.75	0.16
50% NPK +Lime	1.56	0.56	1.54	0.72	0.15
50% NPK+FYM+ Lime	1.54	0.53	1.81	0.80	0.16

Maharastra

Pooled data of three consecutive years on pH, total acidity, exchangeable acidity, exchangeable Al and lime requirement of acid soil reveal that application of lime @ 1/10th LR reduced soil pH by about 0.6 units in farmers practice and recommended level of fertilizer application. The incorporation of lime in farmers practice as well as soil test based fertilizer dose + Lime resulted in significant reduction in total and exchangeable acidity and exchangeable Al^{+3} content of soil after the harvest of groundnut crop (Table 21).

Table 21: Effect of lime on soil acidity contributing parameters after harvest of groundnut.

Treatments	Parameters				
	pH	Total acidity {Cmol (p^+) kg^{-1}}	Exch. Acidity{Cmol (p^+)kg^{-1}}	Exch. Al {Cmol (p^+)kg^{-1}}	Lime requirement (t ha^{-1})
FP	5.27	3.8	0.36	0.17	12.1
FP + Lime	5.89	2.1	0.15	0.05	8.4
100% NPK	5.29	3.8	0.33	0.16	12.0
100% NPK + Lime	5.92	2.1	0.14	0.05	8.3

Pooled data of three years experiments.

It was also evident that application of lime did not result in significant changes in organic carbon, available nitrogen, phosphorus and potassium content of soil (Table 22).

Table 22: Effect of liming on organic carbon and nutrient status of soil after harvest of groundnut.

Treatments	O.C. (g kg^{-1})	Avail. N (kg ha^{-1})	Avail. P_2O_5 (kg ha^{-1})	Avail. K_2O (kg ha^{-1})	Each. Ca {Cmol (p^+) kg^{-1}}	Exch. Mg {Cmol (p^+) kg^{-1}}
FP	15.5	301	10.2	283	2.67	1.10
FP + Lime	16.0	309	14.1	307	9.23	2.88
100% NPK	16.0	312	11.8	300	3.02	1.36
100% NPK + Lime	16.5	322	15.2	328	9.92	3.15

Pooled data of three years expts.

Meghalaya

The collected maize grain and straw samples from field trials were analysed for the N, P and K removal from soil. Addition of lime equivalent to 1/10th LR, considerably increased the uptake of major nutrients (Table 23).

Table 23: Removal of total N, P and K (kg ha^{-1}) by maize at Mawpan and Umrai maidan villages.

Treatments	Mawpan			Umrai maidan		
	N	P	K	N	P	K
FP	21.1	6.9	21.3	23.2	6.9	22.8
FP + Lime	35.9	10.6	31.9	37.0	10.0	34.0
100% NPK	57.3	15.5	47.3	63.8	17.1	55.7
100% NPK + Lime	76.5	21.1	62.9	90.1	24.7	74.8

Pooled data of 10 expts.

Odisha

The soils under groundnut (sole) were strongly acidic (pH 5.1). The average LR by Woodruff Buffer method was 8 t ha^{-1}. However, lime application was recommended @ 20% LR i.e., 1.6 t ha^{-1}. The paper mill sludge ($CaCO_3$ equivalent 70 per cent) was applied as liming material @ 2.28 t ha^{-1} in furrow below the seed rows behind the plough, at the time of sowing of crops. Treatments receiving lime application, maintained higher soil pH whereas unlimed soils turned more acidic (Table- 24). Amount of NH_4^+ + NO_3^- -N in soil increased in lime and NPK treated soils, which compared to the initial status of nitrogen may be due to the atmospheric N_2 fixing capacity of both the leguminous crops (groundnut and pigeonpea) in cropping sequence. Phosphorus fertilized soil maintained higher available P status than that in unlimed soil. Soil in farmer's practice without and with lime exhibited lower available P status probably due to removal by crops. Irrespective of potash application in different treatments, due to large removal by crops and some leaching through coarse textured soil, its status decreased drastically (Table- 24). Liming of soil also accelerated its loss by its replacement from exchange complex by added Ca through liming. Liming increased Ca status in soil, whereas in unlimed soil, the Ca status decreased compared to the initial value. The exchange acidity of the soil was 0.48 Cmol (p^+) kg^{-1}. Lime application neutralized the acidity and reduced it in soil. In unlimed soil, removal of Ca by crops, leaching loss of Ca aggravated the acidity problem and increased exchange acidity of soil.

Table 24: Effect of liming on soil health in groundnut-pigeonpea cropping system.

Treatments	pH	OC (gkg^{-1})	N	P	K	Ca Cmol (p^+) kg^{-1}	Exch. Acidity Cmol (p^+)kg^{-1}
				($kgha^{-1}$)			
FP	5.0	3.1	205	10	152	1.06	0.64
FP + Lime	6.0	3.7	250	9	139	1.52	0.33
RP	4.9	3.7	250	19	128	0.97	0.68
RP + L	5.9	4.0	312	15	119	1.36	0.23
Initial	5.2	4.4	198	12	224	1.21	0.48

In the second set of experiment, impact of liming was studied with combination of FYM and fertilizer. Uptake of P was significantly higher with 50% NPK + Lime + FYM (22.0 kg ha^{-1}) followed by that under full dose of fertilizer (20.3 kg ha^{-1}). Application of recommended dose of fertilizer or half of it integrated with lime resulted in comparable amount of K uptake, which was significantly higher compared to that under farmer's practice. Integrating FYM application along with lime enhanced K uptake by crops. The removal of calcium and magnesium ranged from 59 to 134 and 18 to 54 kg ha^{-1}, respectively under farmer's practice and half of recommended fertilizer integrated with lime (Table-25). Lime either alone or with FYM, when integrated with 50 per cent NPK resulted in significantly higher Ca and Mg uptake compared to that under farmer's practice and application of full dose of chemical fertilizers without lime.

Table 25: Effect of liming, FYM and fertilizer on nutrient uptake (kg ha^{1}) in groundnut – pigeonpea cropping system.

Treatments	Nutrients				
	N	P	K	Ca	Mg
FP	180	11.7	147	59	18
FP + Lime	310	20.3	232	106	33
100% NPK	310	18.5	223	110	41
100% NPK + Lime	376	22.0	250	134	54

West Bengal

Among the five LR methods tested, the amount of lime estimated through SMP method, brought the soil pH nearest to optimum , while the same with other methods such as Peech, Adams and Evan, Woodruff and Yuan could bring soil pH to 7.7, 7.6, 6.6 and 6.8, respectively (Table- 26). Results, thus indicate that the amount of lime applied estimating through SMP method could bring the soils towards neutrality suggesting its superiority over other methods tested.

Table 26: Changes in pHw values of soils on liming on the basis of lime requirement estimated by different methods.

Soil/Method	Liming Methods					
	Initial	SMP	Peech	Adams & Evans	Woodruff	Yuan
Matikata	5.2	6.8	7.8	7.6	6.5	6.6
Baranachina	5.2	6.8	7.8	7.7	6.8	6.9
Bilshi	4.7	7.1	7.8	7.5	6.9	7.0
Sonarpur	5.0	7.1	7.8	7.6	6.8	7.0
Purbamallikpara	5.0	7.3	7.9	7.6	6.8	7.2
Chopra	4.3	7.0	7.7	7.5	6.5	6.9
Itahar	4.8	7.0	7.6	7.5	6.8	6.9
Hemtabad	5.4	7.0	7.5	7.5	6.8	6.9
Hutmura	5.7	7.0	7.9	7.7	6.8	6.9
Hura	5.4	7.0	7.7	7.7	6.6	6.7
Bankura	5.3	6.8	7.5	7.5	6.9	7.0
Taldangra	6.0	7.1	7.6	7.5	6.9	7.0
Sonamukhi	5.4	7.3	8.0	7.7	6.9	7.1
Mithekhali	4.5	6.9	7.3	7.5	6.3	6.4
Tangrakhali	4.5	6.9	7.3	7.7	6.5	6.6
Chhoto Dumki	4.0	6.7	7.4	7.6	6.2	6.8
Ithhola	4.9	6.8	7.4	7.7	6.4	6.7
Merigang	4.9	6.7	7.7	7.7	6.6	6.7
Mean	5.0	7.0	7.7	7.6	6.6	6.8

To test the relative efficiency of basic slag as a liming material, eight experiments were conducted with mustard crop. After recording the straw and grain yield, plant samples were analysed for N, P and K following standard methods. Results (Table- 27)

Table 27: Effect of liming on the content of N, P and K in mustard plants variety- Sita

Location and name of farmer	Treatments	N, P and K content (%) in mustard plants		
		N	P	K
Jhargram (Pukuria anchal)	RD	0.39	0.13	0.71
Subodh Mahato	RD + Lime	0.50	0.33	1.09
Jhargram, (Dahijuri anchal)	RD	0.50	0.15	1.50
Binpur-1Nirod Singh	RD + Lime	0.69	0.29	2.08
Jhargra (Dahijuri anchal)	RD	0.53	0.36	1.30
Binpur-1Sheikh Asimuddin	RD + Lime	0.92	0.54	1.55
Purulia (Gopladih)Dhiren Mahato	RD	0.45	0.19	1.68
	RD + Lime	1.07	0.55	1.97

howed that liming resulted in significant increases in N, P and K content of the crop.

Soil samples collected after the harvest of mustard were analysed for different parameters viz., pH, EC, org C and available N, P and K. Besides these, total acidity, exchange acidity, exchangeable aluminium and total potential acidity were also analysed to estimate the change that occurred upon liming. Results (Table 28) show that application of lime caused significant increase in pH, orgC, available N, P and K in all the locations. Results (Table 29) also show that total acidity, exchange acidity, exchangeable aluminium and total potential acidity decreased upon liming.

Table 28: Changes in soil properties on liming.

Location and name of farmer	Treatments	pH (1:2.5)	EC mmhos	OC %	Available N, P and K ($kg\ ha^{-1}$)		
					N	P	K
Jhargram (Pukuria anchal)	RD	4.6	0.15	0.21	164.56	25.64	114.61
Subodh Mahato	RD + Lime	5.1	0.18	0.33	181.29	31.44	154.23
Jhargram, (Dahijuri anchal)	RD	4.9	0.14	0.29	129.92	17.63	170.61
Binpur-1Nirod Singh	RD + Lime	5.8	0.16	0.39	181.80	33.64	260.22
Jhargra (Dahijuri anchal) Binpur-1	RD	5.0	0.12	0.21	168.79	35.29	204.59
Sheikh Asimuddin	RD + Lime	5.2	0.17	0.24	229.52	54.79	282.24
Purulia (Gopladih)	RD	5.5	0.20	0.34	147.89	18.05	213.55
Dhiren Mahato	RD + Lime	6.1	0.29	0.35	186.12	29.33	305.24

Table 29: Effect of liming on the forms of soil acidity after harvest of mustard crop.

Location and name of farmer	Treatments	Acid parameters{ Cmol (p^{+}) kg^{-1}} Total acidity	Exchange acidity	Exchangea-ble aluminium	Total potential acidity
Jhargram (Pukuria anchal)	RD	2.65	2.04	1.54	5.67
Subodh Mahato	RD + Lime	1.54	0.96	0.88	4.27
Jhargram, (Dahijuri anchal)	RD	1.78	0.79	0.54	5.34
Binpur-1Nirod Singh	RD + Lime	0.81	0.25	0.33	3.98
Jhargra (Dahijuri anchal)	RD	1.77	0.63	0.63	5.33
Binpur-1Sheikh Asimuddin	RD + Lime	1.48	0.35	0.30	4.00
Purulia (Gopladih)	RD	1.80	0.50	0.63	4.67
Dhiren Mahato	RD + Lime	1.49	0.25	0.36	3.03

Screening of acid tolerant genotypes of crops

Promising varieties of pulses and oilseed crops commonly grown by farmers in acid soils, were grown in presence and in absence of lime, to identity relatively acid tolerant genotypes/varieties.

Assam

Yield data of rapeseed genotypes tolerant to soil acidity are presented in table- 30. The low yield of the crop may be attributed to typical upland condition and rain fed situation. Amongst the genotypes, *TS 29* and *TS 36* were promising as liming showed 19.26 and 19.05 per cent response in yield, respectively. In general, liming brought about a significant effect on yield and per cent response of lime varied from 1.31 to 19.26 per cent among the six genotypes of rape seed. The local variety of rape seed i.e. *Sonmukhi* performed equally in respect of yield under limed (4.03 q ha^{-1}) and unlimed (4.57 q ha^{-1}) condition.

Among the six cultivars of green gram, *ML 131* and *Pratap* showed higher response to lime in acid soil. The response of different cultivars was *ML 131* (74.96%) > *Pratap* (52.7%) > *AAU 34* (28.99%) > *GS 24* (24.89%) > *Sonmogu* (9.06%) > *K 851* (3.87%). *K 851* showed almost similar yield in lime (7.78 q ha^{-1}) and unlimed (7.49 q ha^{-1}) condition and can be recommended for resource poor farmers, who can not afford to invest on lime application in acid soils.

Table 30: Varietal screening of highly profitable cultivars of Rapeseed and Green gram in acid soils region of Assam.

Cultivars	Rape seed * (qha^{-1})			Green gram ** (qha^{-1})			
	No lime	Lime	Per cent response	Cultivars	No lime	Lime	Per cent response
M-27	4.65	5.29	13.76	Pratap	7.35	11.23	52.70
TS 29	4.57	5.45	19.26	K 851	7.49	7.78	3.87
TS 36	4.83	5.75	19.05	ML 131	6.59	11.53	74.96
Varuna	4.27	4.67	9.37	AAU 34	7.67	9.89	28.99
Pusa Kalyani	4.50	4.86	8.00	GS 24	6.95	8.68	24.89
Sonmukhi	4.57	4.63	1.31	Sonmogu	7.84	8.55	9.06

*One year on -station trial**Pooled data of two year on- station trials.

Himachal Pradesh

Comparative performance of eight soybean cultivars/genotypes was studied in acid soil condition of Himachal Pradesh under limed and unlimed condition. Response of lime in different cultivars of soybean varied from 12.88 to 37.18 per cent. *Shivalik* showed the highest response of lime (37.18%) followed by *Lee* (34.45%) cultivar of soybean, while minimum response of lime (Table 10) was recorded in *Harasoya* (12.88%). The increased yield with lime application was mainly due to increasing number of pod/plant or bolder seed size compared to that under un-limed condition.

Comparative performance of ten Brassica cultivars/genotype was studied in acid soil. (Table- 31). Perusal of data reveal that *NUDB 26/11*, *HPN-3*, *TERI-9903* and *CAN 130* performed very well under un-limed situations compared to rest of the cultivars . The yield of *HPN-3* and *TERI-9903*, however, decreased under limed situations. The yield potential of both cultivars was around 5 to 6 q ha^{-1}. Per cent response on yield due to liming varied from -45.27 (*HPN 3*) to 130.24 (*GSE 1*) among the cultivars tested for Brassica in acid soils.

Table 31: Varietal screening of Soybean and *Brassica* cultivars in acidic soil conditions of Himachal Pradesh.

	Yield (qha^{-1})						
	Soybean**			Brassica*			
Cultivars	No lime	Lime response	Per cent	Cultivars	No lime	Lime	Per cent response
Bragg	16.55	20.26	22.42	GSE 3A	4.76	5.52	15.97
Harasoya	15.06	17.00	12.88	GSE 1	2.48	5.71	130.24
P-10	16.38	19.07	16.42	NUDB 26/11	6.29	9.84	56.44
Lee	13.12	17.64	34.45	BPN-1	3.75	5.52	47.20
Palamsoya	13.15	16.59	26.16	HPN – 3	6.03	3.30	-45.27
P-16	16.30	20.35	24.85	NRCN-1	2.41	4.38	81.74
Shivalik	13.69	18.78	37.18	TERI 985	2.60	5.21	100.38
Pb – 1	12.26	14.81	20.80	CAN 130	5.33	8.76	64.35
-	-	-	-	ORK- 1	4.70	5.59	18.94
-	-	-	-	TERI –9903	5.40	3.24	-40.00

*On the basis of one year data

**Pooled data of 3 year

Jharkhand

On-farm trials were conducted with three pea cultivars viz., *Arkel*, *BR-12* and *Swarnrekha* to judge the performance of lime in acidic soil (pH <5.5) on pea yield (Table- 32). Per cent response to lime was highest with *Arkel* followed by *BR-12* and *Swarnrekha*. *Swarnrekha* cultivar of pea had the highest yield (11.80 q ha^{-1})

under unlimed condition and response to lime was low (3.89%). This variety was suitable for resource poor farmers of the acid soil region.

Table 32: Response of some Pea cultivars under limed and unlimed condition in acidic soils of Dumka

Varieties	Grain Yield (qha^{-1})		
	Unlimed	Limed	Per centresponse
Arkel	4.40	6.00	36.36
BR–12	9.90	10.60	7.70
Swarnrekha	11.80	12.20	3.39

One year on-farm trial

Response of lime varied from 32.72 to 95.61 per cent in different cultivars of Black gram in trials conducted continuously for three years in acidic soils (pH < 5.5) of Ranchi. BU-1 showed the highest response (95.01%) of lime application followed by TU-44-1 (63.35%) and PU-19 (57.85%), while lowest response (32.72%) was recorded in KU-301 (Table- 33).

Table 33: Pooled yield (qha^{-1}) and response of Blackgram cultivars under lime and no lime situation in acid soils of Ranchi

Cultivars	No lime	Lime	Per cent response
TU 94-2	3.83	5.99	56.40
TU 99-18-2	3.87	5.42	40.05
PU–19	4.46	7.04	57.85
BU–1	3.19	6.24	95.61
KU–301	3.24	4.30	32.72
T–9	4.40	6.52	48.18
*TU–44–1***	1.91	3.12	63.35

*Pooled data of 3 years

** one year data of on-farm trial.

Kerala

Lime application significantly increased the grain yield in all eight cultivars of cowpea. Response to lime application varied from 5.67 to 47.52 per cent among the cultivars tested (Table- 34). In unlimed condition *Subhra*, *Chharodi*, and *C-152* gave higher yield of cowpea grain (4.50, 4.36 and 4.07 q ha^{-1}) as compared to others. In limed condition, trend of yield was different, *Chharodi* gave the highest yield (5.68 q ha^{-1}) followed by *C-152* (5.43 q ha^{-1}) and *Subhra* (5.24

q ha^{-1}). Yield of *V-16* was on par in both unlimed (3.00 q ha^{-1}) and limed (3.17 q ha^{-1}) conditions.

Lime application in acid soil increased the vegetable cowpea yield in all the varieties except *TC-99-1*. The variety, *TC-99-1* was found to be acid tolerant and very high yielding and did not respond to lime application (Table- 34). All other varieties recorded increase in yield due to lime application and responses varied from 5.77 to 49.40 per cent. The local variety (*Local-2*) recorded 49.40% and *PTB-1* recorded 35.45% increase in vegetable cowpea yield due to lime application. Among the limed plots, *PTB-1* recorded the highest yield of 6.19 t ha^{-1}.

Table 34: Varietal differences in cowpea grown in acid soils of Kerala.

	Grain cowpea (qha^{-1})				*Vegetable cowpea (qha^{-1})		
Cultivars*	No. lime	Lime	Per cent response	Cultivars*	No. lime	Lime	Per cent response
Chharodi	4.36	5.68	30.27	TC-99-1	7.17	5.70	-20.50
Subhra	4.50	5.24	16.44	V-585	2.64	3.15	19.32
Krishnamani	3.96	4.58	15.66	Local-1	3.70	4.23	14.32
Kankamani	3.13	4.24	35.46	PTB-1	4.57	6.19	35.45
GC – 3	3.23	4.73	46.44	Pusa Komal	3.82	4.92	28.80
V – 240	2.82	4.16	47.52	Anaswara	2.75	3.31	20.36
V – 16	3.00	3.17	5.67	Bhagya-lekshmi	3.81	4.03	5.77
C - 152	4.07	5.43	33.41	Local-2	3.32	4.96	49.40

* One year data of on-station trial

Maharastra

Response due to liming varied from 12.90 to 20.32 per cent among six cultivars of groundnut (Table- 35). In unlimed situation, highest yield was recorded in *M-13* (27.31 q ha^{-1}) followed by *Konkan Gourav* (26.75 q ha^{-1}), *Konkan Tapora* (25.73 q ha^{-1}), *TG-26* (24.85 q ha^{-1}), *Phule Pragati* (22.85 q ha^{-1}) and *SB-XI* (20.75 q ha^{-1}). The dry pod yield and response due to lime in case of *TG-26*, & *Konkan Topora* were significantly higher over *SB-XI* and *Phule Pragati*.

Table 35: Yield of cultivars of groundnut (qha^{-1}) in acid soils of Maharastra

Cultivars*	No lime	Lime	Percent response
SB – XI	20.75	23.98	15.57
Phule pragati	22.85	26.29	15.05
TG – 26	24.85	29.90	20.32
Konkan Tapora	25.73	30.35	17.96
Konkan Gaurav	26.75	30.20	12.90
M – 13	27.31	31.47	15.23

* Pooled data of 3 year on- station trials.

Meghalaya

In acidic soil of Meghalaya, five varieties of soybean were tested for their response to lime under acidic soil condition. Response of different varieties to lime varied from 17.87 (*Ankur*) to 33.54 (*MACS 694*) per cent. Among the varieties, *TS 80-1* can be considered comparatively tolerant to soil acidity(Table 36).

Among the varieties of groundnut tested, the highest response of lime was found in *JL-24* cultivar (124.38%) followed by *ICGS-76* (116.38%), *VRI-2*(109.78%), *OG-52*(104.82%), *ICGS-44*(99.32%) and *K-134* (86.33%). In unlimed condition, yield was highest in JL-27

Table 36: Varietal screening of Soybean, Frenchbean, Groundnut and Mustard in acid soils of Meghalaya.

	Soybean (qha^{-1})				French bean (qha^{-1})		
Cultivars	Unlimed	Limed	Per cent response	Cultivars	Unlimed	Limed	Per cent response
Ankur	16.37	21.86	33.54	Amber	9.75	11.84	21.44
MACS 694	14.10	16.62	17.87	Arka komal	9.02	10.30	14.19
TS 80-1	17.86	23.35	30.74	HUR –14	6.85	9.40	37.23
MAUS 98-2	15.86	19.79	24.78	HUR –15	4.18	4.60	10.05
MAUS 695	15.94	19.83	24.40	HUR –139	5.52	8.55	54.89
-	-	-	-	HUR –168	6.20	9.80	58.07
	Groundnut (qha^{-1})				Mustard (qha^{-1})		
ICGS – 76	11.17	24.17	116.38	Panchavati	0.47	4.00	751
JL –24	13.00	29.17	124.38	TS –36	0.67	3.50	422
OG –52	10.17	20.83	104.82	SEJ –2	0.30	3.77	1156
ICGS-44	8.78	17.50	99.32	TS –38	0.67	3.70	303
VRI-2	10.33	21.67	109.78	TS –46	0.30	3.30	1000
K -134	11.85	22.08	86.33	PT -303	0.46	3.56	674

*One year data of on- station trial

cultivar (29.17 q ha^{-1}). *K-134* and *ICGS-44* were lower responsive cultivars of groundnut to lime application.

French bean in acidic soil of Meghalaya was responsive to liming (Table- 36). Response of lime in the six varieties varied from 10.05 to 58.07 per cent. Least response of lime was with *HUR-15* and highest with *HUR-168*.

Odisha

All the varieties of groundnut (Table- 37) showed good response of lime on pod yield and responses varied from 33 to 51 per cent. In unlimed condition *Smruti* (13.8 q ha^{-1}) and *Somanath* (13.2 q ha^{-1}) recorded the highest yield, while in limed condition, *Smruti* and *ICGV 95296* showed better performance than others with 19.7 and 19.3 q ha^{-1} pod yield, respectively.

In another trial, *AK-12-24* did not show response of lime application due to its acid tolerant nature. This cultivar was a good yielder (24.8 q ha^{-1}) under unlimed condition (Table- 38). *TAG-24* recorded highest response (40%) to lime followed by *TG-3* and IG*-26,* respectively with 39 and 27 per cent yield response over unlimed condition. In a trial with a series of *KCM* varieties of pigeon pea, *KCM 33* responded (120.4%) the best and *UPAS 120* responded the least (33.1%), among the ten cultivars. In unlimed condition, *UPAS - 120* cultivar of pigeon pea showed better performance with 16.9 q ha^{-1} yield. Response of lime in *KCM-33* and *KCM-303 varieties of pigeonpea under acidic soils,* was more than 100 per cent, while other varieties also showed a good yield response due to liming (33.1 to 97.6 per cent) .

Table 37: Pod yield (q/ha) of groundnut genotypes as influenced by liming.

2003 – 04				2004 – 05			
Variety	Lo	L	Response (%)	Variety	Lo	L	Response (%)
Gangapuri	12.0	16.5	36	TMV –2	20.8	24.7	19
Somanath	13.2	17.7	34	TAG –24	18.2	25.5	40
DH-45	11.9	17.1	43	AK-12-24	24.8	25.3	2
Smruti	13.8	19.7	43	TG –3	15.7	21.8	39
ICGV- 95296	11.9	19.3	62	IG –26	18.8	23.9	27
ICGV- 95249	9.8	16.4	33	NFP-M-D	20.6	24.3	18
ICGV- 95299	12.7	18.1	43	Smruti	24.2	27.6	14
BARC –4 –3	11.6	17.5	51	Kushali	18.3	21.5	18
Mean	12.1	17.8	-	Mean	20.2	24.3	-

Table 38 : Response of pigeon pea varieties to lime application in acid soil

Variety	Seed yield of pigeon pea			
	Nolime	Lime	Mean	Response (%)
KCM -154	10.8	15.2	13.0	40.7
KCM -350	8.4	16.6	12.5	97.6
KCM -33	4.9	10.8	7.9	120.4
KCM -316	14.7	18.9	16.8	29.0
KCM -609	13.4	20.7	17.1	54.5
KCM -356	14.8	20.1	17.5	35.8
KCM -247	8.1	15.6	11.9	92.5
KCM -252	14.0	21.7	17.8	56.1
KCM -303	10.0	20.1	15.1	101.0
UPAS- 120	16.9	22.5	19.7	33.1
Mean	11.6	18.2	-	-

West Bengal

The varieties of mustard responded well to lime application in acid soils of West Bengal. In unlimed condition, *Sanjukta* cultivar of mustard recorded highest yield of 8.88 q ha^{-1} followed by *Bhagirathi* with 8.56 q ha^{-1} yield (Table- 39).

Table 39: Screening and response of mustard cultivars to liming in acidic red laterite soils of West Bengal.

Cultivars	Yield (q/ha)		
	Unlimed	Limed	Response (%)
B -85*	2.20	7.00	218
Varuna*	3.30	8.00	142
Pusa Bold*	4.50	9.50	111
Sarma*	3.30	8.00	142
Kranti*	2.50	8.30	232
Sarma**	6.82	10.83	59
Sita***	7.40	12.71	72
Sanjukta***	8.88	14.71	66
Bhagirathi***	8.56	13.87	62

* Data of one exp. With $1/10^{th}$ LR.

** Pooled data of 7 expts. with $1/5^{th}$ LR.

*** Pooled data of 11 expts. with $1/5^{th}$ LR.

Three wheat cultivars viz., *Rajlaxmi*, *PBW-343* and *K-9107* were tested in limed and unlimed situation in acidic red lateritic soils of West Bengal. Response of lime over unlimed situation was 62.23, 57.42 and 42.37 per cent, respectively in *K-9107*, *PBW-343* and *Rajlaxmi* cultivar of wheat when grown on acidic soils (Table- 40).

Table 40: Screening and response of Wheat cultivars to liming in acidic red lateritic soils of West Bengal

Cultivars	Yield (q/ha)		
	Unlimed	Limed	Response (%)
Rajlaxmi	15.67	22.31	42.37
PBW – 343	17.31	27.25	57.42
K – 9107	16.47	26.72	62.23

* Pooled data of 7 expts. with 1/5th LR.

Acid tolerance and response to lime application in different crops and varietities

On the basis of acid tolerance and lime responsiveness, the following varieties were screened to be tolerant to soil acidity (Table- 41).

Table 41: Crop varieties relatively tolerant to soil acidity (pH < 5.5) in different regions.

Acid Soil Region / State	Crop	Varieties
Assam	Rape seed	Varuna , Sonmukhi
	Summer green gram	K 851, Sonmugu
Himachal Pradesh	Soybean	Bragg, Pb I, Harasona
	Gobhi sarson	ONK I, Hoyala, HPN –3
Jharkhand	Blackgram	KU – 301
Kerala	Vegetable cowpea	Bhagyalakshmi
	Cowpea (Bush type)	V – 16
Meghalaya	French bean	HUR - 15
Orissa	Groundnut	Smruti
	Pigeonpea	UPAS -120
West Bengal	Mustard	Sanjukta, Pusa bold
	Wheat	K – 9107, PBW - 343

References

Anonymous (2006) Consoildated report of ICAR Network Project on Soil characterization and Resource Management of acid Soil Regions for increasing productivity (2000-2005), Nodal Centre, Birsa Agricultural University, Ranchi, pp1-53.

Sharma, P.D. and Sarkar, A.K. (2005) Managing acid soils for enhancing productivity,Technical Bulletin. 23. Indian Council of Agricultural Research, Natural Resource Management Division, KAB II, New Delhi. pp 1-24.

6

Acid Soils of North-Eastern India

Mausumi Raychaudhuri and S. Raychaudhuri

The north eastern hill region covers a large share of acid soils of India occupying 14.95 m ha which is a major constraint for sustainable crop production. Acid soil limits the availability of calcium and phosphorus in soil, lowers base saturation and reduces microbial activities in soil adversely affecting the crop growth and yield. About 95 % of the soils of this region, except Nagaland (about 70 %) are acidic in reaction. About 84, 77, 76, 60, 57 and 47 per cent soils of Arunachal Pradesh, Manipur, Meghalaya, Mizoram, Sikkim and Tripura, respectively have soil pH below 5.0 and are considered strongly acidic (Panda, 1998). The soils are highly degraded because of heavy rainfall and removal of top soils. Land slides are also contributing towards land degradation. About 16 % of the area is affected by water erosion (www.india stat.com). About 53.8, 28.2, 42.6, 53.9, 89.2, 60, 33 and 59.9 per cent area are degraded in Arunachal Pradesh, Assam, Manipur, Meghalaya, Mizoram, Nagaland, Sikkim and Tripura, respectively. Including Assam, north-east constitutes 80 % of the total area under acid soils of the country. The land use pattern of north-east region is presented in table 1 and area, production, yield of foodgrains and fertilizer consumption are presented in Table 2.

Table 1: Land Use Pattern in North Eastern Region (area in '000 ha)

	Assam	Arunachal Pradesh	Manipur	Meghalaya	Nagaland	Tripura	Mizoram	Sikkim	Total NE Region	All India
Total Geographical Area	7844	8374	2233	2243	1658	1049	2108	710	26219	328726
Forests	1853	5154	1693	948	863	606	1594	319	13030	69635
Area not available for cultivation	2626	64	27	226	98	134	133	250	3558	43324
Fallow Lands	186	110	0	215	163	2	231	35	942	24857
Net Area sown	2753	211	236	284	316	280	95	107	4282	141364
Total cropped area	3984	276	236	337	402	295	95	118	5743	195104
Cropping Intensity %	144.7	130.6	100	118.7	127.2	105.5	100	109.8	117.1	138
Area available for shifting cultivation.	839.1	308.8	1201.4	208.6	522.5	605	376.1	-	4061.5	3514.2

*Provisional data, '0' relates to the area below 500 hectares

Source: Directorate of Economics and Statistics, Department of Agriculture and Cooperation.

Other than soil acidity and associated problems, shifting cultivation, steep slope leading to high soil erosion, injudicious fertilizer use, low nutrient status, low fertilizer use efficiency, slow rate of adoption of modern technology, unavailability of economically viable liming materials, monocropping due to moisture stress in *rabi* season and lack of drainage and irrigation facility leading towards flood and drought condition are some of the management issues that have contributed towards the low productivity of this region.

Table 2: Total foodgrains production (2010-11) and fertilizer consumption (2009-10) of Northeast region

State	Area ('000 ha)	Production ('000 tonnes)	Yield (kg/ha)	Fertilizer Consumption ('000 tonnes)
Arunachal Pradesh	200.7	333.7	1663	0.82
Assam	2766.5	4876.5	1763	242.27
Manipur	264.1	592.7	2244	12.04
Meghalaya	132.6	239.0	1803	3.68
Mizoram	53.6	66.8	1246	5.47
Nagaland	298.8	568.3	1902	0.94
Sikkim	76.2	110.3	1448	0
Tripura	275.4	712.4	2587	14.2
NE Zone	4067.9	7499.8	1832	279.4
All India	126770.6	244779.7	1931	26486.44

Source: Directorate of Economics and Statistics, Department of Agriculture and Cooperation.

Classification of Soils

The north eastern hill region is in the warm perhumid agro-climatic region and classified into three sub-regions viz., i) Himalayan hills comprising Sikkim and Darjeeling hills of West Bengal ii) North eastern hills consisting Arunachal Pradesh, Meghalaya, Nagaland, Karbi Anglong and North Cachar hills of Assam and iii) Southern hills and valleys comprising Manipur, Tripura and Mizoram. The National Bureau of soil survey and land use survey (NBSS&LUP) in collaboration with State Governments have developed the soil resource map of the region on 1: 250,000 scale. The soils of these regions are classified as Inceptisol (45 %), Entisol (28.2 %), Ultisol (14.8 %), Alfisol (4.7 %), Mollisols (0.6 %) and Histosol (patches in Tripura). Majority of the soils come under Inceptisol covering 80 % in Tripura, 45.7% in Meghalaya, 38.4 % in Manipur, and 37.3 % in Arunachal Pradesh 33 % in Sikkim (Table 3).

Table 3: Major soil type of North- Eastern Hill Region

Soil Order	% Geographical area of the State						
	Arunachal Pradesh	Manipur	Meghalaya	Mizoram	Nagaland	Sikkim	Tripura
Inceptisols	37.3	38.4	45.7	37.3	76.0	33.4	80.0
Entisols	35.6	23.1	10.7	21.5	4.0	43.0	8.0
Ultisols	14.2	36.4	40.0	-	17.2	-	7.0
Alfisols	0.3	0.2	3.6	2.6	4.8	-	5.0
Mollisols	-	-	-	38.6	-	23.6	-
Misc. Land	12.6	1.9	-	-	-	-	-

Source: Velayutham, M and Bhattacharya, T. 2000.

The acid soils of this region show variation in their mineralogical, physical, chemical and biological properties owing to the variation in climate, vegetation, parent materials and degree of interaction among the active factors of soil formation.The great groups prevalent in Assam region are: Inceptisols-Paleudalfs, Epiaquepts; Alfisols- Eutrudepts, Dystrudepts; Entisols- Fluvaquents, Udorthents. Soil of hill top /upper hill slopes are shallow to deep, loamy skeletal to fine loamy and excessively drained with severe to very severe erosion hazards. Most of these soils have structural or coloured B horizon and are belong to *Typic/ Humic Dystrudepts* while the *Typic/ Lithic Udortents* show either eroded soils or the soils at initiation of development with A/C horizons. Soils of the valleys are very deep, fine loamy drained with very slight erosion. They have structural B horizon and are classified as *Humic Endoaquepts* (Singh et al 2005).

The great groups prevalent in Meghalaya region are: Ultisols-Kandihumults,Haplohumults, kandiudults; Inceptisols-Hapludalfs; Alfisols- Epiaquepts, Eutrudepts, Dystrudepts; Entisols- Fluvaquents, Udorthents.The great groups prevalent in North Eastern region are: Entisols- Fluvaquaents, Udifluvent, Udorthents; Inceptisols- Eutrudepts, Dystrudepts; Alfisols-Paleustalfs, Kanhapludalfs, hapludalfs; Ultisols- Kandihumults, kanhaplo humults, palchumults, Haplohumults, kandiudults, kanhapludalts, paleudalts, hapludults. (Staff, NBSS&LUP, 2002). NBSS&LUP identified 107 soil series in Assam, 12 series in Meghalaya and 48 soil series in Tripura. In Tripura 18 series are grouped under the order Ultisols (Kandihumults, kandiudults , paleudults, palchumults, Haplohumults, hapludults, kandiudults,), 5 in Entisols (Fluvaquents, Endoaquents, Udorthents), 24 in

Inceptisols (Endoquepts, Dystrudepts) and 1 in Alfisols (kandiudalfs). Wide variations are observed in the subgroup with Typic, Aeric, Humic, Fluvaquents, Oxic, Rhodic, Aeric, Lithic, Aquic, Oxy-aquic (Chhailengta-I) and Ruptic-Ultic,(Ganganagar) and soil texture varied from coarse loamy, fine loamy to loamy (Bhattacharya et al 2004).

The parent materials from which the soil has been derived mostly govern the distribution of clay minerals in the soils. Chakraborty *et al.* (1979) reported distribution of light and heavy mineral from 3 pedons developed under different climatic and topographic positions in the southern bank of the Brahmaputra river in Assam and indicated mixed mineralogy dominated by the resistant minerals, like quartz, zircon and weatherable minerals like mica, feldspar, chlorite etc. Dominance of illite has been reported in the alluvial soils of Assam (Chakraborty,1982). Soils of Nagaland and Cherapunji developed on the shale and siltstone with bands of weathered, rather soft and micaceous, medium grained yellowish-grey wackes and siliceous limestone are dominant in illite followed by kaolinite (Prasad *et al., 1980). A*lluvial soils of Pasighat, Tripura and Assam are dominant in illite and kaolinite (Prasad *et al., 1980;* Chakraborty *et al.,* 1982). The soils of Upper Shillong Nayabunglow and Burnihat developed on quartzite are dominant in kaolinite and illite (Prasad *et al.,* 1980). The highly leached hill Ultisols of Manipur were found dominant in kaolinite (Mausumi Raychaudhuri 1995).

Soil acidity

Laterisation, podsolisation, intense leaching and accumulation of undecomposed organic matter under marshy conditions have contributed towards acid soil development. In addition, soil acidification due to prolonged use of acid forming fertilizers and heavy leaching are also contributory factors. The occurrence of acid soils in this region is presented in Table 4 and Fig 1.

Table 4: Occurrence of acid soil in North Eastern Hill Region (Area in '000 ha)

State	Strongly acidic (pH < 4.5)	Moderately acidic (pH 4.5-5.5)	Slightly acidic (pH 5.5 - 6.5)	Total
Arunachal Pradesh	4.78	1.74	0.27	6.79
Assam	0.02	2.31	2.33	4.66
Meghalaya	-	1.19	1.05	2.24
Manipur	0.43	1.44	0.32	2.19
Mizoram	-	1.27	0.78	2.05
Nagaland	0.12	1.48	0.05	1.65
Tripura	0.06	0.75	0.24	1.05
Sikkim	0.28	0.32	-	0.60
Total	5.69	10.50	5.04	21.23

Sources: NBSS & LUP, Nagpur

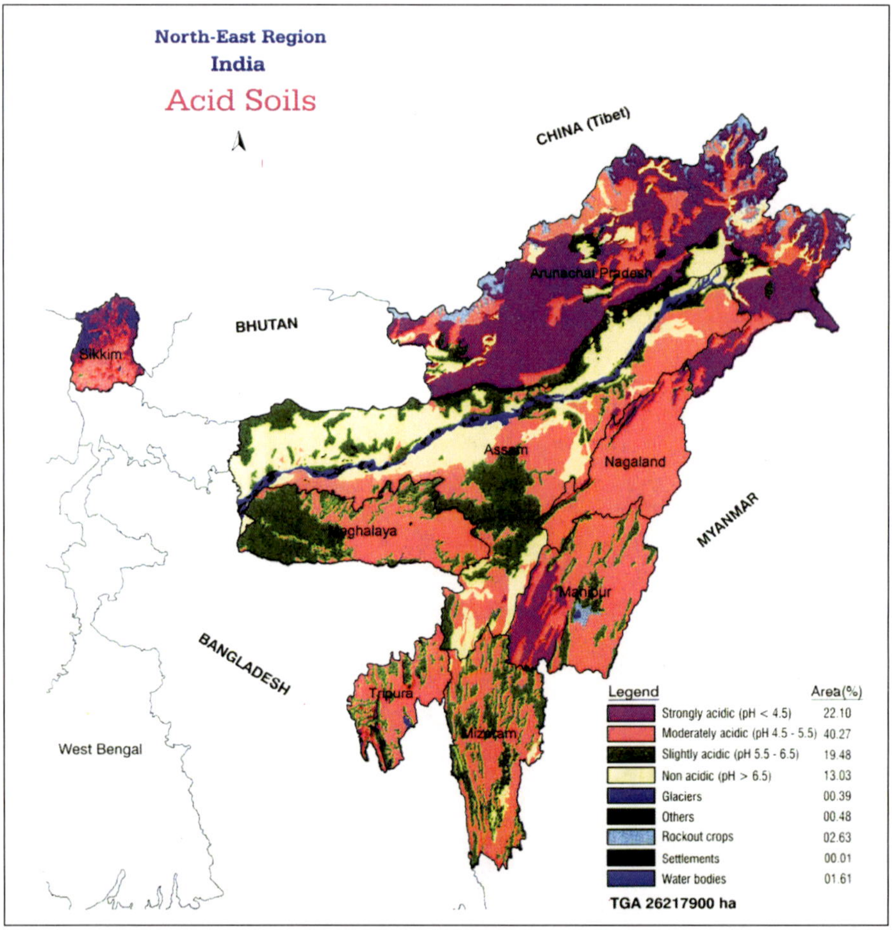

Fig 1: Occurrence and extent of acid soils of North-east India

Heavy rainfall, altitude and slope made the hill and tilla soils of this region more acidic than the valley soils. Several researchers (Singh and Datta 1988, Lahiri and Chakravorty 1989, Gangopadhyay et al 1990) reported decrease in pH and clay content and increase in sand CEC and organic matter with increase in altitude. Total acidity of Manipur valley soils ranged from 3.45 to 18.67 cmol (p^+) kg^{-1} and the hill soils ranged from 10.2-28.8 cmol (p^+) kg^{-1} (Mausumi Raychaudhuri et al 2000). Sannigrahi and Pandey (2000) also reported higher amount of total acidity in the hill soils than the valley soils.

Different Forms of Acidity

Total acidity of soil comprises of pH dependent acidity and exchangeable acidity due to H^+ and Al^{3+}. The pH dependent acidity contributes more towards soil acidity than exchangeable acidity due to high organic matter content and presence of oxides of Al and Fe in this region. The relatively low contribution of exchangeable acidity towards total acidity is also reported by several workers (Pratt and Blair, 1962; Mclean *et al.*, 1964). Sharma *et al.*, 1990 reported that the electrostatically bound H^+ and Al^{3+} comprised 21 and 79 % of exchange complex, respectively, whereas pH dependent and exchange acidity accounted for 71 % of the total acidity of the acid soils of India. The rest of the unaccounted acidity was due to hydrolysis of Fe and Mn on the exchange sites of the soil . In Assam, old alluvial soils are more acidic than soils close to Brahmaputra and pH ranges from 4.5 to 5.0. In soils of Toklai (Assam) and Aizawl (Mizoram), contents of exchange H^+, Al^{3+} and extractable Al^{3+} were 0.15, 0.08 and 0.70, 0.57 1.8, and 2.4 c mol (p^+) kg^{-1}, respectively (Sharma *et al.,* 1990). The pH dependent acidity comprising 60.1 and 50.0 per cent of total acidity was 7.1 and 4.7 c mol (p^+) kg^{-1} in Tocklai and Aizawl, respectively.

Kailash Kumar *et al.* 1995 reported that the total and pH dependent acidities of Manipur Hill Soils ranged from 10.2 to 28.8 and from 10.1 to 24.8 cmol (p^+) kg^{-1}, respectively. The pH dependent acidity was the dominant component of the soil acidity and contributed to 92.2 per cent of the total acidity, while the exchangeable acidity constituted only minor fraction (7.8 per cent) of the total acidity. The average contribution of EB-Al^{3+} and EB-H^+ acidity to the total exchange acidity was 60.7 and 39.3 per cent, respectively. Exchangeable $Al^{3+}$was more prominent in the soils with pH 4.5-5.5 where the exchangeable acidity was dominant. Land use pattern also affected various forms of soil acidities in such soils. The soils under shifting cultivation exhibited maximum acidities of either kind followed by forest cover and terrace cultivation. Major contributing factors for different forms of acidities in these soils were exchangeable Al^+, exteractableAl^+ , exchangeable H^+ and organic matter.

Cation exchange capacity (CEC) of hill soils is low. Different components of CEC of Manipur hill soils revealed that Neutral salt CEC ranged from 3.8 to 27.7 cmol (p^+) kg^{-1} and constituted 81 per cent of the total CEC of these soils (Kailash Kumar et al.1995). The total CEC which is the sum of neutral salt CEC and pH dependent CEC showed an average value of 15.3cmol (p^+) kg^{-1}. The pH dependent CEC contributed 19.7 per cent towards total CEC and ranged between 1.0 and 6.8 cmol (p^+) kg^{-1}. Weakly dissociated acid groups of organic matter contributed more towards pH dependent CEC (71.4 per cent) than sesquioxide coatings on the clay mineral surfaces or partially neutralized complexes of Al and Fe in the interlayers (28.6 per cent). Organic matter showed significant positive correlation with total CEC, neutral salt CEC, total pH dependent CEC and pH dependent CEC due to weakly associated acid groups of organic matter and accounted for 40.2, 27.9, 47,2 and 48.0 per cent variation in them, respectively.

Nair and Chamuah (1993) studied in detail the soil acidity in Meghalaya soils. The pH, exchangeable and extractable Al^{3+} varied from 4.7 to 5.4, 0.04 to 3.53 and 0.06 to 5.44 c mole (p+) kg:', respectively (Ram *et al.,* 1987). The pH of soils of Meghalaya were negatively correlated (r = - 0.76) with exchangeable Al^{3+} content in the soil (Prasad *et al.,* 1978). Laskar *et al.,* 1983 reported that about 90 percent of soils have pH below 5.6. Exchangeable Al^{3+} do not exceed 0.5 c mol (p^+) kg^{-1} in 95 per cent soils but Fe^{3+} and H^+ seem to have their contribution in soil acidity. Upland soils were having more saturation of H^+ and Al^{3+} as compared to lowlands. Both available Fe and Mn contents were present in high amounts in lowlands compared to uplands (Datta and Ram, 1993). So, iron toxicity is reported to be fairly widespread in lowlands comprising about 40.05 per cent of gross cultivable area in Tripura.

High organic matter content

Organic carbon content of the soils in the region is high, except some parts of Assam and Tripura owing to low decomposition rate.

The pH of soils collected from lowland valley and foot hills showed significant negative correlation (r = 0.273*) with soil organic matter content. The pH dependent acidity formed the major portion of acidity, whereas exchange acidity constituted a minor fraction of the total acidity. The weak acids formed during organic matter decomposition due to the presence of number of functional groups containing H^+ ions contributed towards pH dependent acidity. The organic matter content contributes more towards pH dependent acidity compared to oxides of Al and Fe. Patiram and Prasad 1991 and Kumar 1995 reported that the CEC of soils of this region was highly pH dependent.

Acid soils can be broadly categorized into two groups viz.,i) Acid soils with high organic matter content > 0.75 percent and ii) acid soils with low organic matter content <0.75 percent. The north eastern hill soils are dominated by soils with high amount of organic matter content and most of its characteristics viz., forms of acidity, forms of macro, secondary and micro nutrients in soil and their availability to plants are governed by organic matter content in the soil.

High Al^{3+} and Fe^{3+}content

The Al^{3+} replaces Ca^{2+} and Mg^{2+} (dominating bases), which are leached out due to heavy precipitation giving rise to the acidic characteristics of the soil. Prasad et al 1985 reported that Al^{3+} occupied 0.4 to 40.3 percent of the cation exchange capacity of the East Khasi Hill soils of Meghalaya.

The relative abundance of different forms of Al [cmol (p^+) kg^{-1}] in the hill soils is in the order exchangeable (0-2.38) < extractable (0.03-7.56) < crystalline (1.4-16.0) < amorphous (22.2-33.2) < organic bound (20.8-108.5) (Mausumi Raychaudhuri 2005) and the amount of exchangeable Al ranged from nil to 7.1 cmol (p^+) kg^{-1} (Prasad et al., 1985, Kumar and Raychaudhuri 1997, Patiram 1990, Das et al 1996 and Sen et al. 1992). The acid hill soils of this region have similar Al status. Organically bound form is the major fraction of Al in these soils due to the high organic carbon content (14 – 49 g/ kg) in the soil. Prasad et al 1985 also reported that Al bound to organic amorphous soil constituents were more in higher altitudes and they are correlated positively with soil organic carbon. The exchangeable Al^{3+} had significant positive correlation with total exchangeable acidity. The EB-Al^{3+} constituted the major portion of the exchangeable acidity in most of the soils. Soils where pH range is < 5.0, iron also behaves like aluminium and contributes hydrogen ions to the system. The total Fe content of these soils ranges from 2.76 to 5.62 per cent. Amorphous and organically bound portions contribute more than 65 per cent to the total iron content and DTPA extractable Fe is a minor fraction of the total Fe in these soils. Aluminium sensitive crops such as wheat, barley, soybean etc., get damaged when Aluminium saturation of cation exchange capacity exceeds 10 per cent (Patiram 1991). Upland soils have more saturation of H^+ and Al^{3+} as compared to lowlands. The lowlands are waterlogged due to lack of proper drainage. Singh et al 1992 reported iron toxicity in Meghalaya soils having soil pH <5.5, active Fe>1.25 % organic matter >3% with pH usually remaining below 6 after continuous submergence of soil. About 90% of the acid soils of Tripura with pH below 5.6 contain Fe^{3+} and H^+ that contribute towards soil acidity. Occurrence of Fe and Mn in high amounts in lowland soils was reported by Datta and Ram 1993. Widespread iron toxicity in lowlands covering about 40.05 per cent of gross cultivable area in Tripura was also reported.

Impact of Shifting Cultivation

Present *Jhum* practice with a reduced cycle of 3-5 years is an extravagant and unscientific form of land use that increases soil acidity and poses serious environmental, social and ecological problem. It leads to heavy loss of organic matter from the soil that suffers from erosion hazards. Kailash Kumar et al 1994 reported that total acidity ranges from 12.1 - 28.8 cmol (p^+) kg^{-1} in shifting cultivated areas compared to 11.0-21.5 cmol(p)kg-1 in the terraced cultivated areas. Studies at Manipur reveal that the extractable [4.06 cmol (p^+) kg^{-1}] and exchangeable [1.32 cmol (p^+) kg^{-1}] forms of Al dominates in these soils compared to that of terraced land [1.48 and 0.47 cmol (p^+) kg^{-1}, respectively] and under forest cover [3.11 and 1.01 cmol (p^+) kg^{-1}, respectively] (Mausumi Raychaudhuri 2005).

The National commission on Agriculture reported (1976) that about 2.7 million hectare of sloppy land of this region has been adversely affected due to shifting cultivation causing an annual loss of 181 metric tonnes of soil. The estimated nutrient loss per annum is about 603, 0.42 and 4.28 hundred tonne of organic carbon, available P and K, respectively. The shortened Jhum cycle, speedy deforestation and soil degradation were major contributory factors for the development of acid soils in this region.

Injudicious Fertilizer Use

The soils are rich in organic carbon content but the availability of N is not adequate due to low mineralization rate and the crops show deficiency symptoms of stunted growth and yellowish green colour. The crops are highly responsive to nitrogenous fertilizers. The total consumption of chemical fertilizers (N, P_2O_5 and K_2O) of this region (8 states) was 255800 tonnes among them Assam recorded highest (213940 tonnes) followed by Manipur (19010 tonnes) and then followed by Tripura (13330 tonnes) during 2007-08. Application of chemical fertilizers in the ratio of 11 : 2.6 : 1 in Manipur, 7.7 : 3.7 : 1 in Meghalaya and 5 : 2.1 : 1 in Arunachal Pradesh compared to that of FAO standard i.e. 4:2:1 as Nitrogenous: Phosphatic : Potassic fertilizers not only increases the acidity of the soil but also induce insect pest incidence. Use of nitrogenous fertilizers alone lowers the soil pH. This is because they do not have a basic cation to counteract the acidity produced during their breakdown to simpler forms. Thus, its injudicious use in these soils is a total disaster at the cost of the ecology of the soil. On the other hand, the total N, P_2O_5 and K_2O consumption of NE Region in the ratio of 2.1 : 1 : 1 shows a great scope to increase crop productivity with higher nutrient use.

Nutrient Status

The texture of hill soils of this region varies from sandy loam to clay loam where as the valley soils vary from sandy clay loam to clayey. Chakravorty et al 1978 reported that the old alluvial soils of the Brahmaputra valley are silty loam to clayey while new alluvial soils are mostly sandy, silt loam or clay loam. The soils are medium to high in organic carbon and the total N content; low to medium in available P content (Table 5) and medium to high in available K content. Almost all the soils of Meghalaya, Tripura, Manipur, Arunachal Pradesh, over 50 % of Nagaland, 40 % of Mizoram and 30 % of Sikkim are rated low in available phosphorus (Prasad et al.1981). Phosphate fixation due to high content of exchangeable Al and Fe is the key factor for P deficiency in the region.

Table 5: Different forms of Phosphorus in NEH Region

State	Range of Phosphorus (ppm)				Source
	Total	Organic	Bray P_1	Bray P_2	
Manipur	349-1993	67-454	-	3.0-26.7	Kailash Kumar and Mausumi Raychaudhuri 1997.
Meghalaya	375-1575	118-718	0.5-15.0	0.5-87.0	Patiram et al. 1993.
Mizoram	154-341	1.1-111	2.0-13.9	-	Singh, O. P. and Datta, B. 1987.
Nagaland	480-1920	----------	-3.1-48.6		Anonymous
Sikkim	376-1650	68-864	2.5 -101.0	-	Patiram et al. 1990a.

Prasad et al 1981 found that the critical limit of organic carbon for higher elevation (>1000) soils of Meghalaya was 1.5 per cent. However, the rate of mineralization is low due to acidic nature of soils irrespective of soil organic matter status. In most of the soils, N and P are the limiting factors for optimum crop yield. Studies on the P sorption and desorption characteristics of the soils of this region (Ram et al 1987, Patiram et al 1990b, and Patiram and Prasad 1991) indicated that the solubility of P in soil solution is governed by oxides of Al and Fe. The deficiency of K occurs on highly degraded soil where the topsoil is completely eroded due to heavy leaching or landslides. The Scientists of the National Bureau of Soil Survey and Land Use planning have estimated that 42 % of the area in NEH region covering Meghalaya, Manipur, Nagaland and Tripura and eastern part of Arunachal Pradesh is affected by water erosion. Sikkim and parts of Arunachal Pradesh are heavily damaged both by water erosion and landslides. The soils, which are strongly acidic in nature, are not favourable for crop growth, thus, left barren leading to excessive erosion and degradation.

Das et al 2011 observed that 40% soils in Dibrugarh district, 36.8% in Sibsagar, 9% in Jorhat and 20% soils in Gilaghat district of Assam are deficient in sulphur and need sulphur fertilization for sustainable crop production.

Micronutrient Status

Very little information is available about the micronutrient status of this region. The DTPA extractable Zn, Cu and Mn in soil ranges from 0.72-3.28, 0.6-2.8 and 3.0-162 ppm, respectively. Most of the soils of Sikkim are high in available Fe, Cu, Zn and Mn content. The distributions of the available status of these elements in different hill soils are presented in table 6. The availability of some of the micronutrients especially B and Mo are also affected by the acidity and often adversely affects crop yields.

Table 6: Available micronutrient status of NEH soils

State	Micronutrients (ppm)				Source
	Fe	Cu	Zn	Mn	
ArunachalPradesh	7.0-73.6	0.14-4.07	0.11-1.12	1.6-64.6	Anonymous, 1980
Manipur	17.8-243.5	0.4-7.6	0.4-5.8	27.5-248.7	Mausumi Raychaudhuri and S. Raychaudhuri 2006
Meghalaya	21-77	0.06-3.64	0.03-3.3	0.8-110.0	Patiram & Prasad, R. N. 1984.
Nagaland	10-92	0.2-2.5	4.5-55.6	10.5-34.7	Ghosh, G. & Ghosh, S. K. 1982
Sikkim	6.8-138.9	0.3-5.8	2.1-11.8	3.8-132.1	Patiram et al., 2000.
Tripura	1.8-61.8	0.2-2.2	2.1-4.0	11.4-33.6	Datta, M. & Ram, M. 1993.

Soil Management

The management of acid soil should aim to improve the production potential either by counteracting soil acidity through the addition of amendments, meeting the deficiencies through balanced fertilization with an integrated approach or by manipulating the agricultural practices through a farming system approach. Organic farming, Paddy cum fish culture, Integrated farming system, 3 tier farming system are some of the technological options to conserve soil and prevent erosion and associated loss of nutrients. But, due to small land holdings, low adoption rates and less awareness, soils are becoming more acidic day by day owing to natural and human induced factors. Amelioration of acid soils by liming material is the most suitable option for increasing the production potential of hill soils. There is about 5810 million tones of lime deposits in NEH region (Table 7) and the quality is cement grade containing about 50% of CaO. Most of the crops grown in this region are highly responsive to lime increasing the yield from 30 to 150 per cent. The lime requirement after every 3rd crop @ 1 t/ ha as estimated for the north eastern region soils with pH below 5 is 3.75 m tones, which can be easily obtained from its own reserve for 2324 years or more. This needs attention of the administrators, researchers, industrialists and extension workers to popularize liming among the farmers and also to frame policies so as to make lime available to the farmers at a cheaper cost.

Table 7: Availability of limestone and lime requirement of North-east Region

States	Total acid soil (,000 ha)	Net cropped area ('000 ha)	Soils below pH 5 require amendment ('000 ha)	Lime (CaO) reqd (Mt) @ 1t/Ca CO_3/ha	Limestone deposits (Mt)
Arunachal Pradesh	1955	200	1642	1.642	383
Assam	612	2774	153	0.153	963
Manipur	481	223	370	0.37	21
Meghalaya	1030	212	783	0.783	4308
Mizoram	1050	97	630	0.63	-
Nagaland	127	309	55	0.028	135
Sikkim	76	112	43	0.043	-
Tripura	203	280	95	0.095	-
Total	5534	4207	3771	3.745	5810

Compiled from data available in www.indiastat.com

Lime requirement of acid soils are estimated by two chemical ways. One way is to apply lime to increase the soil pH to 7.0 that will enhance macro and secondary nutrient availability, symbiotic and non-symbiotic biological activity, and will reduce the toxic effect of micronutrients. The other way is to add lime to neutralise the exchangeable AI content in the soil that may be uneconomical. Average per cent contribution of different forms of Al to the total Al was 0.27, 0.81, 4.68, 3.76 and 1.00 for exchangeable, extractable, organic bound, amorphous and crystalline forms, respectively that was found in hill soils of Manipur. The entire acid hill soils of this region have the similar Al status because of their high organic matter content. Application of lime to neutralize the exchangeable Al has been found suitable for Sikkim (Pradhan and Khera, 1976) and Meghalaya (Prasad *et al.,* 1982).

Dutta *et al.* (1983) have reported high values of lime requirement (3.2-27.2 tonne/ha) for the soils of Nagaland. Lime requirement by Woodruff's method for the soils of Tripura was estimated as 1.35 to 4.95 tonne/ha (Laskar *et al.,* 1983). The lime requirement of Sikkim soils by Shoemaker's, Woddruff's and Hunter's method range between 4.4 to 17.7, 2.9 to 10.1 and 0.5 to 2.0 tonne/ha, respectively (Gupta and Prasad, 1983). Lime requirement methods were standardized for Manipur hill soils by Kailash Kumar *et al.*(1997). The methods of Shoemaker *et al.* (1961), Woodruff (1948), Brown and Cisco (1984) and Pratt and Blair (1962) showed high degree of correlation with the standard LR by the method of Bhumbla and Mclean (1965). The LR determined by Shoemaker *et al.*, Woodruff, Brown and Cisco , Pratt and Blair and Bhumbla and Mclean showed significant correlation with all forms of soil acidity. Variations in the LR by these methods accounted more by exchangeable acidity than pH dependent acidity. On the basis of R^2 values Brown and Cisco buffer method was found to be the most promising for the determination of LR in these soils.

Bora (1967), reported that lime requirement decreased in the order: *Bheel* soil> lateritic soil > old alluvial hill soils > new alluvial soils in representative soil samples of Assam, Meghalaya and Mizoram. Lime requirement of rice growing soils of Sibsagar district of Assam were significantly correlated with clay CEC, pH of soils (Borthakur and Majumdar 1967).

Crop response to Liming

Liming increases the availability of most of the nutrients viz. nitrogen, phosphorus, calcium and magnesium to the plant and has beneficial effect on these type of soils in increasing soil productivity (Mandal et al 1966, Prasad et al. 1983, Datta and Gupta 1984, Gupta et al. 1989 and Patiram 1991) and application of 0.25 LR (Lime Requirement) based on buffer method was recommended for optimum yield of crops. Maize, the second important crop grown in this region after paddy was found highly responsive to lime. Application of lime significantly increased the maize yield in Meghalaya (Prasad et al, 1978), Sikkim (Patiram 1991), Nagaland (Datta et al 1983) and Manipur (Raychaudhuri and Kumar 2002) (Table 8). Liming beyond 0.25 LR level decreased maize yield significantly under controlled condition (Mausumi Raychaudhuri, 1997). Pressmud, an industrial waste of sugar mill factory contains 40% of CaO,14%OC,0.7% N, 0.3%Pand 0.4%K is an alternative for lime which increased the maize yield with increased doses without showing any adverse effect (Goswami et al,1976; Datta and Gupta 1983) but decreased yield when applied in higher doses (Raychaudhuri and Raychaudhuri) . Prasad *et al.* (1983) observed that application of lime equivalent to 25% of lime requirement showed direct and residual effect on the grain yields of maize in the soils of East Khasi Hills.

Liming showed an antagonistic effect on the availability of soil K and K uptake by the plants beyond 0.25 LR of lime application (Mausumi Raychaudhuri and Sanyal 1999) in organic matter rich acid hill Ultisol. This is because with the presence of high aluminium content in different forms and organic matter the soils are not expected to follow the normal Gapon exchange behavior i.e. exchanges involving K^+ vs Ca^{2+} + Mg^{2+} ions. Mausumi Raychaudhuri and Kumar 2002 recommended increase in amount of K fertilization with increased level of lime beyond 0.25 LR to maintain the availability of soil potassium to maize and soybean under terraced cultivation.

Table 8: Effect of liming on crop yield grown in different States of North-east Region

State	Crops	Grain yield(q/ha)		Increase over control (%)
		Without lime	With lime	
Manipur	Maize	13.60	34.70	255
	Pigeon Pea	11.90	28.00	235
Meghalaya	Maize	10.35	43.30	319
	Soybean	16.00	30.00	87
	Ragi	18.00	19.00	38
	Rice	18.00	19.50	8
	Wheat	4.25	7.80	83
	Pea	0.46	3.38	634
	Lentil	1.20	3.58	198
	Mustard	0.55	2.80	409
	Bengal gram	1.60	3.75	134
Nagaland*	Maize	30.33	49.22	62
	Soybean	10.97	17.78	62
	Rice	59.39	59.57	0.03
	Wheat	11.52	24.47	112
	Pea	17.23	29.25	70
Tripura	Maize	7.98	13.65	71
	Soybean	4.38	9.35	113
	Groundnut	10.70	14.03	31
	Moong	2.43	3.72	53

*2 tonnes pressmud/ha was used as liming material
Source: Compiled from Annual Reports of ICAR Research Complex for NEH Region

Liming enhances legume – Rhizobium symbiosis by increasing the Ca level in the soils which enhances the root penetration of leguminous crops into deeper layer and also induces normal distribution of nodules on the tap root and lateral root by *Rhizobium* (Mausumi Raychaudhuri et al. 1997). Ca^{2+} starvation reduced greatly the growth of *Rhizobium*, number of nodules and nodule weight. This is because a major 38 KD_a bacterial protein was regulated by the availability of Ca^{2+} but this protein did not bind Ca^{2+} itself (Balatti et al 1991). Liming in association with P fertilization (Mausumi Raychaudhuri et al 1998) not only increased the yield, nitrogen fixation, nodule weight and nutrient uptake by soybean, but also increased the P use efficiency. Liming increased the efficiency of dual inoculation of *Rhizobium* and PSM in acid hill Ultisol and in combination increased the pod yield of groundnut by 75.4 percent (Mausumi Raychaudhuri et al. 2003)

Liming increased the nutrient uptake by the crops especially N and P uptake. The per cent increase in N, P and K uptake by the grains of maize and followed soybean trend in i.e. N uptake>P uptake> K uptake. This is because N and P concentration in grains of both crops increased while the K concentration decreased with increasing lime levels (Mausumi Raychaudhuri and Kailash Kumar, 2002). Bishnoi (1994) reported increase in N uptake by soybean on liming due to increase in the available soil N content. Datta and Gupta (1984) reported increase in the available P content of the soil and P uptake by maize with liming. Sudhir (1983) reported antagonistic effect of Ca on K uptake by groundnut (Arachis hypogaea L.).

Mausumi Raychaudhuri et al 1997 reported that liming increased the N uptake by the soybean grains significantly. *Rhizobium* inoculation (TAL 102) also increased the N uptake significantly. Lime and *Rhizobium* interaction was also significant as inoculation increased the N uptake in the limed soils significantly but the response was very low and insignificant in unlimed soils. This may be attributed to the fact that *Rhizobium* inoculation in limed and unlimed soils increased the *Rhizobium* population (Raychaudhuri and Raychaudhuri 2008). But the increase was maximum in limed soils which resulted in more nodule number, nodule weight and nitrogen fixation than the unlimed soils. Moreover, *Rhizobium* inoculation in limed and unlimed soils increased the organic matter contain of the soil which also contributed towards the N uptake of the grains.

Liming alone and in combination with *Rhizobium* inoculation increased the rhizobial population remarkably (Raychaudhuri and Raychaudhuri 2008). Moreover, if *Rhizobium* treated seeds are lime pelleted, that makes the micro environment for *Rhizobium* more conducive to multiply, reducing the initial shock for environmental difference for rhizobia.

Adsorption–desorption behaviour of P and Zn on liming was studied by Venkatesh *et al.* (2005). Phosphate adsorption was described successfully by Freundlich and Langmuir adsorption isotherms. Lime application increased P adsorption in both Alfosols and *Entisols*. Lime application had little impact on adsorption maxima while bonding energy increased with increasing lime levels. The Freundlich coefficient, 1/n decreased with lime application while Freundlich K did not show any definite trend. Phosphorus supply parameter of soil increased with increasing P levels and decreased with lime application in both *Alfisols* as well as *Entisols.* The Langmuir adsorption maxima, bonding energy and P- buffering capacity were positively correlated with amorphous Al, free Fe oxides and clay content (Majumdar et al 2005).

Adsorption of Zn significantly increased with different levels of Zn and lime. *Entisols* recorded more Zn adsorption than *Alfisols.* Lime application @ 0.5 LR and above resulted in more than 97 per cent of Zn adsorption as compared to un-limed condition. Supply parameter of Zn increased with Zn levels in all the soils with a maximum value at 10 $\mu g\ mL^{-1}$ added Zn. All the lime levels decreased Zn supply parameter except 0.25 LR in Alfisols. Desorption of Zn also decreased with liming and Alfisols recorded more desorption than Entisols. Supply parameter and desorption showed significant negative correlation with soil pH.

Liming increases the P and N use efficiency in the acid soils. This is due to the increases in the available P and N content in soil as a result of liming. Symbiotic N uptake by the soybeans grains increased to maximum with 0.50 LR and further increased when coupled with 26 kg P per ha followed by 0.25 LR and 26kgPha^{-1} (Table 9). Per cent utilization of the applied P and PUE also increased P fertilization and liming (Table 9).

Considering the high cost of lime, residual effect of lime has been studied and the lime was found effective even after fourth year of its application in maize (cereals) for first two years then followed by soybean (legumes) for next 2 years (Mausumi Raychaudhuri and Kumar 2002, Raychaudhuri and Raychaudhuri 2009). To minimize the amount of lime, furrow liming @ 250, 500 and 1000 kg/ha $CaCO_3$ has been recommended for crops (Patiram 1994, Mausumi Raychaudhuri and Raychaudhuri 2008). Liming @ 0.25

LR (Lime Requirement) for cereals through broadcasting or 500 kg/ha lime in furrows for legumes are recommended under permanent terraced cultivation in the acid hill soils of north eastern region (Table 10).

Table 9: Fertilizer use efficiencies in terms of nutrient uptake by the grains

Treatments	Symbiotic N uptake (kg/ha)		Fertilizer P uptake (kg/ha)		Apparent recovery of fertilizer P (%)		Phosphorus use efficiency (PUE)	
	1994	1995	1994	1995	1994	1995	1994	1995
0.25LR	32.4	29.5	3.85	2.07	-	-	-	-
0.50LR	49.5	49.1	5.30	4.36	-	-	-	-
P_{13}	18.9	17.2	1.72	2.22	13.2	17.0	19.2	18.5
P_{26}	50.2	46.4	4.92	4.88	18.9	18.5	27.3	24.2
0.25LR +P_{13}	40.9	39.0	4.74	3.55	36.5	27.3	44.6	36.9
0.25LR +P_{26}	78.4	69.3	8.50	6.23	32.7	23.9	43.1	35.0
0.50LR +P_{13}	58.2	55.9	6.16	5.45	47.4	41.9	61.5	54.6
0.50LR +P_{26}	98.8	98.9	10.34	9.74	39.8	37.5	52.3	47.7

Source: Mausumi Raychaudhuri et al 2007

Table 10: Nutrient management in oilseeds and pulses

Treatments	Soyabean (q/ha)	% increase over control	Black gram (q/ha)	% increase over control	Groundnut (q/ha)	% increase over control
Control	10.8	-	6.8	-	8.3	-
Furrow lime @500kg/ha	19.1	76.9	12.6	85.3	12.5	51.2
Farmyard manure @ 10 tonnes/ha	18.7	73.1	13.1	92.6	15.5	87.7
Lime+Farmyard manure	27.8	157.4	16.8	147.1	17.7	114.3

Source: Mausumi Raychaudhuri 2005.

Industrial wastes like lime sludge from paper mills, pressmud cake produced through carbonation process of sugar mills, cement kiln wastes can be used depending on the calcium content and availability. Datta and Gupta 1983 reported increase in yield of maize-wheat crop sequence with application of 2 t/ha press mud along with fertilizers in an acid soil of Nagaland. The neutralization capacity of some of the liming materials in terms of calcium carbonate equivalent are presented in Table 11.

However, liming is not recommended for lowland paddy grown under water logged condition.

Table 11: Calcium carbonate equivalent of some naturally occurring liming materials

Liming Material	Typical CCE (%)	Tons Required to be Equivalent to 1.0 Ton of $CaCO_3$
Calcite (pure)	100	1.0
Calcitic limestone	75 to 100	1.3 to 1.0
Dolomitic limestone	75 to 108	1.3 to 0.9
Aragonite	95 to 100	1.1 to 1.0
Hydrated lime $Ca(OH)_2$	120 to 136	0.8 to 0.7
Wood ash	30 to 70	3.3 to 1.4

Nutrient management

N recommendation for optimum yield of local rice varied from 40-60 kg N/ha and for improved cultivars,it varied from 60-100 kg/ha (Singh and Patiram 1987, Gupta and Laskar 1988, Patiram and Rai 1988, Patel and Singh, 1990 and Hussain and Sharma 1991). For upland rice 30 kg N/ ha (Patiram *et al* 1987) and for potato 120 kg N/ha (Sahota and Grewal 1984, Sharma and Awasthi, 1989, Patiram and Prasad, 1985) is recommended. Higher dose of N @ 80-100 kg N / ha and lower dose of N @ 20 – 40 kg/ha is recommended for cereals like maize, wheat, finger millet and oilseeds/pulses respectively (Laskar *et al* 1983, Patiram and Rai, 1988). Urea is widely used nitrogenous fertilizer in this region, which is subjected to heavy losses due to leaching, run off and denitrification processes. This is a major constraint for efficient use of native as well as applied N. In general, percent recovery of applied N is only about 10-15 per cent in this region and major portion is lost by the above processes. The use of slow release nitrogenous fertilizers viz., rock phosphate coated urea, urea super granule in place of urea is recommended.

Azotobacter, *Azospirillum* and phosphate solubilizing microorganisms (*Bacillus polymyxa*, *Bacillus megatherium*, *Pseudomonas striata* etc) are presently being used in large scale. Blue green algae (BGA) are long been recognized as potential Biofertilizer for flooded rice, but due to their stringent water management requirement, they are yet to be commercialized. Some other microorganisms such as VAM fungi, plant growth promoting rhizobacteria (PGPR, mainly fluorescent pseudomonads) and *Acetobacter diazotrophicus* are fast coming up as new biofertilizers.

Organic inputs like, Azolla @ 10 t /ha, suitable strains of *Azotobacter, Azospirillum*, Blue green algae, green manure and farm yard manure increased the N use efficiency and can supplement 15-30 kg N/ha to wet land rice grown in this soil. Incorporation of paddy straw @ 5 t/ha is also beneficial in mobilizing the nutrients for plant uptake (Table 12). Kumar and Raychaudhuri 2000 reported an increase of 20.8 percent in lowland paddy with *Azospirillum* over control (30.6 q/ha) and further increase when coupled with half dose of NPK. Thus, application of N fertilizers through a combination of organic (50 %) and inorganic (50 %) means is recommended for this region.

Table 12: Nutrient management in paddy

Treatmant	Yield (q/ha)	Per cent increase over control
NPK(60kg N,40 kg P_2O_5 and 30 kg K_2O)	56.2	52.3
Azospirillum+PSM+NPK	57.2	55.0
Paddy straw @5 tonne/ha+NPK	58.7	59.1
Paddy straw @5 tonne/ha+ 1/2NPK	55.5	50.4
Azospirillum+PSM+ 1/2NPK	52.6	42.5
Azolla compost + 40 kg N through urea	39.85	97.8
Farmyard manure@10 tonne/ha +40 kg N through urea	38.70	92.1
Azollafresh(10 tonne/ha) +40 kg N through urea	43.65	116.6
80 kg N through urea (3 splits)	39.35	95.3
Green manure(Sunnhemp) + 40 kg N through urea	37.50	86.1
Azolla inoculation (1 tonne/ha) + 40 kg N through urea	37.00	83.6

Source: Mausumi Raychaudhuri 2002.

The native symbiotic and nonsymbiotic N fixers population are limited under moderately and strongly acidic condition. Seed inoculation with TAL 102 strain of *Rhizobium japonicum* to soybean (cv. Pusa 16) increased the *Rhizobium* in the soil by 12.4 fold and increased the number of effective nodules and the total N content of the soil by 5 per cent. P nutrition to the soil @ 26 kg P ha^{-1} increased the *Rhizobium* population by 2.6 fold and also increased the nodule weight significantly. Inoculation along with P nutrition increased the rhizobial population by 54, 4.4 and 20.8 times over control, inoculation and nutrition, respectively.

Yadav and Raychaudhuri 2001 compiled reports of more than 250 demonstrations on biofertilizers (N fixers and P solubilisers) conducted in Assam, Manipur and Sikkim by RBDC, Imphal, GOI without any fertilizer input as basal dose and a small quantity of urea provided uniformly as top dressing. They reported that paddy responded to inoculation under different conditions and recorded 2 - 43 per cent increase in yield over untreated control. Biofertilizer not only supplemented 15-30 kg N ha^{-1}, but also increased yield over recommended NPK and control. Hence, the low pH or anaerobic conditions due to submergence were not the limiting factors for building up of root-*Azospirillum* association in rice fields and use of biofertilizers proved to be economical, particularly, when farmers can not afford chemical fertilizers.

The soils are deficient in available P and the crops are highly responsive to phosphatic fertilizers. Experimental results indicate that the application of 50-80, 120, 50-80, 80 and 60 kg P_2O_5 /ha is beneficial for rice, potato, wheat, maize, soybean and french bean, respectively. Mixture of rock phosphate and SSP (3:1) has been found to be more efficient in Tripura and Manipur in crops (Mausumi Raychaudhuri et al 2000). The initial P requirement is met by SSP and then gradually by slow release of P from rock phosphate. Liming increases the availability of phosphorus to the plant. Per cent utilization of the applied P and PUE is increased with liming (Mausumi Raychaudhuri et al 1998). Suitable strains of P solubilising microorganisms viz. *Bacillus polymixa, Pseudomonas striata* were also found effective when applied in combination with lime, organic and inorganic fertilizers. Application of 60-80, 40-50 and 30-40 kg/ha K_2O/ha are recommended for wet land paddy, maize and oilseeds/pulses based cropping systems, respectively for K deficient soils. But, recommendations on the basis of the soil test values are more efficient and balanced than the general recommendations.

Singh and Boordoloi, 2011 reported that variation in soil and crop management practices under different land use system plays a recessive role in governing the soil fertility status. The land use system that is moderately disturbed and regularly replenished with nutrient sources including house hold generated organic wastes; maintain soil fertility at a fairly satisfactory level. On the other hand, the land use system where the soil is excessively disturbed suffers worst from severe nutrient imbalances due to loss of one or more nutrients. The moderately disturbed orchard soils, which receive continuous application of high analysis fertilizers with little use of organic inputs, also show severe imbalances including micronutrient deficiencies. Surface soil in a general term maintained higher level of fertility than the sub surface soil. The logical trend of correlation between the soil properties more or less existed with explainable deviations at the surface layer, but was extremely erratic in most instances in case of the sub-surface layer.

Balanced Fertilization

Crops grown in north-east acid soils suffer due to deficiency of nitrogen, phosphorus, calcium, magnesium, molybdenum and boron and toxicity of iron and manganese (Panda 1987). Integrated plant nutrient management involving chemical, organic and biological sources have proved to be ideal for sustaining soil health.

Experimental results suggest that *Azospirillum* (C_2 strain) and Phosphatika (*Pseudomonas striata*) alone and in combination increased the grain yield by 20.8, 14.6 and 22.8, respectively over control. In addition to half and full recommended doses of N, P, K, the yield increase by 33 and 50 per cent respectively over control. Farm yard manure (FYM) @ 10 t/ha alongwith biofertilizers (*Azospirillum + Azotobacter*) increased the yield of maize by 144 per cent. When coupled with half and full recommended doses of N, P, K, the yield further increased. Inter-cropping of paddy and maize with groundnut/soybean/ beans are advisable in low fertility soils. The effect of nutrient application in the form of INM on the yield of maize, black gram and soybean is presented in Table 13. Other than the combination of farmyard manure and suitable strains of biofertilizer a combination of farmyard manure along with furrow liming @ 500 kg/ ha has been recommended for oilseeds and pulses.

Organic inputs in addition to retaining soil fertility, increases moisture and cation holding capacity of the soil and improves soil structure. Experimental data revealed that the soil pH increased from 4.8 to 5.63 with lime @ 500 kg/ha and further to 5.81 when combined with FYM @ 10t/ha, thus providing a congenial environment for the nutrients to be available to the plants. The exchangeable Ca + Mg content increased significantly with FYM @ 10t/ha, Lime @ 500 kg/ha and Lime + FYM from 2.27 to a maximum of 3.58 cmol (p^+) kg^{-1} with lime. FYM and lime alone and in combination increased the organic matter content in the soil significantly by 28.3, 17.7 and 32.3 per cent, respectively over control (1.19 percent). Average quality FYM contains 0.4-0.6% N, 0.2-0.3 % P_2O_5 and 0.6-0.8 % K_2O i.e. 12 –17 kg nutrients per tonne.

Lime @ 500 kg/ha + FYM @ 10t/ha was the best treatment as it increased the pH, available nutritional status and the yield to a maximum (Mausumi Raychaudhuri 2005).

Table 13: Per cent increase in grain yield of crops under integrated nutrient management

Treatments	Maize		Soybean		Blackgram	
	Yield (q/ha)	% increase over control	Yield (q/ha)	% increase over control	Yield (q/ha)	% increase over control
Control	15.9	-	8.5	-	6.7	-
FYM@ 5 tonne/ha + Biofertilizer	22.1	39.0	18.8	121.2	11.5	71.6
FYM@10 tonne/ha + Biofertilizer	38.8	144.0	27.7	225.9	13.6	103.0
FYM@ 5 tonne/ha + Biofertilizer + ½ NPK	30.7	93.1	27.2	220.0	12.6	88.1
FYM@ 5 tonne/ha + Biofertilizer + NPK	35.2	121.4	28.3	232.9	13.3	98.5
FYM@ 10 tonne/ha + Biofertilizer+1/2 NPK	47.1	196.2	30.8	262.4	15.6	132.8
FYM@ 10 tonne/ha + Biofertilizer + NPK	52.1	227.7	31.4	269.4	17.2	156.7
CD (P=0.05)	10.3	-	2.88	-	4.59	-

Source: Mausumi Raychaudhuri 2002.

As agro-ecosystems are open and dynamic systems, balanced fertilization can only aim at nutrient supplementation in order to assist crop performance and at the same time to minimize losses. The balance of nutrients in soil ecosystem was described by the following equation (Mausumi Raychaudhuri 2005)

$$NR_{th} = \Sigma th\ (NP_t + NA_{\Delta t} - NU_{\Delta t} - NL_{\Delta t})$$

Where, NR is the soil inorganic and organic nutrients at time (th), NA is the soil inorganic and organic nutrients added or returned to the soil during the time interval Δt. NP is the soil organic and inorganic nutrients present at time t. NU is the uptake of nutrients by the harvested product during the time interval t. NL is the inorganic and organic nutrients lost during the time interval Δt, t is the time of sowing and th is the harvesting time and Δt is the time inerval between t and th. The equation simply states that if nutrients removed are greater than additions, the reservoir of nutrients remaining within the soil pool will decline. Thus, on the basis of the fertility status of the soil, a judicious application of nutrients in the form of organic and inorganic in a balanced way to overcome the deficit in the soil is recommended for sustainable crop production in this region.

Future Thrust

The north eastern region soils are acidic in nature. A flowchart is being presented in Fig 2. showing possible options for sustainable management of acid soils of north. The pH can be adjusted by practising paddy based cropping system in lowlands and ameliorating uplands for maize based cropping system. Liming is the best option for its amelioration. Any substance containing 75 per cent calcium and 90 percent $CaCO_3$ neutralization equivalent can be used as liming material. North east has good reserve of lime that can be judiciously exploited for better crop production. The most common material used for liming agricultural soils is finely ground limestone that has all the desired characteristics of an agricultural liming material. Imposition of rules and regulations on lime quality and standardization of liming material based on size and neutralization effect is required for its effective utilization in ameliorating acid soils.

Integrated farming system is recommended for valley as well as hill ecosystem for sustainable crop production. Acid tolerant suitable varieties of paddy, maize, oilseeds, pulses, horticultural crops may be developed, screened and evaluated for increased crop production. Integrated nutrient management coupling organic and inorganic nutrient sources in right proportions is one of the best options found for sustainable crop production in this region. Agroforestry based farming system with multipurpose trees in the upper part of the hills with strong acidity may be encouraged. Aquaculture based farming system and paddy based farming system are recommended for lowland areas.

Cropping intensity is very low in northeast region. Mostly mono cropping is being practiced due to moisture stress in *rabi* and summer season. Incorporation of lime and organic matter improves moisture content in the soil and increases cropping intensity. Zero tillage is also found promising in the valley region with sandy clay loam soils and can be extended for oilseed and pulse production.

Organic farming with a holistic approach is a sustainable option in the hills. Most of the hill agriculture is organic by default due to zero fertilizer and pesticide application. Therefore, organic farming may be promoted by covering more and more area with proper certification that can fetch premium price to the product which in turn will improve the livelihood of the farmers of this region.

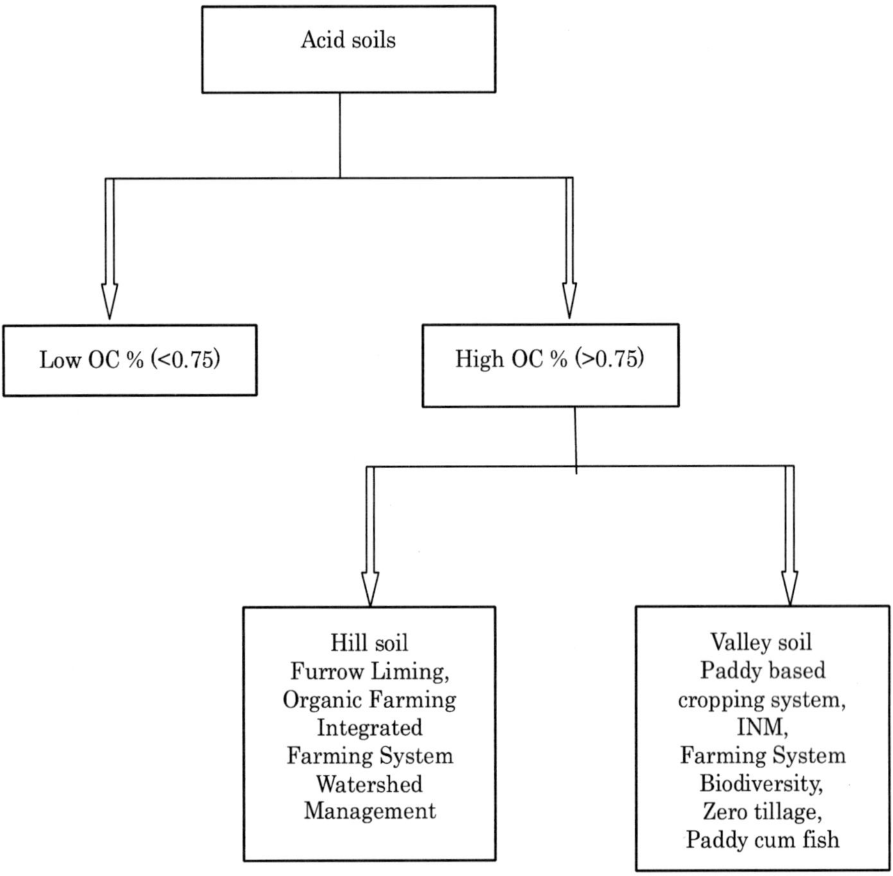

Fig 2: Management of Acid Soils

Climate Change is a global issue and its impact on agriculture is of concern. As per the IPCC Report there is increase in temperature by 2^0 to 4^0 C and variation in rainfall and its uneven distribution. The north east has a high reserve of organic matter, with increase in temperature, the availability of nutrients through will increase, which in turn, will increase the crop production. Research on the impact of climatic variability on natural resources and crop behavior needs to be assessed and possible adaptive measures are needed to mitigate the climatic aberrations

mineralisation. Apparently, it shows climatic variability has both positive and negative impact on agriculture in northeast region.

Technologies are many but their applications are very few. The research and education wings of central and state government and autonomous body viz., Indian Council of Agricultural Research, Department of Soil Conservation, Agriculture, Horticulture, Fishery, and Veterinary Sciences, State Agricultural Universities, Central Agricultural University are devoted in developing technologies to combat the adverse effect of acidity in northeast agriculture. Still a huge gap exists between technology developed and its extension to the beneficiaries. Linkages between extension officials and researcher need to be strengthened for better understanding of the problem and the solution developed. Extensive training programme to bring awareness among the farmers about the developed technologies with proper demonstration needs to be organized. Gender issues need to be addressed for active participation of women in agriculture. The NGOs should also work hand in hand with the farmers and the researchers for deeper percolation of the available technologies.

References

Anonymous, 1980, Annual Report, ICAR Research Complex for NEH Region, Umiam, Meghalaya.

Balatti, P. A., Krishnan H.B and Pueppke S.G (1991) Calcium regulates growth of Rhizobium fredii and its ability to nodulate soybean cultivar peking. *Canadian Journal of Microbiology* 37: 542–548.

Bhattacharyya, T. Sarkar, D., Dubey, P.N., Ray, S.K., Gangopadhyay, S.K., Baruah, U., Sehgal, J. 2004. Soil Series of Tripura, NBSS Publ. No.111 ,National Bureau of Soil Survey & Land Use Planning, Nagpur, p115.

Bhumbla, D.R. and Mclean, E.O. (1965). Aluminum in soils VI. Changes inpH-dependent acidity, CEC and extractable Al with additions of lime to acid sulfate soils. *Proceedings of the Soil Science Society of America*, 29:370-374.

Borthakur and Majumdar 1967. A Laboratory study on lime requirement of some acid soils of Sibsager district of Assam. *Journal of Assam Soil Science Society*. 19:1-4

Brown J. R. and Cisco J. R. (1984). An improved Woodruff buffer for estimation of lime requirements.*Soil Science Society of America Journal*. 48 : 587-591.

Chakravorty, D.N., Sehgal, J. L. and Dev, G. 1978. Influences of climate and topography on the pedogenesis of alluvium-derived soils of Assam. *Indian Journal of Agricultural Chemistry* 11: 77-97

Das, N. K., Basumatary, A. and Borkotoki B (2011) .Interrelationship of Forms of Sulphur with its Availability Indices and Soil Properties in Entisols of Assam. *Journal of the Indian Society of Soil Science* 59, 134-140.

Das, T. H., Sarkar, D. and Sehgal, J. 1996. Soil acidity on steep slope of Sikkim at higher altitude. *Journal of The Indian Society of Soil Science* 44: 319-321.

Datta, M and Gupta, R. K. 1983. Utilization of pressmud as amendment of acid soil in Nagaland. *Journal ofthe Indian Society of Soil Science* 31: 511-516.

Datta, M and Gupta, R. K. 1984. An appraisal of inorganic phosphorus forms in acid soils of Nagaland. *Journal of the Indian Society of Soil Science.* 32: 511-516.

Datta, M. and Ram, M. 1993. Status of micro nutrients in some soil series of Tripura. *Journal of the Indian Society of Soil Science.* 41: 776-777.

Datta, M.,Gupta, R.K and Prasad, R.N.(1983 b) Bulletin, ICAR Research Complex for NEH region,22.

FAO, 1958; Efficient Use of Fertilizers, Ed: Vladinir Ignatieff & Harlod J. Page, FAO Agricultural Studies No. 43

Gangopadhyay, S. K., Das, P. K., Mukhopadhyay, S. S. and Bansal, S. K. 1990. Altitudinal pattern of soil characteristics under forest vegetation in Eastern Himalayan Region. *Journal of the Indian Society of Soil Science.* 38:93.

Ghosh, G. and Ghosh, S. K. 1982. Micronutrients in some soils of Nagaland. *Indian Agriculturist.* 26: 91-99.

Goswami, N. N., Leelavati, C. R and Dravid, M.S.(1976) *Bulletin of the Indian Society of Soil Science* 11:238.

Gupta, R. K., Singh, R. D., Rai, R. N. and Prasad, R. N.1989. Effect of lime and micronutrients on yield and uptake of micronutrients by wheat on acid soils. *Journal of the Indian Society of Soil Science* 37: 70-73.

Gupta, S. K. and Laskar, S. 1998. Economic use of nitrogen and wetland rice cultivation in Tripura. *Journal of Hill Research.* 1: 92-97.

Hussain, S. M. and Sharma, U. C. 1991. Response of rice (Oryza sativa) to nitrogen fertilizer in acidic soil of Nagaland. *Indian Journal of Agricultural Sciences.* 61: 662-664.

Kailash Kumar and Mausumi Raychaudhuri 1997. Phosphorus fractions as affected by different land uses in acid hill soils of Manipur. *Journal of the Indian Society of Soil Science.* 45: 574-577.

Kailash Kumar and Mausumi Raychaudhuri 2000. Integrated Nutrient Management for Rice in Manipur. Silver Jubilee Souvenir, ICAR Research Complex for NEH Region, Manipur Centre. 22-25.

Kailash Kumar, K. V. P. Rao and L. J. Singh (1995). Forms of acidity in some acid *Inceptisols* under different land uses in Manipur. *Journal of the Indian Society of Soil Science,* 43:338-342.

Kailash Kumar, L.J.Singh and K.V.P.Rao (1995).Components of pH dependent CEC in acid soils of Manipur. *Journal of the Indian Society of Soil Science.*, 43:263-264.

Kailash Kumar, L.J.Singh and K.V.P.Rao (1997).Evaluation of lime requirement methods for acid soils of Manipur. *Journal of the Indian Society of Soil Science.*, 45:404-406.

Kumar Kailash, and Mausumi Raychaudhuri. 1997, Phosphorus fractions as affected by different land uses in acid hill soils of Manipur. *Journal of the. Indian Society of Soil Sci ence* 45: 574-577.

Lahiri, T and Chakravorty, S. K. 1989. Characteristics of soils of Sikkim at various altitudes. *Journal of the Indian Society of Soil Science.* 37: 451-454.

Laskar, S., Dadhwal, , K. S. and Prasad, R. N. 1983, Soils of Tripura and their fertility management. *Research Bulletin* No. 23., ICAR Research Complex for NEH Region, Umiam, Meghalaya.

Lyngdoh, J. C. and Shukla, L. M. 1993. Fertility status of some Alfisols. *Journal of the Indian Society of Soil Science* 41: 707-709.

Majumdar, B. Kailash Kumar, Venkatesh, M. S. and Patiram (2005). Effect of liming on phosphorus adsorption and desorption behaviour of acidic *Alfisols* and *Entisols* of Meghalaya. *Journal of the Indian Society of Soil Science* 53:188-193.

Mandal, S. C., Sinha, H, Prasad, C. R. and Ali, M. A. 1966. Studies on liming of acidic red loam soils. *Journal of the Indian Society of Soil Science* 14:127-131.

Nair, K. M. and Chamuah, G. S. 1993. Exchangeable Aluminium in Soils of Meghalaya and Management of Al^{3+} Related Productivity Constraints. *Journal of the Indian Society of Soil Science* 41: 331-334.

Panda N. 1987. Acidic soils of eastern India, their chemistry and management. *Journal of the Indian Society of Soil Science* 35: 568-581.

Panda, N. 1998. Soils of North-Eastern Region and their management for sustainable production. Paper presented at the Symp. on ICAR Research Complex for NEH Region, Umiam, Meghalaya.

Patel, C.S and Singh, J. 1990, Efficacy of form, timing and application methodof urea and urea based fertilizer in wetland rice (*Oryza sativa*) on Histosols and Haplaquent soil under mid altitude of north eastern hills region. *Indian Journal of Agricultural Sciences.* 60: 169-173

Patiram 1990, Forms of aluminium in some acid soils of Sikkim. *Journal of the Indian Society of Soil Science.* 38: 652-655.

Patiram 1991. Liming of acid soils and crop production in Sikkim. *Journal of Hill Research* 4: 6-12.

Patiram and Prasad, R. N. 1984. Forms of potassium in the soils of East Khasi hills of Meghalaya. *Journal of the Indian Society of Soil Science* 32: 168-171.

Patiram and Prasad, R. N. 1985. Efficacy of time of potassium application in wetland rice on Haplaquent of Meghalaya. *Indian Journal of Agricultural Scences* 55: 338-341

Patiram and Prasad, R. N. 1991. Growth of maize and phosphorus uptake in relation to phosphorus adsorption characteristics of acid soils. *Journal of the Indian Society of Soil Science* 39: 302-307.

Patiram and Rai, R. N. 1988. Response of maize rice and wheat to fertilizer nitrogen in Sikkim. *Journal of Hill Research* 2: 18-20.

Patiram, Prasad, R. N. and Munna Ram 1993. Inorganic phosphorus fractions and their relation to available phosphorus indices in acid soils of Meghalaya. *Journal of the Indian Society of Soil Science* 41: 430-433.

Patiram, Prasad, R. N. and Singh, R. P. 1987. Response of rainfed direct seeded rice to potassium on an Alfisols of Meghalaya. *Annals of Agricultural Research* 8: 214-220.

Patiram, Rai, R. N. & Prasad, R. N. 1990a. Form of soil phosphorus and suitable extractants for available phosphorus in acid soils of Sikkim. *Journal of the Indian Society of Soil Science* 38: 237-242.

Patiram, Rai, R. N. & Prasad, R. N. 1990b. Phospate adsorption by acid soils from different altitude. *Journal of the Indian Society of Soil Science* 38: 602-608.

Patiram, Upadhyay, R. C. Singh, C. S. and Ram, M. 2000. Micronutrient cation status of mandarin orchards (*Citrus reticulate Blanco*) orchards of Sikkim. *Journal of the Indian Society of Soil Science* 48: 246-249.

Prasad, R. N., Barooah, R. C. and Patiram. 1978. Annual Report, ICAR Research Complex for NEH Region, Indian Council of Agricultural Research, Meghalaya.

Prasad, R. N., Patiram and Munna Ram 1985. Forms of Al in soils of East Khasi Hills, Meghalaya. *Journal of the Indian Society of Soil Science* 33: 523-527.

Prasad, R. N., Patiram, Barooah, R. C. and Munna Ram 1981. Soil Fertility in North Eastern Hill Region, *Research Bulletin* No. 9.

Prasad, R. N., Patiram, Barooah, R.C. and Ram M. 1983. Direct and residual effect of liming on yield of maize and uptake of nutrients in acid soils of Meghalaya. *Journal of the Indian Society of Soil Science* 31: 233-235

Pratt,P. F. and Blair, F. L.. (1962). Buffer method for estimating lime and sulfur applications for pH control of soils. *Soil Science* 93: 329 -3 32.

Ram, M., Prasad, R. N. and Patiram 1987. Studies on phosphate adsorption and phosphate fixation in Alfisols and Entisols occurring different altitudes of Meghalaya. *Journal of the Indian Society of Soil Science* 31: 207-216.

Raychaudhuri Mausumi 1997. Some aspects of Potassium dynamics in soils. PhD Thesis. Calcutta University.

Raychaudhuri Mausumi 2002. Nutrients management of acid soils of Manipur. *Indian Farming.* 52 (6): 13-15.

Raychaudhuri Mausumi 2005. Nutrients management of acid soils of the north-east India. *Indian Farming.* 55: 22-29.

Raychaudhuri Mausumi and Kailash Kumar. 2002. Direct and Residual effect of lime on soil characteristics, yield and nutrient uptake of maize (Zea mays) and soybean (Glycine max) in an acid hill soil of Manipur. *Indian Journal of Agricultural Sciences.* 72: 67-69

Raychaudhuri Mausumi and Raychaudhuri, S. 2008. Managing symbiotic N-fixation in acid soils. *Indian Farming.* 57: 25-28.

Raychaudhuri Mausumi and Raychaudhuri, S. 2008. *Rhizobium* inoculation and furrow liming interaction on the yield of and nutrient uptake by groundnut (Arachis hypogeaea) in an acid hill Ultisol. *Journal of the Indian Society of Soil Science* 56: 76-79.

Raychaudhuri Mausumi and Raychaudhuri,S 2006. Annual Report. ICAR Research Complex for NEH Region, Umiam, Meghalaya.

Raychaudhuri Mausumi and Sanyal, S. K. 1999. Forms of potassium and the Quantity – Intensity Parameters of an acid hill Ultisol after liming. *Journal of the Indian Society of Soil Science* 47: 229-234.

Raychaudhuri Mausumi, Kailash Kumar and S. Raychaudhuri. 1997. Lime-Rhizobium interaction on Soybean in Kanhaplohumult of Manipur Hills. *Journal of the Indian Society of Soil Science* 45: 739-742

Raychaudhuri Mausumi, Kailash Kumar and S. Raychaudhuri. 1998. Response of soybean to lime and phosphoruson an Ultisol of Manipur. *Journal of the Indian Society of Soil Science* 46: 628-632.

Raychaudhuri Mausumi, S. V. Ngachan, S. Raychaudhuri and Singh, A. L. 2003. Yield Response of groundnut (*Arachis hypogaea*) to dual inoculation and liming of an acid hill Ultisol of Manipur. *Indian Journal of Agricultural Science,* 73:86-88.

Raychaudhuri Mausumi, Suresh, V. R. and Lokeswar Singh (eds). 2000. Agricultural Technologies for Manipur. Published by S.V. Ngachan, Joint Director, ICAR Research Complex for NEH Region, Manipur Centre, Imphal.

Raychaudhuri, Mausumi and Raychaudhuri, S. 2009. Potassium dynamics and its management in acid soils. In Proceedings of the IPI-OUAT-IPNI International Symposium on Potassium Role and Benefits in Improving Nutrient Management for Food Production, Quality and Reduced Environmental Damages. Bhubaneswar.

Sahota, T. S. and Grewal, J. S. 1984, Manurial and fertilizer requirement of potato in Meghalaya. *Fertiliser News* 29: 27-32.

Sannigrahi, A. K. and Pandey, R. 2000. Nature of acidity in some hilly soils of Arunachal Pradesh. *Annals of Agricultural Research.* 21(3): 438-440.

Sen, T. K., Nayak, D. C., Singh, R. S., Dubey, P. N., Majhi, A. K., Chamuah, G. S. and Sehgal, J. 1992. Pedology and edaphology of benchmark acid soils of north eastern India. *Journal of the Indian Society of Soil Science* 45: 782-790

Sharma, U. C. and Awasthi, R. P. 1989. Response of potato to fertilization in north eastern hill states. *Journal of Hill Research.* 2: 25-33.

Shoemaker, H. E., Mclean, E. O. and Pratt, P. F. (1961). Buffer methods for determining lime requirement of soils with appreciable amounts of extractable aluminum. *Proceedings of the Soil Science Society of America*25: 274-277.

Singh, A.K. and Bordoloi, L.J. 2011.Study in Soil Fertility Variation as Influenced by Land Use System and Soil Depth Interaction in Acid Hill soils of Nagaland. *Journal of the Indian Society of Soil Science* 59:198-204.

Singh, B. P., Das, M.,Ram, M., Dwivedi, B. S. and Prasad R. N. 1992. Characterisation of Fe-toxic soils and affected plants in Meghalaya. *Indian Journal of Agricultural Sciences* 62: 326-328

Singh, O. P. and Datta B. 1988. Organic carbon and nitrogen status of some soils of Mizoram occurring at different altitudes. *Journal of the Indian Society of Soil Science* 36:414-420

Singh, O. P. and Datta, B. 1987. Phosphorus status of some hill soils of Mizoram in relation to pedogenic properties *Journal of the Indian Society of Soil Science* 34: 700-705

Singh, R.P. and Patiram 1987. Response of wetland rice to applied potassium on Haplaquent of Meghalaya. *Indian Journal of Agricultural Sciences* 57:398-340.

Singh, R.S., Baruah, Utpal, Sarkar, Dipak, Butte, P.S., and Gajbhiye, K..S. 2005. Soil science of Meghalaya.NBSS Publ.121 NBSS&LUP, Nagpur, p.86.

Staff, NBSS&LUP (2002). Soils of India. NBSS Pub.94, National Bureau of Soil Survey and Land Planning, Nagpur, India, pp.130+11 sheets map.

Velayutham, M and Bhattacharya, T. 2000. Soil resource management. In Yadav, J. S. P. and Singh G. B. (eds.) Natural Resource Management for Agriculture production in India, pp. 1-35, Alfa Printers, New Delhi.

Venkatesh, M. S. Kailash Kumar, B. Majumdar and Patiram (2005). Sorption and desorption of Zinc as influenced by limimg in some acid soils of Meghalaya. *Journal of the Indian. Society of Soil Scence* 53:319-323.

Woodruff, C. M. (1948). Testing soils for lime requirement by means of a buffered solution and the glass electrode. *Soil Science* 66:53.

Yadav, A. K. and Raychaudhuri S. 2001. Performance of inoculants in farmers' field of North-east. In Three decades of Biofertilizer and Organic farming in North-east India. [Eds. Yadav, A. K. and Raychaudhuri S.). Published by North-eastern Region of Biofertilizers, SML, AAU, Jorhat. Pp 99-110.]

7

Acid Soils of Jharkhand

A.K. Sarkar

Jharkhand state lies between 21°58'2" to 25°8'32" North latitude and 83°19'05"to 87°55'03" East longitude covering an area of about 7.97 m.ha. and accounts for 2.4 per cent of the total geographical area (TGA) of the country. It is bound on the east by West Bengal, on the west by Chhattisgarh, on the north by Bihar and on the south by Odisha. Reliable information on the location, extent and quality of soil and land resources is the first requirement in planning for the sustainable management of land resources. The components of land i.e., soils, climate, water, nutrient and biota are organised into an eco-system, which provides a variety of services that are essential for the maintenance of the life support system and the productive capacity of the environment.

The state is primarily rainfed and crop production largely depends on onset,progress and withdrawal of monsoon. Failure of HATHIA rain has adverse effects on the reproductive phase of rainy season crops as well as establishment of winter season crops.High intensity rainfall coupled with light soils and undulating topography results in soil erosion, run off and loss of plant nutrients.Jharkhand is a part of the eastern plateau and hill region (Agroclimatic region VII) of India. Based on rainfall,temperature, terrain and soil,the state has been divided into three agroclimatic sub zones i.e.Central and North-eastern plateau ,Western plateau and South-eastern plateau (Fig.1). The distribution of districts in these 3 zones and the important features of these zones are presented in Table 1 and 2.

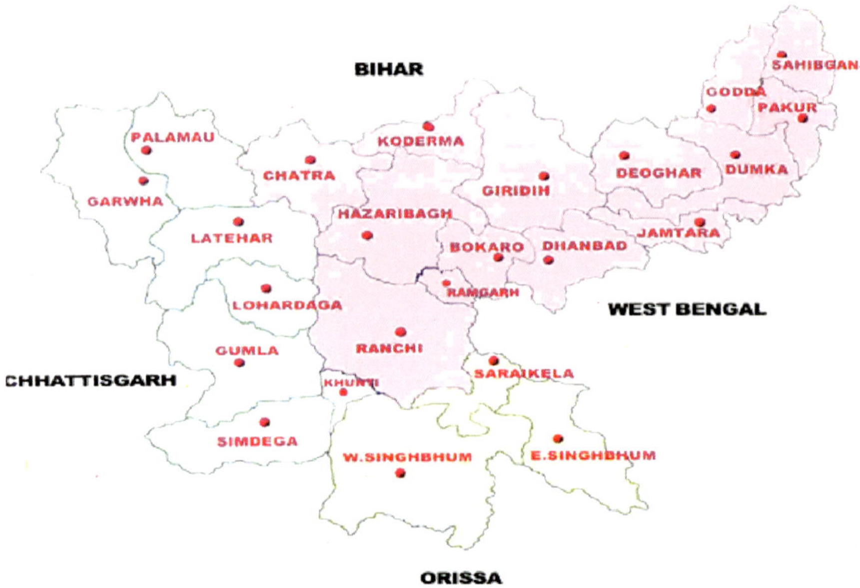

Fig.1: Agro Climatic Regions of Jharkhand

The Chotanagpur plateau is composed of ancient rocks of diverse origin. The rocks of the area are of igneous origin, while the later ones are metamorphic and sedimentary. It is made up of archean gneiss and schists. Chotanagpur granite occurs in two zones, one zone includes SanthalParganas, Hazaribagh, Palamau and the other includes Ranchi, which occurs as intrusive into the country schistose rock. The land utilization pattern of the state is given in the table 3. Out of 7.97 m.ha area in the state, only 22.7% (1.81 m.ha) is under cultivation, out of which only 11.3% area is irrigated.

Table 1: Agro-climatic sub zone: Jharkhand

- Sub-Zone IV: consists of 14 districts covering almost 50% of the geographical area of the state
- IVA: Santhal Parganas consists of 6 districts
- IVB: North Chotanagpur and South Chotanagpur
- Sub-Zone V: Consists of 7 districts with highly drought prone areas
- VA: Palmau and Garhwa: Normal Soils
- VB: Latehar, Simdega, Gumla, Chatra and Lohardaga (5 districts)
- Sub-zone VI: 3 districts (Kolhan region): East Singbhum, West Singbhum and Saraikela-Kharsawan

Table 2: Agricultural & climatic features of agro-climatic sub-zones

	Sub-zone IV	Sub-zone V	Sub-zone VI
Geographical area (Lakh ha.)	35.18	30.95	15.57
Irrigation	11.40	12.60	7.80
Land put to non-agricultural use in (Lakh ha.)	3.08	1.80	1.95
Barren and uncultivated land (Lakh ha.)	2.60	2.00	1.13
Cultivable waste Land (lakh ha.)	1.20	1.00	0.55
Net area sown (Lakh ha.)	7.12	8.00	2.94
Total cropped area (Lakh ha.)	7.95	8.94	3.79
Cropping intensity (%)	112	112	129
Annual Rainfall (mm)	1270	1226	1400

Table 3: Land Use pattern of Jharkhand State

S No.	Total area classified	Area (m ha)	% of TGA
1.	Forest	2.33	29.2
2.	Net sown area	1.81	22.7
3.	Barren and unculturable waste	0.57	7.2
4.	Non agricultural use	0.79	9.9
5.	Orchards& Pasture	0.20	2.5
6.	Culturable wasteland	0.28	3.5
7.	Current and other fallow	1.99	25.0
	Total	7.97	100.0

Source: Fertilizer and Agriculture Statistics, Eastern Region (2003-2004)

2.0 Soils and their fertility

Soil is the base for terrestrial life supporting system. The farming situation is based on the type of land. The land located in the upper slope is called TANR land and that in the lower slope is called DON land. In between the two, there are medium land, which have intermediate properties of upland and lowland.The soils of upland and medium land are red and lateritic,highly permeable, coarse textured, shallow depth, low base saturation, acidic soil reaction, low cation exchange capacity and poor water retention capacity. These soils are generally deficient in available Phosphorus,poor in organic matter and available nitrogen. Lowland soils are wet and often suffer from excess moisture due to heavier texture,poor drainage with near neutral soil pH.Important features of the soils under different land situations are presented in Table 4.

Table 4: General characteristics of upland, medium land and low land

Characteristics	Upland land	Medium land	Lowland
Colour	Red or Brown red	Yellow or yellowish	Grey or Greyish
Texture	Light textured (loam)	Medium texture (Clay loam)	Heavy textured
Drainage	Well drained	Moderately drained	Poorly drained
Soil reaction	Low pH	Moderately acidic	Neutral pH
Soil fertility	Poor in organic carbon Ca, Mg, N P&S	Poor in organic carbon N, Ca, Mg	Medium in N & organic carbon

It is estimated that out of the total cropped area of 22 lakh hectares in the state, about 10 lakh hectares are acidic in reaction.Out of this ,about 4 lakh hectares of land are highly acidic(pH less than 5.5).These soils mainly from the districts of Bokaro,Dhanbad,east singhbhum, Giridih, Gumla, Hazaribagh, Jamtara, Lohardaga, Ranchi,Saraikela,simdega and west singhbhum pose major threats to crop production (Table 5.). The extent of Phosphorus deficiency in acidic soils reveal its widespread occurrence,which has an adverse impact on crops .

Table 5: Districts with more than 50% area affected by soil acidity

District	Cultivable land '000 ha	Acidic soil pH <5.5 '000 ha	% land acidic	Deficiency of P (%)
Bokaro	163.5	113.6	69.5	66.0
Dhanbad	139.1	83.4	60.0	68.8
East Singhbhum	174.4	125.9	72.7	88.7
Giridih	295.3	165.7	56.1	79.0
Gumla	414.7	287.8	69.4	80.4
Hazaribagh	243.4	128.3	52.7	57.8
Jamtara	137.5	88.4	64.3	73.1
Lohardaga	95.4	68.3	71.6	15.6
Ranchi	524.1	383.6	72.3	40.0
Saraikela	144.9	97.2	67.1	94.9
Simdega	293.0	214.4	73.2	90.5
West Singhbhum	382.1	280.8	73.5	95.3

Based on the analysis of 12,823 soil samples collected from the districts of the state using G.I.S,the district level soil fertility maps were developed by the NBSS&LUP,Regional centre,Kolkata and Birsa Agril. University,Ranchi.Some of the salient results are discussed below:

Three soil orders namely Entisols, Inceptisols and Alfisols were observed in different districts of Jharkhand. Alfisols were the dominant soils covering 54.0 per cent of TGA followed by Inceptisols (24.4 %) and Entisols (19.4 %). Vertisol is only present in Pakur district to an extent of 0.6 per cent of TGA of the district.

Soils of the state, in general, are low to very low in available phosphorus and sulphur, medium in available nitrogen & potassium status and deficient in available boron. Available macro and micronutrients in different categories of soil fertility are as follows:

- The organic carbon content of soils of about 47 per cent of TGA are low to medium (<0.5% to 0.75%)
- Soils of majority area (about 70% of TGA) of the state have medium status of available nitrogen (280 to 560 kg/ha) and about 19.6 per cent area have low available N content (<280 kg/ha).
- Soils of about 66 per cent area are low (<10 kg/ha) in available phosphorous content while about 28 per cent area are found to be medium (10 – 25 kg/ha) in available P content.
- Most of the soils (about 51% of TGA) have medium potassium content and about 18 per cent are low (below 108 kg/ha) in available potassium content.
- Soils of about 38 per cent area are low (<10 mg/kg) whereas soils of about 31 per cent are medium (10-20 mg/kg) in available sulphur content.

The DTPA extractable micronutrient status revealed that based upon the critical limits:

- All the soils are sufficient in iron (>4.5 mg/kg) and manganese(>2.0 mg/kg)
- Soils of about 7 per cent area are deficient in available zinc (<0.5 mg/kg)
- Soils of about 4 per cent area are deficient in available copper (<0.2 mg/kg).
- Soils of about 45 per cent of the state are found to be deficient in available boron (<0.5 mg/kg).

3.0 Crops and cropping pattern

Selection of crops relatively tolerant to soil acidity is an effective strategy to address soil related constraints. Soil reaction preference for different crops (Mandal*et al.*1975) presented in table 6 reveal that transplanted rice tolerates soil acidity to a considerable extent because water logging raises the soil pH. Minor millets, potato, mesta, niger, mustard are suitable for acid soils. But maize, wheat, pulses, oilseeds, sorghum do not grow well in acid soils unless the soils are limed.

Table 6 : Preference of different crops to soil pH

Crops	Ideal pH range
Cereals	
Maize,Sorghum,wheat,Barley	6.0 to 7.5
Millets	5.0 to 6.5
Oats	5.0 to 7.7
Legumes	
Field bean,Soybean,pea,lentil	5.5 to 7.0

Rainwater harvesting and recharge of groundwater is a priority area for ensuring irrigation for dry season crops.Existing water bodies(ponds,ditches,wells,water storage structures)need renovation for harvesting of rainwater and its subsequent use for providing irrigation to crops.Thus,crop diversification especially through intercropping(maize ,pulses,oilseeds,vegetable based)in rainfed uplands and mediumlands and sequence cropping (rice based)in lowland is essential for increasing crop productivity in acid soil regions (Sarkar 2013).Some such diversification possible in different land situations in about 10 lakh hectares of the state(Verma etal.2007) is presented below(Table 7)

Table 7: Efficient Cropping Systems for Different Land Situations:
Upland : Rainfed ecosystem having water retention capacity of 180-200 mm/metre soil depth

Existing	Diversification Suggested
*Rice/Maize-Fallow *Rice/Gundli-Fallow *Pigeonpea/Blackgram-Fallow *Fallow-Niger/Horsegram	*Pigeonpea + Groundnut (1:2) *Pigeonpea + Blackgram (1:2) *Pigeonpea + Rice (1:3) *Pigeonpea + Maize (1:1) *Maize + Blackgram (1:2) *Maize + Soybean (1:2) *Blackgram-Safflower/Linseed/Lentil/ Chickpea/Niger/Horsegram *Maize-Safflower/Linseed/Lentil/Chickpea *Soybean-Safflower/Linseed/Lentil/ Chickpea
Medium land : It is very potential land situation having limited access to adequate irrigation facilities for sequential cropping	
Existing	Diversification Suggested
• Rice/Maize/Chickpea/Lentil/ Pea • Rice/Maize-Barley/Mustard /Linseed • Rice/Maize-Potato/Wheat	• Maize + Blackgram-Wheat • Maize-Early potato-Late Wheat • Maize-Early potato-Late Pea (green pods) • Maize-Wheat-Greengram • Maize-Potato-Green gram • Maize-Potato-Onion • Maize-Toria-Wheat • Maize-Early Potato/Toria-Late Wheat – Green gram • Maize-Wheat + Lentil (4:2) • Maize-Wheat + Mustard (8:2) • Maize – Lentil + Mustard (5:1) • Maize-Potato + Wheat (1:1) • Rice-Wheat-Green gram/Black gram • Rice-Potato-Green gram/Black gram • Rice-Early Potato-Late Wheat • Rice-Toria-Late Wheat • Rice-Early Potato-Late Peas • Rice-Early Potato-Onion • Rice-Berseem • DeenanathGrass-Berseem-Maize+Cowpea • Rice-Early Potato/Toria-Wheat

Besides, field crop based systems, fruit based land use systems and rice based *utera* cropping and adoption of intercropping systems have shown promise especially in acid soils of Jharkhand.

4.0 Acid Soil Management

Pioneering work on management of acid soils was carried out at Ranchi by Mandal and his coworkers during the past five decades. These workers have identified liming as the most important input for amelioration of acid soils(Mandal *et al.*1966; Mandal *et al.*1975; Mathur, 1980: Mathur *et al.*1985). Liming improved the physical,chemical and biological properties of acid soils,which resulted in increased availability of plant nutrients, enhanced Nitrogen fixation by micro-organisms. This, in turn brought about significant yield improvements in crops grown in such soils.Considering major soil related constraints in acid soils of eastern India, studies on their management were carried out by several workers.These have been well documented in several publications (Panda *et al.* 1991; Panda *et al.* 1996; Panda *et al.* 2009; Sarkar *et al.* 1994; Sarkar *et al.* 2007; Rattan *et al.* 2009). Current thinking on acid soil management is to integrate the nutrient management and soil ameliorationwith water management, conservation technologies using a system approach of crop production to tackle the food and nutritional security of the population. This is important for livelihood security of the farmers, especially of rainfed regions with small and marginal land holdings (Anonymous, 2012; Sarkar, 2005; Sarkar, 2013). Liming is cost-effective and can increase the crop productivity levels at least two-fold provided, locally available, low cost liming materials are available to the farmers at sowing time. Method of lime application and extent of low pH soils (Fig.2 &3) have been shown below.

5.0 Lime as an amendment for Acid soils

Systematic work was carried out on the use lime as an amendment with an objective to reduce soil acidity and to bring the unavailable forms of nutrients in acid soils to available form. The investigations at Kanke (Ranchi) were focused on (i) response of crops to liming (ii) lime requirement (iii) quality of liming materials (iv) fineness of liming materials (v) frequency of liming (vi) methods of liming application and, (vii) crop rotation in relation to liming. It was found that all the liming materials, when added on equivalent lime (CaO or CaO + MgO) basis were more or less equally effective in increasing the yield of crops (Mandal *et al.*, 1966, Mandal and Jha, 1970). The minimum fineness of the liming material should be such that it passes through 10 mesh sieve, but in the case of basic slag, materials which pass through 30 to 40 mesh sieve are recommended for application as a soil amendment.

Subsequently,attempts were made to reduce the dose of lime to one fourth to one half of the lime requirement and apply in split doses.Lime provides Ca to soil,which is in short supply in acidic

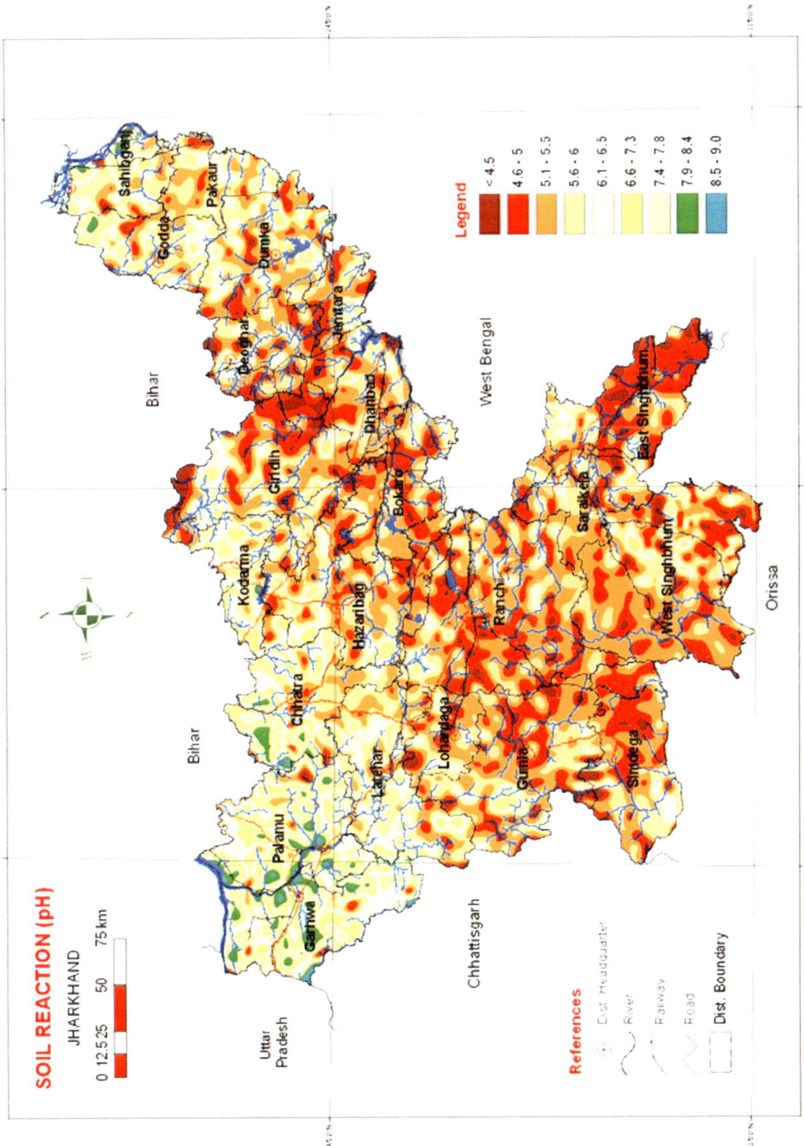

Fig. 2: Soil reaction (pH)

Fig. 3: Lime Application by farmers

soils due to removal by growing crops and leaching due to high rainfall in slopy terrain.

Mathur et al. (1983) applied carbonate and sulphate salts of Ca and Mg at the rate of 2 and 4 q/ha in furrows like a fertilizer nutrient at the time of sowing of crop and observed an increase in the yield of Soybean from 19 to 32 % and in Groundnut from 21 to 48% in acid sandy loam soil (pH 5.6) of Ranchi. The improvement in crop yield due to gypsum, irrespective of fall in pH and increase in exch.Ca^{2+}status indicated that it was the effect of calcium nutrition rather than of pH . Subsequent studies made by Mathur et al. (1985) revealed that by application of 3-4 q of $CaCO_3$/ha in furrows below the seed at sowing could increase the yield of groundnut, soybean, gram and barley ranging from 20 to 36 per cent. Liming at this rate increased the pH as well as calcium saturation of soils by 37 to 50 per cent (Table 8), which confirmed with the required calcium saturation of 40 percent in kaolinite rich soils.

Table 8: pH,exchangeable Ca saturation of Soils as affected by lime application

Lime	Groundnut			Soybean			Gram		
(Q/ha)	pH	Exch. Ca**	Ca saturation(%)	pH	Exch. Ca**	Ca saturation(%)	pH	Exch. Ca**	Ca sat. (%)
Control	5.2	3.3	38.4	5.1	3.2	37.2	5.5	3.2	37.2
As per LR*	6.9	5.3	61.6	6.7	3.5	40.7	6.9	4.7	54.6
2	5.6	3.4	39.5	5.4	3.4	39.5	5.7	3.4	59.5
3	5.6	3.5	40.7	5.6	3.7	43.0	5.8	3.6	41.9
4	5.7	3.6	41.9	6.0	4.1	47.7	6.0	3.9	45.3
	6.0	4.5	50.0	6.1	4.3	50.0	5.2	4.2	48.8

*Lime requirement

**Cmol(P^+)kg^{-1}

Calcium as a plant nutrient needs more attention than lime application in acid soils.Mathur et al. (1985) showed that $CaCO_3$ applied @2to4q/ha in furrows at the time of sowing improved the crop yields of several crops in acid soils of Ranchi(Table 9).

Table 9: Grain yield(q/ha) of Crops with lime application in Acid soils

Lime (q/ha)	Groundnut		Soybean		Gram		Barley	
	Yield	% increase	yield	% increase	yield	% increase	yield	% increase
Control	17.5	–	20.3	–	13.1	–	24.3	–
As per LR*	22.7	29.9	28.1	38.4	18.4	40.1	35.6	46.2
2 (furrow)	20.8	19.0	24.8	22.2	16.0	22.1	30.1	23.6
3 (furrow)	20.8	19.0	27.3	34.6	16.4	25.4	32.0	31.6
4 (furrow)	20.7	18.5	26.3	29.3	17.2	31.5	33.8	38.8
5 (furrow)	21.3	21.9	26.1	28.6	16.7	27.2	33.7	38.6
C.D.at 5%	1.2		1.4		1.3		1.4	

*L.R.- Lime requirement of soils growing groundnut,soybean,gram and barley were 46.5, 55.0,28.5, and 28.5 q/ha,respectively.

On the basis of lime response, crops like pigeon pea, soybean and cotton has been classified as high responsive and chickpea, lentil, peas, groundnut, maize and sorghum as medium responsive and small millet, rice, potato etc. as low responsive.

Mathur *et al.,* (1989) compared lime application at the rate of 1/10, 1/15 and 1/20th of lime requirement (LR) applied in Rhizosphere at sowing time for six years with LR dose applied as broadcast only once in the beginning. The mean yield of urdbean, soybean, groundnut, Lentil and others crops grown with 1/20 LR dose of lime was at par with 1 LR dose. After six years exchangeable Ca^{2+} was found to be the highest in soils receiving 1/10 LR dose of lime and was more profitable than LR dose (Table 10).The uptake of Phosphorus by crops was greatly enhanced by liming.The return not only increased with time at all levels of split doses of lime but also the gap between full dose and lower doses of lime narrowed down,showing thereby more profitability at lower doses in most crop rotations.

Table 10: Grain yield of crops due to liming (q ha^{-1}) in acidic soils of Jharkhand

Treatments	Crops						
	Moong	Groundnut	Urad	Soybean	Lentil	Pea	Gram
No Lime	2.8	12.0	5.8	13.3	11.4	11.3	2.2
1 LR(Broadcast)	6.1	16.7	12.1	22.0	17.7	17.6	5.1
1/10 LR(Furrow)	6.0	17.2	11.5	22.0	17.6	16.6	6.1
1/15 LR(Furrow)	5.8	16.0	11.3	21.7	17.1	16.2	5.9
1/20 LR(Furrow)	5.3	15.2	11.1	20.4	15.8	15.6	5.7
CD (P=0.05)	0.03	1.6	1.8	19.8	1.8	1.7	1.8

6.0 Basic slag as a liming material

Indian basic slag is poor in P but has high amounts of lime and trace elements. Studies have been made in the country on the utilisation of certain Indian basic slags relatively rich in P (7-8% P_2O_5) like the ones obtained from steel melting shop no. 2 of TISCO. All steel slags contain high amounts of lime (30-45% CaO), which is as efficient as $CaCO_3$in neutralsing soil acidity. Use of basic slag in agriculture has been suggested as a P-carrier and as an amendment for acid soils(Mandal and Jha, 1970). The beneficial role of basic slag in increasing crop production has been well established in the acidic soils of Chotanagpur plateau of Bihar on maize, wheat, gram and groundnut with a number of steel slags containing no P or different amounts of P in comparison to lime (Mandal and Jha, 1970).

Basic slag may be used @ 2 to 4 quintal per hectare for pulses, oilseeds, fruits, flowers and vegetables annually. Tata Steel Industries, Jamshedpur (Jharkhand) generates about 6 lakh tones of basic slag annually and that could be used in acid soils. Organic manures applied with basic slag helps in increasing its solubility and hence its effectiveness.There are reports of heavy metal (Cd, Cr, Ni etc.) addition with basic slag application. Fe and Al oxides present in high amounts in red and lateritic soils are very good scavengers for heavy metals. They bind these toxic heavy metals tightly and thus prevent them to remain in soil in plant available form.There is also hardly any possibility of these metal ions to escape into soil solution as such and to percolate through soil profile to contaminate underground aquifers (Bhat et al. 2010)

Field experiments in ten state govt. farm of Jharkhand state were conducted during *Kharif* 2010 through Agricultural Technology Mangement Agency (ATMA), Ranchi, Gumla, Lohardaga, Simdega, West Singhbhum, Saraikella, East Singhbhum, Bokaro, Dhanbad and Dumka.

The basic soil characteristics of identified experimental field are presented in table 11.

Table 11: Basic Soil characteristics of selected field locations in different districts of Jharkhand

Districts	Location	Soil parameters				
		Soil pH	Organic carbon(%)	Available N(Kg/ha)	Available P_2O_5 (Kg/ha)	Available K_2O (Kg/ha)
Ranchi	S.M.Farm Bundu	5.4	0.39	240.0	21.7	147.3
Gumla	Farm	5.2	0.38	235.0	20.4	151.2
Lohardaga	Farm	5.2	0.42	255.0	23.5	169.5
Simdega	S.M.FarmSaldega	5.5	0.39	240.0	22.6	163.5
W. Singhbhum	KVK FarmJagarnathpur	5.5	0.35	220.0	20.8	180.6
Saraikella	KVK FarmSaraikella	5.3	0.32	205.0	18.5	124.5
E. Singhbhum-I	Vill.-Gorgora Block-Jamsedpur	4.8	0.27	180.0	22.7	103.8
E. Singhbhum-II	S.M.FarmPatamda	5.1	0.50	295.0	27.9	108.2
Bokaro	S.M.FarmKasmar	5.1	0.27	180.0	16.9	103.2
Dhanbad	Farm	5.4	0.41	250.0	17.8	175.7
Dumka	Asbani farm,Raneshwar	5.3	0.34	215.0	19.5	140.6

The treatments details for the field experiments were as follows:

T_1: Basic slag (BS)@ 4 quintal /ha in furrow (Provided by Tata steel for testing)

T_2: Basic slag (BS) + NPK (100 % of recommended dose of selected crop)

T_3: Basic slag (BS) + 50 % NPK (50 % Recommended dose of selected crop)

T_4: NPK (100 % of recommended dose of selected crop)

T_5: Farmers practice (FP)

Basic slag analysed 48 % CaO, 2 % Total P and was ground to pass through 80 mesh sieve.

Results indicated that the grain yield of Wheat increased from 39.50 to 50.26 qha^{-1} , Gram from 8.15 to 12.56 qha^{-1} , Mustard from 5.90 to 8.10qha^{-1} and Rai from 6.42 to 8.56 qha^{-1}with application of basic slag as compared to NPK alone & farmers practices (no basic slag). The highest grain yield of Wheat, Gram, Mustard and Rai was obtained with Basic slag (BS) + 100 % NPK followed by Basic slag (BS) + 50 % NPK then NPK alone, basic slag alone and the lowest with farmers practices treatments (Table 12). Similar results were also reported by Sharma and Sarkar (2005) in acid soils of India with other liming materials such as lime or dolomite.

Table 12: Effect of basic slag & fertilizers on crop yield (q/ha) in different district of Jharkhand

Districts	Crops	Treatments				
		BS	BS+NPK	BS+50%NPK	NPK	FP
Ranchi	Gram	10.25	12.64	11.76	11.95	
	(RGS- 963)	(15.6)*	(42.5)	(32.6)	(34.7)	8.87
Gumla	Wheat	45.15	50.26	48.21	49.12	
	(Sonalika)	(11.2)	(23.8)	(18.7)	(21.0)	40.60
Lohardaga	Gram	10.95	12.23	11.37	11.55	
	(BG 372)	(20.2)	(34.2)	(24.8)	(26.8)	9.11
Simdega	Gram	11.20	12.56	11.30	11.43	
	(PG-114)	(18.5)	(32.9)	(19.6)	(20.9)	9.45
W. Singhbhum	Mustard	6.80	8.10	7.85	7.96	
	(Shivani)	(15.2)	(37.3)	(33.0)	(34.9)	5.90
Saraikella	Wheat	44.50	48.42	46.25	47.35	
	(K-9107)	(12.6)	(22.6)	(17.1)	(19.9)	39.50
E. Singhbhum-I	Gram	10.70	12.12	11.25	11.42	
	(BG-372)	(16.2)	(31.6)	(22.1)	(24.0)	9.21
E. Singhbhum-II	Rai	7.15	8.56	8.16	8.30	
	(A 404)	(11.4)	(33.3)	(27.1)	(29.3)	6.42
Bokaro	Wheat	39.75	41.80	40.20	41.23	
	(Sonalika)	(18.8)	(25.0)	(20.2)	(23.3)	33.45
Dhanbad	Gram	9.17	10.56	9.43	9.55	
	(PG-114)	(12.5)	(29.6)	(15.7)	(17.2)	8.15
Dumka	Gram	9.35	10.65	9.87	10.15	
	(PG-114)	(11.0)	(26.5)	(17.2)	(20.5)	8.42

*Figure in parenthesis indicate per cent yield increase over farmers practice.

The per cent yield increase over farmers practice was from 11.2 to 23.8 in Wheat, from 8.15 to 42.5 in Gram, from 5.9 to 8.1 in Mustard and 6.42 to 8.56 in Rai indicating Gram and Wheat are more lime responsive crop as compared to Mustard and Rai. Results also indicate that basic slag/ LD slag application @ 4.0 qha-1 improved grain yield response in Wheat by 11.2 & 12.6 per cent, in Gram by 11.0 to 20.2 per cent, in Mustard by 15.2 per cent and in Rai by 11.4 per cent. Similar response was recorded when basic slag/ LD slag + 100 % NPK and Basic slag (BS) + 50 % NPK was compared with recommended dose of NPK (Table 12).

7.0 Lessons from long term studies in acid soils

The field experiment underway at Ranchi in an acidic red loam soil (pH 5.5)with maize-wheat crop sequence since 1955 has provided insight on management of acid soils for higher crop productivity. Crop yield over the past four decades (Table 13) reveal a decline in N,NP,and NPK plots.Basal application of lime with thse inorganic sources of plant nutrients resulted in sustainable crop yields with high profits from input use.The reason is the deterioration of soil fertility in terms of lowering of soil pH,lowering of base saturation and availability of secondary and micronutrients with fertilizer use without taking care to ameliorate the soil acidity.This could be remedied with liming + nutrient use as per crop requirements.This practice improves the nutrient use efficiency by crops and the profitability from nutrient use for crop production.

Table 13: Grain yield (kg/ha) of maize & wheat grown in rotation from 1956 to 2005 (50 years) as affected by fertilizers and lime

Treatments	Year					Mean value
	1956to 1965	1966 to 1975	1976 to 1985	1986 to 1995	1996 to 2005	
Maize (Kharif)						
Unfertilized	1638	914	525	523	492	716
N	1638	1029	027	030	078	421
NP	2174	2436	296	095	724	1033
NPK	2637	2830	626	145	952	1309
Lime + NPK	2402	3038	3970	4806	4506	3893
Wheat (Rabi)						
Unfertilized	519	609	626	976	802	721
N	559	291	NIL	017	124	150
NP	1608	2174	1007	435	1607	1380
NPK	1738	2277	1707	916	1708	1695
Lime + NPK	1426	2728	3169	4984	3854	3402

8.0 Need for lime and P application in Acid soils

Acid soils, ingeneral, are deficient in available Phosphorus. This is primarily due to the fixation of added Phosphorus in soils.The use efficiency of added P fertilisers is very low in acid soils due to these reasons. This has led to low crop productivity levels (<2-3t/ha) of crops, such as rice, wheat, maize etc. as compared to the desired level of about 5 t ha^{-1}. This calls for needed changes in the nutrient use policy to make it more rational and scientific.Re-emphasizing the need of lime and phosphorus application in crops grown in acid soils can bring about desirable changes in the crop production scenario in the region.

Addition of P-fertiliser in acid soils results in a precipitation reaction between exch.Al and added P resulting in the formation of a highly insoluble Al-P compound.Liming such soils to pH above 6.0 reduces exch.Al- ion concentration and increases availability of fertiliser P for crop production.Results over the years from Long Term Fertiliser Trials in acid soils of India reveal significant yield enhancement with P application in Soybean-wheat, Maize-wheat and Finger millet-maize crop sequences(Table14).

Relative response of N,P and K in Maize and Wheat crops in acid soils(pH-5.5) of Ranchi (Jharkhand) over the past 50 years of Permanent Manurial Trial confirm that - i) Response of P to crops in acid soils is higher than that of N and K. ii)Response of P is more in *rabi* crops than that in case of *kharif* over the years. iii)Extent of response to P application in maize was 15 kg grain $kg^{-1}P$, while in case of wheat,it was 32 kg grain $kg^{-1}P$ (Table 15).

Recent studies under IPNI-India programme at Ranchi (Paleustalf, pH - 5.6, P - 12.6 kg ha^{-1}, and organic C- 0.45%) indicate that for a 5tha^{-1} maize, P uptake was about 25 kg ha^{-1}. Omission of P from the crop fertilization schedule reduced the P uptake by Maize to 16 kg ha^{-1} in the same experiment. This was associated with yield reduction of Maize by about 40% in P-omission plots (Table16).

Results, thus, suggest the need for a focussed attention on the P fertiliser use for mitigating its deficiency in acid soils of eastern India.This will not only make agriculture more profitable but will also ensure better utilization of resources with adequate food for masses.

Table 14: Average grain yield of crops (t ha^{-1}) over the years in long-term experiments on Acid Soils of India

Centre	Crops	Control	N	NP	NPK
Palampur(1973-98)	Maize	0.3	0.6	2.0	3.2
	Wheat	0.3	0.4	1.8	2.5
Ranchi(1973-99)	Soybean	0.7	0.4	0.9	1.6
	Wheat	0.8	0.4	2.3	2.6
Bangalore(1986-99)	Fingermillet	0.6	0.9	1.2	4.3
	Maize	0.3	0.5	0.8	2.2

Ref. Sarkar and Singh (2002)

Table 15: Response of nutrients, manure and amendment in maize – wheat sequence in acid soils during last five decades under Permanent Manurial Trial, Ranchi.

Treatments	1956-69	1970-79	1980-89	1990-99	1999-07	Mean
Maize						
N (kg grain/ kg N)	4.50	0.63	-4.48	-4.53	-3.33	-2.72
P(kg grain/kg P)	27.75	53.83	4.55	1.67	15.72	15.44
K(kg grain/ kg K)	10.83	10.66	5.48	0.85	9.96	6.24
Lime (kg grain/kg lime)	-	-	7.76	5.47	5.37	5.40
FYM (kg grain/ quintal FYM)	4.35	6.45	6.93	14.10	14.50	9.95
Wheat						
N (kg grain/ kg N)	-0.02	-3.95	-6.74	-8.11	-6.27	-5.31
P(kg grain/kg P)	52.78	77.46	17.21	16.49	41.59	32.36
K(kg grain/ kg K)	3.49	3.94	14.36	3.78	3.46	5.35
Lime (kg grain/kg lime)	—	—	5.84	4.36	2.95	4.05
FYM (kg grain/ quintal FYM)	4.05	3.88	5.78	10.91	9.16	6.89

Sarkar and Singh(2002)

Table 16: Yield and nutrient uptake of Maize in nutrient omission and fertilized plots (Pooled data of 2009 and 2010)*

Treatments	Grain yield (t ha^{-1})	Nutrient uptake (kg ha^{-1})		
		N	P	K
NPK (250:120:120)	5.38	145.0	25.6	97.2
(-N)	1.22	34.6	7.4	35.5
(-P)	3.48	92.0	16.3	67.9
(-K)	4.13	98.9	16.7	63.6
SSNM (200:90:100)	5.67	132.0	22.9	108.3
CD (P = 0.05)	0.77	17.8	3.1	19.8

*IPNI(India Programme)personal communication.

9.0 Status of Boron and Molybdenum

Boron deficiency occurs in highly leached sandy soils in high rainfall areas, acidic soils, calcareous soils & soils low in organic matter status. Crops such as Cauliflower, Cabbage, Turnip, Legumes, Rapeseed are highly sensitive to Boron deficiency while Broccoli, Carrot, Raddish, Tomato, Spinach, Lettuce, & Cotton are medium sensitive crops. Application of Boron containing fertilizers such as Borax, Solubor, Boric acid, Granubor, Boronated SSP @ 2 to 3 kg B ha^{-1} can control Boron deficiency and improve quality of produce. Majority of farmers are unaware about the role of Boron in plant nutrition and benefits which can be derived from application of fertilizers containing Boron. To ensure adequate and correct use of Boron in crops, farmers need proven and practical information either directly or through farm advisory services, soil and plant testing laboratories, training programmes, mass media and other sources.

Boron is required in different amounts for plants. Monocotyledons generally require less B than dicotyledons. Roots of monocots have lower capacity to absorb than roots of dicotyledons. Cruciferous plants generally require highest amount of B. Boron is relatively immobile in plants, therefore the early casualties of deficiency are observed in the young growing parts of the plant.

Quality of crops especially pulses, vegetables & oilseeds depend to a large extent on the supply of plant nutrients deficient in a particular soil. Thus, in soils with poor organic matter status having pH values less than 5.5 need application of Boron, lime and organic manures for proper crop quality. Browning and brown rot of cauliflower, cracking of litchi fruits, heart rot of beet, water soaked areas in stem and fruits are some of the disorders commonly observed in Boron deficiency, reducing the cost of produce to less than 50% .

Deficiency Vs. Toxicity of Boron

The range between deficient and toxic levels of Boron is narrow in soils. Hence, the farmers must be careful in the rate of B fertilizer application. Sensitive crops to excess Boron are : soybean, kidneybean, grapes, peach, peanuts etc.

In field experiment with rate of Boron application in soil low in available B (0.3 mg kg^{-1}) with groundnut it was observed that Boron application to crops beyond 3kg/ha resulted in reduction in pod yield of groundnut and B uptake by pods (Table 17)

Table 17: Effect of Boron application in Groundnut

Boron levels(kg B ha^{-1})	Pod yield(q ha^{-1})	Boron uptake by pods(g ha^{-1})
0	11.4	4.3
1.5	13.2	12.9
3.0	15.3	26.8
4.5	11.9	24.0
CD (P = 0.05)	1.7	4.4

Farming practices which may cause Boron deficiency

- Irrigation with alkaline water very low in Boron and high in calcium
- Cropping under drought conditions
- Intensive cultivation of crops
- Cropping in soils low in available phosphorus
- Cropping in low organic matter soils
- Over liming & high soil pH

Status of available Boron in soils of Jharkhand reveal that about 45% area under cultivated soils are deficient in available Boron. Priority for Boron application in soils is apparent from the results presented especially for the districts of Palamau, Latehar, east Singhbhum, Lohardagga&Saraikela(Table 18)

Deficient symptoms of boron in crops

Since B is relatively immobile in plants, the early casualties of B deficiency are the young growing parts of the plant. The general symptoms are the death of apical (growing) point which results in the regeneration of numerous auxiliary buds producing a bushy growth and slight thickening of leaves. Root growth is stunted and flowers do not form. Its inadequacy is often associated with sterility and malformation of reproductive organs.

Managing Boron deficient soils

After detecting boron deficiency, its remedial measure is important for improving the crop productivity in B deficient soils. Remedial measures include suitable sources, time, rate & method of application of B containing fertilizers.

Name of B containing material	Chemical formula	B content (%)	Amounts (kg) of Material required to supply 1 kg B
Borax (Sodium tetraborate deca hydrate)	$Na_2B_4O_7.10H_2O$	11	10.0
Granubor II (Sodium tetraborate penta hydrate	$Na_2B_4O_7.5H_2O$	14.6	7.0
Solubor	$Na_2B_4O_7.5H_2O$ $Na_2B_{10}O_{16}.10H_2O$	20	5.0
Sodium tetraborate	$Na_2B_4O_7$	20	5.0
Boric acid	H_3BO_3	17	6.0
Boronated NPK	Complex fertilizer NPK+B	0.3-0.5	

Table 18: Distribution on available B content in soil in different districts of Jharkhand.

S.N.	District name	Approx. area ('00 ha)	Area('00 ha) with severe to moderate acidity	% of TGA	Range of HWS-B (ppm)	Area ('00 ha) with low (<0.5 ppm HWS-B)	% of TGA
1.	West Singhbhum	7180	5277	73.5	0.02-7.21	2729	38.0
2.	East Singhbhum	3535	2552	72.2	0.02-0.94	2731	77.3
3.	Saraikela	2721	1828	67.0	0.03-3.03	1496	54.9
4.	Ranchi	7701	5637	73.2	0.02-3.52	3297	42.8
5.	Simdega	3767	2750	73.0	0.01-2.28	1718	45.7
6.	Gumla	5318	3691	69.4	0.02-3.34	2594	48.8
7.	Lohardaga	1492	1068	71.6	0.04-1.07	1066	71.5
8.	Latehar	40	70	15.9	0.02-1.62	3094	84.5
9.	Palamau	5089	229	4.5	0.02-4.18	3414	67.7
10.	Chatra	3807	731	19.2	0.07-4.48	1312	35.4
11.	Hazaribagh	5017	2644	52.7	0.03-7.87	1964	38.9
12.	Koderma	2405	623	25.9	0.02-5.84	576	23.9
13.	Giridih	4939	2771	56.1	0.02-5.23	2303	46.6
14.	Deoghar	2475	953	38.5	0.03-1.94	1117	45.1
15.	Dumka	4410	2139	48.5	0.11-7.15	1204	27.3
16.	Godda	2106	596	28.3	0.05-8.95	521	24.7
17.	Sahebganj	1598	350	21.9	0.07-3.78	615	38.5
18.	Pakur	1805	731	40.5	0.10-7.22	490	27.2
19.	Jamtara	1801	1158	64.3	0.02-6.09	415	23.0
20.	Dhanbad	2087	1252	60.0	0.22-5.9	190	9.1
21.	Bokaro	2862	1989	69.5	0.09-5.03	641	22.5
22.	Garhwa	4036	222	5.5	0.01-2.99	2862	<0.8

Deficiency Symptoms of Boron in Crops

Methods of application

Application of B containing fertilizers should be made only when its deficiency has been established. Injudicious use may result toxicity to plants and soil pollution. Application of B @ 1.0 kg B ha^{-1} is highly beneficial in B deficient soils and it should be practiced in intensively vegetable grown areas. Basal soil application of B was found to be efficient in controlling B deficiency. However, in fruit & vegetable crops foliar spray was superior to soil application. Foliar application of 0.125% boric acid in water with 0.1% surfactant/wetting agent (Teepol) or 0.2% humefactant (calcium chloride) 3 times commencing 10 days after transplanting of seedling at an interval of 8 days in rainy season and 12 days in winter season was found most effective and superior to 0.2% application of boric acid or borax at same interval. Foliar application of borax and boric acid along with surfactants 2-3 times was effective for controlling brown rot of cauliflower in acid soils of Jharkhand. In addition to soil application and foliar spray other methods like seedling dipping and seed soaking to enrich root zone with B can be practiced.

Rate of application of Boron in crops

- Borax (10.5% B), Solubor (19% B), Boric acid (17% B), boronated super phosphate (0.18% B) can be applied in soil as basal @ 1-2 kg B ha^{-1}. Lower rates are preferred to avoid toxicity. Application at 2 year interval based on soil test is recommended.
- Foliar spray of Boron at 0.05-0.15% B is very effective. For both annual and perennial crops, foliar sprays are effective. As B is immobile, frequent sprays are needed. Higher concentration of B solution is toxic.

Response of crops to Boron application and profitability in Jharkhand

During 2004-05 to 2007-08 under the Network Project on "Evaluation of efficacy of granubor II for ameliorating boron deficiency in important field crops of Jharkhand", response of crops to B application was studied in farmers field of Ranchi. Results indicate significant increase in yield of Cauliflower, maize, groundnut, lentiland wheat crops with recommended rates of Boron application with high profits (Sarkar et al. 2010).

Molybdenum

Molybdenum is deficient in acid soils of Jharkhand due to leaching and low pH(Kotur,1994).Systematic studies were carried out in several vegetable crops to study the response of Mo application in acid soils of Ranchi.In a study on the response of Cauliflower under limed and unlimed conditions to Mo reveal interesting results.Mo improved the curd diameter and weight.Soil application of 1 kg.Mo /ha significantly increased the curd diameter ,which was at par with 0.1 per cent Ammonium molybdate as foliar spray.

The lime-Mo interaction was significant in case of curd yield and Mo content in leaf (Table 19)Among the Mo treatments,0.1 per cent foliar application showed the best performance in the presence of lime,which was at par with 1.5 Kg Mo/ha soil application. Between the two levels of soil application,1.5 kg/ha (as Ammonium molybdate) showed a significantly lower curd yield in the absence of lime.In the case of foliar application at 0.2 per cent,the curd yield was significantly reduced,both under unlimed and limed conditions due to excessive accumulation of Mo in leaf (12.2 to 16.5 mg/kg) although no visual toxicity symptoms were noticed on the crop. Results point out that foliar applications were significantly more effective compared to soil application in raising the Mo content of leaf .

Table 19: Effect of lime and molybdenum on curd yield and Mo content in leaf of cauliflower

Lime level	Control	Molybdenum levels			
		Soil application		Foliar application	
		1kg/ha	1.5kg/ha	0.1%	0.2%
		Curd yield (t/ha)			
No lime	12.8	17.6	16.6	17.8	14.0
Lime	15.1	18.2	20.0	20.5	14.5
C.D.(P=0.05)	0.69				
		Molybdenum Content in leaf(mg/kg)			
No lime	0.73	7.56	11.65	8.52	12.20
Lime	1.64	7.63	12.38	12.20	16.39
C.D.(P=0.05)	0.47				

10.0 Research and development needs

The eastern region is rich in natural resources i.e. soil & water. The region constitutes the most fertile plains of the country. Eastern region has about 34 M ha under Inceptisol, 20 M ha under Alfisol and 14.5 M ha under Entisol. Fertiliser consumption in this region is the lowest in the country. 85% farmers of the region are small and marginal. The climate of the region is tropical, hot and humid to sub-humid with high rainfall (Average 1500 mm). Three major river basins in the region are Mahanadi, Brahmani-Baitarani and Subarnarekha. Annual utilizable surface water in the three basins is estimated at 8.8 M ha-m. The region has a good ground water potential. Total utilizable ground water resource for irrigation is about 19.5 M ha-m.

The research and development needs of acid soil regions suggest that the problem should be tackled in an integrated manner. Soil, water, crop management need to be focused as per needs of the farmers (Table 20,21,22).Scientific interventions need to be put in place keeping the soil and water related constraints in view.Finally,an action plan has to be drawn for solving the issue.some of these are discussed below:

10.1 Issues to be addressed

Major issues are soil degradation, soil acidity, poor soil organic matter, secondary and micronutrient deficiencies, inadequate & unreliable surface irrigation system, meagre utilization of ground water resources and soil and water pollution.

Table 20: Soil related constraints in eastern region

S.N.	States	Total degraded land (%)	Acid soils (%) with pH <5.5	Nutrient deficiency problems	Soil & water pollution problems
1.	West Bengal	30.9	40	N, Zn, B, S	As, F
2.	Bihar	—	—	N, P, Zn, B, S	Salinity, Fe
3.	Jharkhand	—	49.0	P, N, B, Ca, S	As, F
4.	Odisha	39.3	80.0	P, N, Ca, B, S	Fe, F
5.	Assam	76.8	50.0	P, N, Ca, B	Fe
6.	Chhatisgarh	53.7	50.0	N, P, K	Fe, F
7.	Eastern UP	52.1	—	Zn, N, P	Salinity

Table 21: Water related constraints in Agriculture

Plain region	Plateau region	Coastal region
1. Groundwater exploitation is low	1. Groundwater exploitation is low	1. Groundwater exploitation is about 30%
2. Large flood prone area (>15%)	2. Probability of drought occurrence is very high	2. Salinity in coastal areas (50%)
3. Waterlogging & salinity in canal command areas	3. Low irrigated area (10-15%)	3. Very low water use efficiency in canal commands.

Table 22: Source of irrigation in eastern States (% of Irrigated area)

State	Canal	Well	Others
West Bengal	23.5	57.6	18.9
Bihar	28.4	54.0	17.6
Assam	87.1	1.2	11.7
Jharkhand	38.9	51.1	10.0
Odisha	64.8	14.8	20.4

10.2 Assessment of farmers need:

1. Farmers need maintenance of existing ponds, wells, tank & their need based rejuvenation.
2. Existing water storage structures available with farmers need improvement for prevention of seepage losses.
3. ITK available with farmers need validation, refinement & promotion.
4. Checking soil erosion by gully plugging, mulching, land leveling & bringing additional area under cultivation.
5. Helping farmers to prepare good quality compost, vermicompost for use in crop production.
6. Desiltation of farm ponds for greater *in-situ* conservation of rainwater.
7. Soil testing & providing soil health cards to farmers.
8. Checking the quality of inputs such as fertilizers, amendments, pesticides from time to time.
9. Managing supply of critical inputs through input dealers such as Borax, lime Dolomite, SSP, Phosphogypsum etc.
10. Regular meeting of Scientists with farmers clubs & visit to their fields & advising them.

10.3 Desired scientific interventions on Soil & Water Management

1. Ensuring availability of lime/dolomite/basic slag/paper mill sludges for managing acid soils of the region. Dose of application is 2 to 4 q ha^{-1} yr^{-1} in lime responsive crops only. Locally available, low cost soil amendments costing Rs.1 to 2 per kg should be used.

2. Quality compost/vermicompost preparation and use @ 3 to 5 t ha^{-1} yr^{-1} in farmers fields needs to be promoted to maintain soil organic matter status.
3. Balanced use of NPK fertilizers as per soil tests & crop requirements is essential for maintenance of soil fertility. Fertilizer use need to be increased per unit of land for higher crop productivity.
4. Secondary and micronutrients deficient in soils should be replenished by soil, foliar or a combination of the two for higher yield and quality of crops.
5. In rainfed areas of eastern India, microwatershed based farming system development is needed to ensure livelihood security of small and marginal farmers.
6. Soil & water quality parameters need continuous monitoring &redressal especially in high intensity cropping System Zones.
7. In coastal saline soils, application of green manure crops like *Sesbania aculeata* or green leaves of *Glyricidia maculata* @ 10 t ha^{-1} in the puddled rice soils with reclamation and fertilization is beneficial.
8. Water management interventions recommended by Water Technology Centre for Eastern Region (ICAR), Bhubaneshwar for different land situations are:

Rainfed uplands

1. Field bunding/graded bunding/contour bunding.
2. Tank cum well system
3. Percolation tank
4. Mahabandh along with conveyance network involving more than 100000 m^3 storage (100 x 200 x 5m).
5. Dugout ponds in rice fields.
6. Check dams/diversion weir on well defined Nala.

Rainfed lowlands

7. Comprehensive drainage system inclusive of field drains, lateral drains and main drain.

8. Subsurface WHS in coastal areas.
9. Very shallow type wells in coastal areas having saline groundwater below 10 m.
10. Conversion of lowlands to aquaculture tanks.
11. Peripheral dykes for saucer shaped waterlogged areas.

Canal irrigated area

12. Renovation of field channel/construction of new field channels.
13. Renovation of drainage system.
14. Raised and sunken bed system for small and marginal farmers.
15. Renovation of minors transferred to Water Users Associations.

Groundwater utilization

16. Groundwater utilization through open dug wells for small and marginal farmers.
17. Manually drilled shallow tube wells in alluvial areas.

Miscellaneous

18. Renovation of existing tanks/ponds on Government/community land.

10.4 Suggested Action Plan for Natural Resource Management in Jharkhand

10.4.1 Small farm diversification

(i) Sequence cropping of pulses and oilseeds in rice fallows, using water saving technologies.

(ii) Intercropping of rice + pigeonpea/Maize + Pigeonpea & Pigeonpea + groundnut with INM package.

(iii) Utilization of excess rain water in medium and lowlands for second crop and fish culture.

(iv) Agri-horticultural system in unbunded uplands

(v) Improved upland rice varieties for aberrant weather conditions.

(vi) INM with green manure crops/grean leaf manuring in transplanted rice.

(vii) Rice-fish-duck/pig farming system.

10. 4.2 Improved rice crop management

1. Promotion of rice with SRI technique in medium and low-lands.
2. Liming to boost crop yield especially pulses, oilseeds, maize, soybean, groundnut etc.
3. INM techniques for transplanted rice involving BGA, Azolla, quality compost & slow release nitrogenous fertilizers.
4. Crop diversification in uplands and medium lands with introduction of QPM or hybrid pigeonpea, soybean, groundnut with appropriate nutrient management strategies.

10.4.3 Managing wastelands & bringing additional areas under cultivation.

The development strategy for the state lies in the fact that poverty is the major issue and the poor almost totally depend on natural resources for their survival.Degradation of natural resources have eroded their life support system and has further intensified the degree and extent of poverty.The overall development of degraded land is bound to necessitate adoption of an integrated approach,which enables to deal with all the forces at work at the same time.Isolated actions treating one or two factors may not be very efficient.

10.4.4 Watershed based natural resource management

(i) Soil and runoff control in a toposequence

(ii) Safe disposal of excess water from lowland rice areas

(iii) Prioritization and identification of critical areas in watershed by remote sensing tools.

(iv) Contingency planning through weather monitoring

(v) Technologies for risk management in watersheds for food and livelihood security

(vi) Improving soil quality in rice based cropping systems.

10.4.5 Training & empowerment of farmers

(i) Soil & water testing

(ii) Input management

(iii) Watershed Management

(iv) Soil& water pollution

(v) Precision farming

(vi) Water saving Technologies.

10.4.6 Revamping the input supply system.

1. Quality composts
2. Micronutrient blended NPK fertilisers
3. Biofertilisers
4. Biopesticides
5. Micro-irrigation
6. Soil Testing Kits
7. Soil amendments (lime, dolomite, basic slag etc.)
8. Secondary & micronutrient sources especially Boron, Phosphogypsum, Molybdenum.

10.4.7 Strengthening of existing institutional arrangements

(i) Soil Testing facilities in all Districts & soil test based nutrient recommendations need to be implemented with proper manpower management.

(ii) A unit of 'Farm Input Supply & Management' need to be established in each sub-region.

(iii) "Agricultural Informatics & Communication" Department should be in place for the State.

(iv) Quality control laboratories need considerable strengthening in terms of manpower & technical expertise.

(v) KVKs & ATMAs of the State must strengthen market linkages.

(vi) Farmer participatory mode of working using a bottom up approach.

10.4.8 Monitoring and evaluation

(i) Monitoring will provide constant feed back on the prevailing situation of the field.

(ii) An in-built information system that can monitor field activities at regular intervals.

(iii) Periodical evaluation is helpful to determine project impact.

(iv) Other evaluation will be designed to measure quantity and quality of people's participation.

(v) Evaluation by accredited research agencies is essential.

(vi) Objective should be to know the effectiveness and possible change in knowledge, skill and attitude of people.

10.5 Policy focus

- Public-Private partnership in areas such as supply & use of soil amendments, micro-irrigation, precision farming.
- Farmers participatory group led approach on soil health management, composting, bio-gas, mulching, residue management & other areas
- Microwatershed based farming system development for rainfed areas for small and marginal farmers.
- Training & Empowerment of farmers.
- Creation of small & medium water harvesting structures in farmers fields
- Improved Agro-advisory service.

References

Anonymous (2012). Sustainable soil,water and Plant Nutrient management for rainfed cropping system.ICAR Niche Area of Excellence Programme. Consolidated report (2007-2012), BAU, Ranchi.pp 1-66.

Bhat, J.A., Kundu, M.C., Hazra, G.C., Santra, G.H., and Mandal, B (2010) rehabilitating acid soils for increasing crop productivity through low cost liming materials. *Sci.Total Environment.*doi.:10.1016

Choudhary,J.S.,sarkar,A.K.,Kumar,Ajay and agarwal, B.K.(2010). Basic Slag as an amendment for acid Soils of Jharkhand for increasing productivity of crops. *Journal of Research,* BAU 22(2)121-127.

Kotur, S.C.(1992). Project report (ICAR) on Micronutrient studies in Vegetable crops,Central Horticultural Experimental Station (ICAR) Ranchi pp1-65.

Kotur, S.C.(1995). Effect of liming and modes of Mo application on the response of cauliflower to Mo in an acid Alfisol. *Journal of the Indian Society of Soil Science* 43(2):296-298.

Mahapatra, P.,singh, R.P.,Singh, B.P.,and Sarkar, A.K.(2007). Long trerm effect of fertiliser,organic manure and amendments on soil health,crop productivity and sustainability.SSAC(BAU). *Technical Bulletin* 4:1-75.

Mandal, S.C. and Jha, K.K. (1970). Agricultural utilization of Basic and other steel slags produced in India. Bull. No.12, Indian Society of Soil Science: 450-453.

Mandal, S.C., Sinha, M.K., and Sinha, H.(1975). In Acid soils of India and liming. *ICAR Technical Bulletin*.75.

Mandal,S.C.,Sinha,H.,Prasad,C.R.,andAli,M.A.(1966). Studies on liming acidic red loam soils.*Journal of the Indian Society of Soil Science*, New Delhi 14:127-131.

Mathur, B.S., Rana, N.K., Sinha, H. and Lal, S.(1985). Lime as an effective source of Fertiliser in acid red loam soil of Bihar. *Journal of the Indian Society of Soil Science* 33:328-332.

Mathur, B.S., Rana, N.K., Sinha, K.P. and Lal, S.(1989). SSAC(BAU)*Research Bulletin* 2/89. BAU, Ranchi.

Mathur, B.S., Sinha, H., Rana, N.K. and Lal, S.(1983). *Journal of the Indian Society of Soil Science* 31:334-340.

Mathur, B.S.(1980). Use of Indigenous rockphosphate as source of Phosphate for acidic soils of Chotanagpur plateau, Bihar.Proceedings Symposium on increasing productivity of acid soils in the eastern region. Fertiliser association of India (eastern region) Calcutta.63-67.

Panda, N., Prasad, R.N., Mukhopadhya, A.K., and Sarkar, A.K. (1991). Managing acid soils for optimum productivity in red,laterite and associated soils of eastern India. *Bulletin of the Indian Society of Soil Science*.15:27-37.

Panda, N., Sarkar, A.K., and Chamuah, G.C. (2009). In chapter on Soil Acidity, "Fundamentals of soil science" (2nd.edition) *Indian Society of Soil Science*, pp 317-328.

Panda, N., sarkar, A.K., and Mathur, B.S. (1996). Chemical degradation leading to soil acidity and its amelioration.*Bulletin of the Indian Society of Soil Science* 17:82-88.

Rattan, R.K., Sarkar, A.K., and Singh, A.K. (2009). Soil and water management for agricultural transformation of eastern india. *Bulletin of the Indian Society of Soil Science* 26:1-120.

Sarkar, A.K. (2005). Managing natural resources for enhancing agricultural productivity in eastern states. *Journal of the Indian Society of Soil Science* 53:435-447.

Sarkar, A.K. and Singh, R.P. (2002). Importance of long-term fertiliser use for sustainable agriculture in Jharkhand. *Fertiliser News* 47(11):107-111.

Sarkar, A.K. and singh, Surendra (2003). Crop response of secondary and micronutrients in acidic soils of India. *Fertiliser News*. 48(4):47-54.

Sarkar, A.K., Jena, D., and Raychaudhuri, Mausumi (2012). Managing acid soils for efficient use of nutrients and water for crop production-challenges and options.Proceedings of the Platinum Jubilee Symposium (Dec.22-25,2009,IARI,New Delhi,editedby.Dr.R.K.rattan). *Indian Society of Soil Science*. pp 79-95.

Sarkar, A.K., Mahapatra, P., and Kumar, A (2007) Management of Macronutrients in Acid Soils. *Bulletin of the Indian Society of Soil Science* 25:10-26.

Sarkar, A.K., Mahapatra, P., Kumar Arvind, and Kumar, Ajay (2010). Boron-for high crop yield,quality and profitable agriculture. *NAE Technical Bulletin* BAU, Ranchi.1:1-11.

Sarkar, A.K., Srivastava,V.C., and Mathur, B.S. (1994) Soil management and rainwater conservation and use in Alfisols under medium rainfall.Bulletin of the Indian Society of Soil Science 16:33-40.

Sarkar, A.K. (2013) Soil Health management in acid Soils.In Soil health Management (ed.Dr.H.L.S.Tandon) Fertiliser Development and Consultative Organisation, New Delhi, India.pp 142-164.

Sarkar, A.K.(2006) Consolidated Report, ICAR Network Project on Soil Characterization and resource management of Acid soil Regions for Increasing Productivity. NodalCentre, BAU, Ranchi. pp 1-53.

Sharma,P.D.,and Sarkar, A.K.(2005) Managing acid soils for enhancing productivity.N.R.M.Division (ICAR) KAB II, New Delhi. *Technical Bulletin* 23:1-24.

Verma,U.N., Pal, S.K.,Thakur, R.,Upasani, R.R. and Singh, M.K. (2007) Cropping Strategies for Rainfed Agro-ecosystem of Jharkhand. Bulletin (agro.)1/2007, BAU, Ranchi. pp 1-70.

8

Acid Soils of Odisha

D. Jena

The state of Odisha located in the Coromondal coast of India has a tropical monsoon climate characterized by high humidity, high rainfall (1497 mm) and short and medium winter. It is the 10th largest and 11th most populous state accounting about 5% of the geographical area and 4% of the population of the country. The geographical area of the state is 15.57 m ha, out of which 0.62 m ha is arable land. Agriculture contributes 28% to state domestic production and employs 64% of the work force.

Rice is the principal food crop of the state, which occupies 50% of the total cultivated area (0.62 m ha) followed by pulses (22%), oilseeds (9%), vegetables (7%) followed by other crops like fruits, spices, sugarcane, jute, etc. (12%). The production and productivity of major crops in the state are much below the National average primarily due to soil acidity, deficiency and toxicity of plant nutrients, low fertilizer consumption per unit of cropped land (63 kg ha^{-1}) and imbalanced nutrient use. Although the state has experienced upward trend in fertilizer use from 7.6 kg ha^{-1} (1980) to 63 kg ha^{-1} (2012), the ratio of N to K_2O fertilizer is still very wide, which declined from 5.9 during 1980 to 4.43 during 2010.

Based on physiography and relief, the state is classified into 4 broad physical divisions namely, northern plateau, central tableland, eastern ghats and coastal plains. The soils of Odisha have been classified under under four orders namely, Inceptisol (7.49 mha), Alfisol (5.62 mha), Entisol (1.53 mha) and Vertisol (0.93 mha). About 70-75% of soils of Odisha are acidic and support important production systems. The extent of acid soils in different districts of Odisha is given in Table 1 (a), (b) and (c).

Table 1(a). Extent of distribution of acid soils (>70%)

Districts	% Distribution	Districts	% Distribution
Cuttack	96.5	Dhenkanal	81.7
Kendrapara	96.0	Puri	81.8
Jagatsingpur	95.0	Rayagada	81.3
Jajpur	95.0	Mayurbhanj	81.0
Koraput	88.1	Nayagarh	80.2
Nawarangpur	87.3	Anugul	79.6
Malkangiri	82.6	Phulbani	74.2
Khurda	82.3		

Table 1(b). Extent of distribution of acid soils (50 -70%)

Districts	% Distribution	Districts	%Distribution
Jharsuguda	69.5	Sambalpur	66.9
Ganjam	66.6	Sundergarh	63.3
Gajapati	62.0	Boudh	60.3
Baragarh	56.7	Deogarh	56.0

Table 1(c). Extent of distribution of acid soils (<50%)

Districts	% Distribution	Districts	%Distribution
Bhadrak	48.9	Sonepur	43.9
Balasore	41.5	Bolangir	39.2
Kalahandi	38.2	Nuapada	32.2
Keonjhar	30.9		

Crop Production Constrains in Acid Soils

The upland acid soils have coarse soil texture with high infiltration rate, low water holding capacity, high permeability, soil crust formation, excessive leaching of nutrients and high bulk density. Seed germination is affected by surface soil crust. Application of organic matter (Biswas and Khosla 1971), tank slit, conservation tillage, contour and strip cropping, intercropping of cereals with legumes or oilseeds and in-situ rainwater harvesting are some of the suitable methods for improving crop production in such soils.

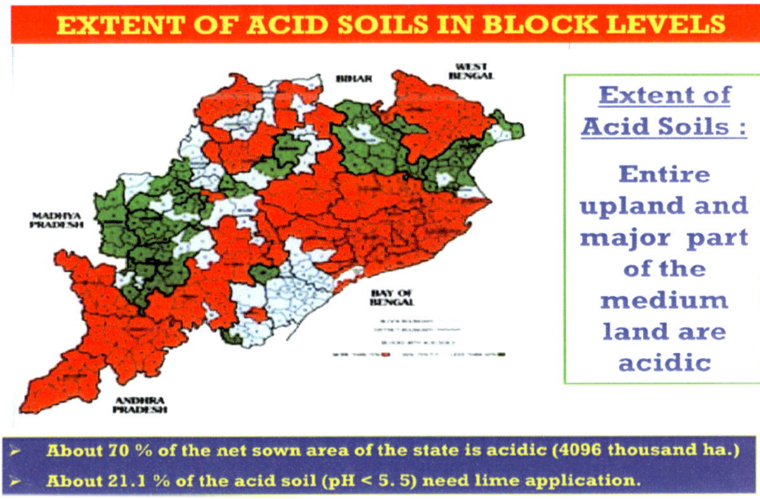

Common problems of acid soils in respect of chemical properties are low pH, low CEC (due to dominance of 1.1 type of clay), low base saturation, high Fe, Al, and Mn, high P fixing capacity, clay fractions consisting of rather low surface active minerals. These problems could be managed by amelioration with liming, which increases soil pH , base status and CEC, inactivates Fe, Al and Mn soil solution, reduce acidity and P fixation in soil (Panda 2007, Misra *et al.* 1989, Sahu & Patnaik 1990, Jena 2008).

Acid soils are deficient in Ca. Exchangeable Mg content of such soils is also poor. Sulphur deficiency is common in upland coarse textured soils. Micronutrients such as B and Mo are often deficient in acid soils. Acid soils of Odisha are generally low in organic carbon and nitrogen status.

Delineation of Acid Soils

More than 70 % soils of Odisha are acidic. Efforts were made to delineate the acid soils of the state. Seven districts of Odisha namely Puri, Cuttack, Khurda, Sambalpur, Nayagarh, Angul and Dhenkanal were considered. About 15 to 20 villages in each of the seven districts were selected for the study. Six soil samples from each village were collected. The longitude and latitudes of each location were recorded using GPS. The samples were analyzed for soil pH in soil water ratio of 1:2.5. The soils were classified into six groups on the basis of pH values such as (1) <4.5, (2) 4.5 to 5.5, (3) 5.5 to 6.5, (4) 6.5 to 7.5, (5) 7.5 to 8.5 and (6)>8.5. Based on the pH values, digitized soil pH maps for each block were prepared using Arc GIS software and are presented in Fig 1.1 to 1.7. A computerised data bank of soil pH was developed. Highly acid soils (pH 4.5 to 5.5) and moderately acid soils (pH 5.5 to 6.5) were of 53346 ha and 74968 ha in Nayagarh districts, 256319 ha and 239851 ha in Cuttack district, 229718 ha and 69373 ha in Khurda district, 306899 ha and 241356 ha in Dhenkanal district, 32194 ha and 155930 ha in Angul district, 384646 ha and 21953 ha in Sambalpur district, respectively. Among these districts, the problem of soil acidity was maximum in Sambalpur district and minimum in Angul district. Thus, a total of 1384399 ha of land in these districts required amelioration through liming (Table 2).

Table 2: Extent (ha) of soil acidity in different districts of Odisha

District	pH range			
	4.50-5.50	5.50-6.50	6.50-7.50	7.50-8.50
Dhenkanal	306899 (53.1)	241356 (41.8)	17575 (3.0)	12163 (2.0)
Cuttack	256319 (50)	239851 (47)	14742.34 (3)	-
Nayagarh	53346 (37)	74968 (52)	16443 (11)	-
Puri	121277 (83)	24304 (16.6)	500 (0.3)	-
Khurda	229718 (75.5)	69373 (23)	4937 (1.6)	-
Angul	32194 (10)	155930(50)	112564 (36)	8328 (3)
Sambalpur	384646 (95)	21953 (5)	-	-

Lime Requirement Of Soils

Lime requirement of a soil refers to the amount of liming material that should be added to raise the pH to a desired level, preferably 6.0-7.0. Several methods have been proposed for estimating lime requirement. More precise methods of determining lime requirement are to measure directly potential or exchange acidity in the system. This is done by using a strongly buffered solution. One of the buffer used was p-nitrophenol buffered at pH 7.0. When this solution is added to acid soils, it reacts and attain a pH value of 7.0. The amount of decrease in pH is directly related to the amount of exchange acidity of the soil.

Tri-ethanolamine (TEA) buffered at pH 8.25 has also been used in measuring exchange acidity. In this method, the TEA and soil are allowed to react and then the suspension is filtered. The filtrate is titrated to some specified pH value and compared with titer of an equivalent amount of TEA solution. The difference is considered to be due to exchange acidity.

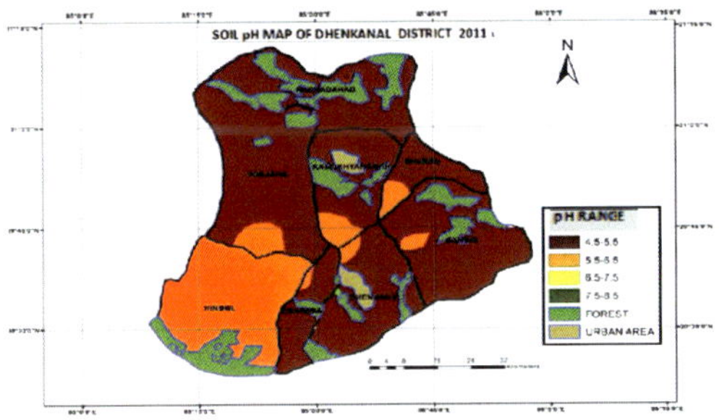

Fig. 1.1. Soil acidity map of Dhenkanal district

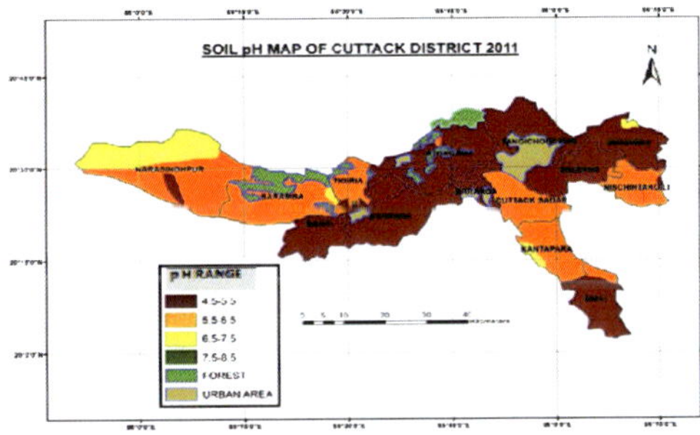

Fig. 1.2. Soil acidity map of Cuttack district

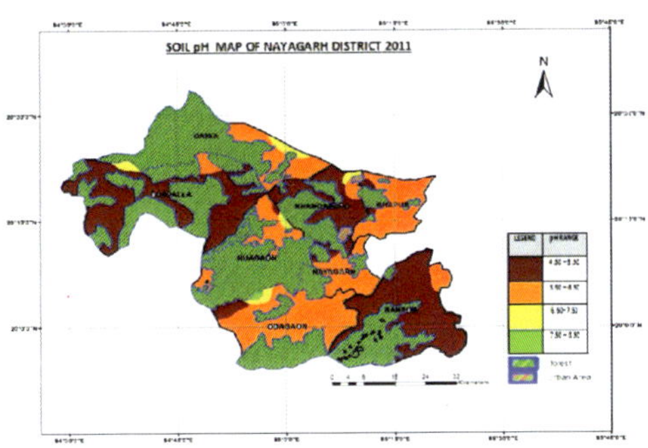

Fig. 1.3. Soil acidity map of Nayagarh district

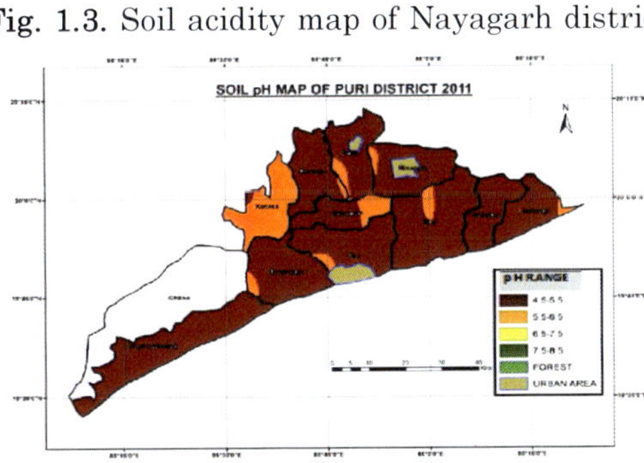

Fig. 1.4. Soil acidity map of Puri district

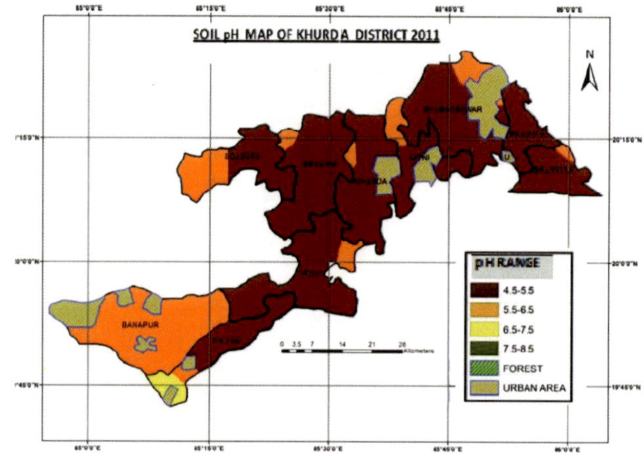

Fig. 1.5. Soil acidity map of Khurda district

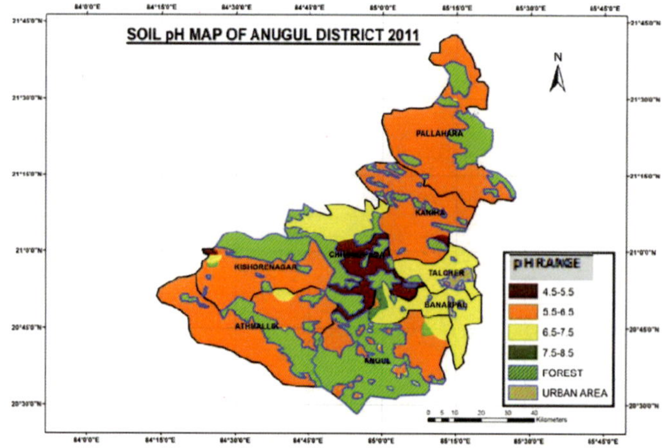

Fig. 1.6. Soil acidity map of Angul district

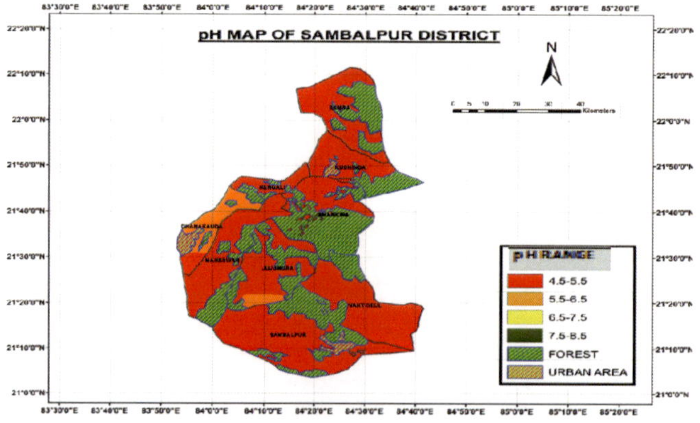

Fig. 1.7. Soil acidity map of Sambalpur district

Each procedure of estimating the lime requirement of soil gives the theoretical amount of Ca that would be needed to raise the pH. When the calculated amount of Ca is added to soils in the field, the pH of the soil is not raised to a level predicted by the testing procedure. Therefore, a "liming factor" has been introduced. A liming factor of 1.5 to 2.0 is commonly used.

Several soil testing laboratories in India use Woodruff's buffer method or SMP method to determine the lime requirement value for ameliorating acid soils. A new method using Ca $(OH)_2$ was developed to find out the lime requirement of acid soils. This method was compared with modified Woodruff's method and Shoemaker's method.

Soil samples of different textural classes were collected from different locations of Odisha. The details of the soil properties are given in Table 3. While some soils were coarse textured, others were medium and or fine textured. Similarly, some soils were of low organic carbon content, while others were of medium to high organic carbon status. All the soils had low exchangeable Al contents.

10 g of soil was treated with 0, 0.5, 1.0, 2.0, 3.0, 4.0, 5.0, 6.0, 7.0, 8.0, 9.0, and 10.0ml of 0.022 M Ca $(OH)_2$ solutions. Distilled water was added to each soil so as to make a soil: water ratio of 1:2.5. After half an hour of intermittent shaking, the pH of the soils was determined with the help of a pH meter.

Effect of Liming on Soil pH

The influence of $Ca(OH)_2$ addition on changes in soil pH are presented in Fig 2. High correlation between pH and calcium hydroxide added exist in sandy loam, loam, silty loam, clay loam and clayey soils. The R^2 values in these soils ranged between 0.88 to 0.99.

Buffering Capacity of Soil

The buffering capacity of the soil was computed from the following equation:

Table 3. Physicochemical properties of the soils

Sl. No.	Name of the soil	Texture	Organic Carbon g/kg	pH	Exch.Al^{3+}	Exch.H^{+}	Exch. acidity	pH dependent acidity	TPA
					C mol(p+) kg^{-1} of soil				
1.	*Typic Ustochrepts*	Loamy Sand	3.53	4.67	0.484	0.176	0.660	2.72	3.38
2.	*Typic Ustochrepts*	Loamy Sand	1.67	5.03	0.264	0.044	0.308	0.472	0.78
3.	*Typic Ustropepts*	Loamy Sand	3.07	4.54	0.484	0.308	0.792	1.808	2.60
4.	*Typic Haplustalfs*	Sandy Loam	2.90	4.69	0.440	0.350	0.704	2.676	3.38
5.	*Typic Haplustalfs*	Loam	2.60	4.72	0.220	0.088	0.308	7.232	7.54
6.	*Typic Ustropepts*	Loam	8.49	5.57	0	0.044	0.044	11.136	11.18
7.	*Typic Ustochrepts*	Silt Loam	3.16	4.86	0.660	0.308	0.968	10.732	11.70
8.	*Typic Ustochrepts*	Silt Loam	4.03	4.41	0.704	0.308	1.012	11.728	12.74
9.	*Typic Ustochrepts*	Clay Loam	4.09	4.51	0.660	0.264	0.924	10.256	11.18
10.	*Typic Ustochrepts*	Clay Loam	3.65	4.93	0.528	0.044	0.572	10.34	10.92
11.	*Typic Ustochrepts*	Clay Loam	2.50	5.08	0	0.132	0.132	8.20	8.330
12.	*Typic Ustochrepts*	Silt Clay Loam	1.34	5.65	0	0.080	0.080	8.36	8.440
13.	*Typic Ustochrepts*	Silt Clay Loam	3.07	4.87	0.220	0.660	0.880	7.54	8.420
14.	*Typic Ustochrepts*	Clay	4.27	4.92	0.484	0.264	0.748	10.29	11.78
15.	*Typic Haplustalfs*	Clay	7.30	4.70	0.396	0.176	0.572	10.18	10.75
16.	*Typic Haplustalfs*	Clay	6.91	4.65	0.264	0.220	0.484	8.530	9.01
17	*Typic Ustochrepts*	Clay	4.80	4.99	0.170	0.220	0.396	8.880	9.27
18	*Typic Ustochrepts*	Clay	4.99	5.38	0	0.190	0.190	9.560	9.75
19	*Typic Ustochrepts*	Clay	5.57	4.71	0.308	0.308	0.616	7.920	8.53

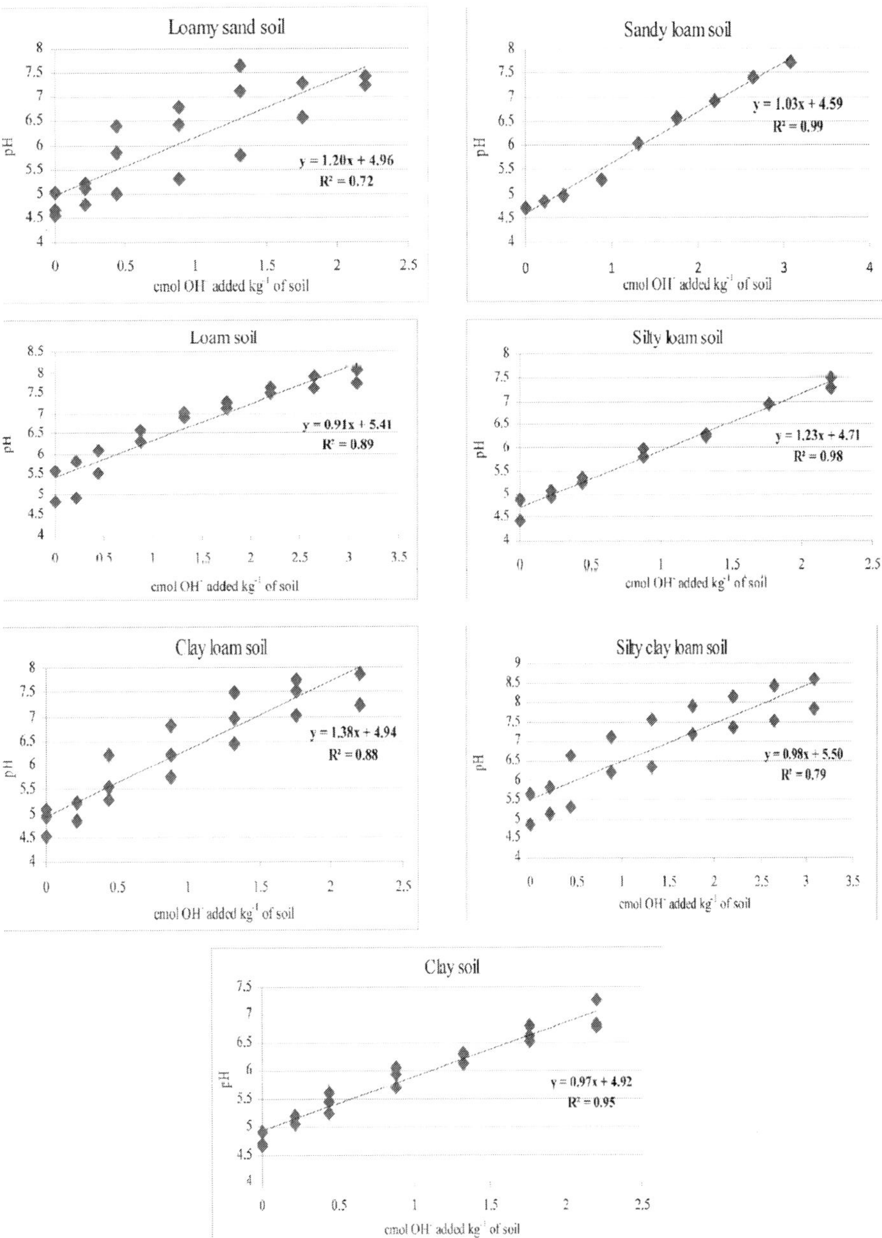

Fig. 2. Effect of levels of lime on pH of soils of different texture

$$BC = \frac{0.44}{pH_4 - pH_2} \qquad \text{Eq. (1)}$$

Where, BC is the buffering capacity (cmol OH^- /kg of soil/pH), pH_2 and pH_4 indicates the pH of the soil when 2 and 4 ml of 0.022 M $Ca(OH)_2$ were added to the soil. The results indicate that buffering capacity varied from 0.51 to 1.08 cmol/kg of soil/pH for different soils. With finer texture of soils, the buffering capacity increased.

Lime Requirement (LR) Determination by $Ca(OH)_2$ Method

The lime requirement (tonnes of $CaCO_3$ ha^{-1}) of the soil was determined by the following equation,

$$LR = \frac{pH_t - pH_0}{pH_4 - pH_2} \qquad \text{Eq. (2)}$$

Where, pH_t was the target pH and pH_0, pH_2 and pH_4 were the pH of the soil, when 0, 2 and 4 ml of 0.022 M Ca $(OH)_2$ solutions were added to the soil. For a target pH of 7, the lime requirements of soils varied from 1.32 to 3.86 tonnes of $CaCO_3$ ha^{-1}. As the buffering capacity of soil increased, the lime requirement of the soil increased.

LR Determination by Modified Woodruff's and Shoemaker's Methods

The lime requirement of these soils was determined following modified Woodruff's method and calcium hydroxide method as per equation (2) so as to raise the soil pH to 7 in both the cases. The lime requirement by modified woodruff method was regressed with that of calcium hydroxide method. The lime requirement value varied from 1.3 to 5.4 tonnes of $CaCO_3 ha^{-1}$ by modified Woodruff's buffer method and 0.56 to 7.87 tones of $CaCO_3$ ha^{-1} by Shoemaker's method. The relationship between these methods was poor and a slope of 1.44 was computed instead of 1.0 in both these methods. This showed that lime requirement determined by modified Woodruff's and Shoemaker's methods were 1.448 times more than that determined by calcium hydroxide method for raising the soil pH to the same value. The relationship between modified Woodruff's buffer method and Shoemaker method was not significant. Calcium hydroxide method was as good as Woodruff's buffer or Shoemaker's method. Further work is required to verify the methodology under field conditions.

Liming – Source And Efficiency

Among the naturally occurring lime sources, calcite, dolomite and stormatolitic lime stones are important. Since calcite and dolomite have industrial use, its application in agriculture is not economical. In India, the total reserve of all categories of lime stone is about 76,446 million tones (mt) out of which 11,562 mt are under proven category, 16,463 mt under probable category and 48,419 mt under possible category (Panda 2007). Odisha has a reserve of 1683 mt. Deposits of about 40 mt of stormatolitic lime stone, a poor grade lime containing 28-32% CaO, 12% MgO and 0.5% P_2O_5 is found in Odisha. Its use in acid soils needs detailed study.

Several industrial wastes such as paper mill sludge (PMS) from paper mills, basic slag from steel industry, preesmud from sugar mills using carbonate process have been tested successfully in acid soil regions of India as amendments, which are eco-friendly. Liming material must be locally available, properly ground and should have high neutralizing value and low cost for use by small and marginal farmers. Although huge amount of basic slag (100 mt per annum) is generated from steel mills located in Bhilai, Rourkela, Bokaro, Durgapur, Burnpur and Jamshedpur, its use is limited due to high grinding cost. On an average, Indian slag contains 1 to 3 % P_2O_5, 24 to 50% CaO and 2 to 10% MgO. The different liming materials available in Odisha are discussed.

(i) Phosphogypsum (PG)

Phosphogypsum is a byproduct of diammonium phosphate factory located at Paradip of Odisha state. When rock phosphate is treated with sulphuric acid, gypsum and phosphoric acid are produced. Phosphogypsum contains 16% S and 20-21% Ca. It also contains about 0.2 to 1.2% phosphorus. Since, Ca in phosphogypsum can leach down faster as compared to lime in light textured soils, sub-soil acidity could be ameliorated with resultant reduction of aluminium toxicity and calcium deficiency. Jena (2008) reported that about 10 million tones of phoshogypsum is dumped around fertilizer industry at Paradip and is an efficient source for correcting S deficiency in crops.

The results of an incubation study revealed that the pH of phosphogypsum – water, pure gypsum –water and soil water suspension on first day was 6.17, 6.17 and 5.34 which increased to 7.27, 7.35 and 5.04 on 30th day of incubation. When Phosphogypsum is applied to soil, the pH dropped by 0.16 unit as compared to normal soil (control) (Fig.3). Application of Phosphogypsum @ 60 kg S ha^{-1} to acid soil reduced the exchange acidity, exch. Al^{3+} and Exch. H^+ from 0.467 to 0.344, 0.2 to 0.1 and 0.267 to 0.244 [cmol(p^+)kg^{-1}], respectively during 30 days incubation period.

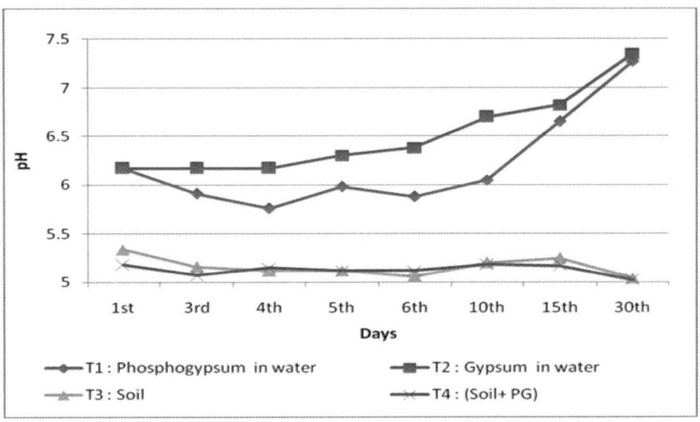

Fig. 3. Effect of Phosphogypsum on pH of acid soil over time

(ii) Paper mill sludge (PMS)

Paper mill sludge is a by-product of paper mill. It is estimated that about 2 lakh metric tones of PMS with 65-79 % neutralizing value and 22 to 35 % Ca content are dumped around four paper mills located in Cuttack, Rayagada, Brajarajnagar and Jeypore of Odisha (Jena, 2008). Efficiency of PMS was studied to find out the optimum dose of lime in reducing the soil acidity by neutralizing exchangeable Al^{3+}. The results revealed that with increase in levels of lime, there was a significant rise in soil pH with decrease in exchange acidity, exchangeable H^+ and Al^{3+}. Application of lime @ 0.2 LR was adequate to bring the pH to a safe level for profitable production of different crops with neutralization of total exchangeable Al^{3+} within first week of application. Jena (2008) studied the effect of different levels of lime (calcite) on pH, exchangeable Al^{+3} and H^+ in an acidic laterite soil of Dhenkanal. Results reveal that application of lime @ 0.2 LR raised the pH from 5.1 to 6.9 (Fig 4) and decreased the exchangeable Al^{3+} from 0.62 to zero cmol (p^+) kg^{-1} within seven days of incubation (Fig 5).

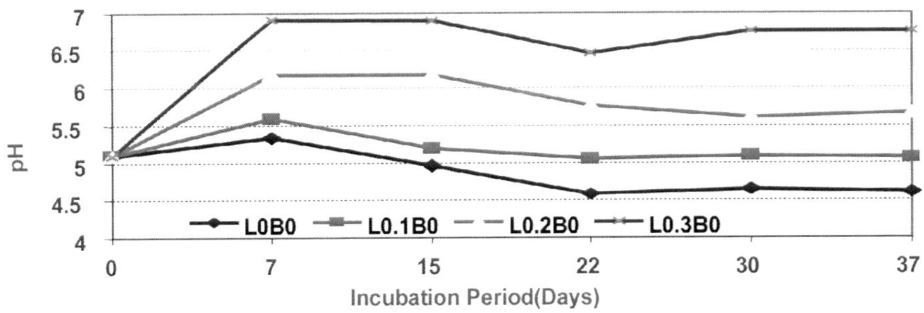

Fig. 4. Effect of lime on soil pH

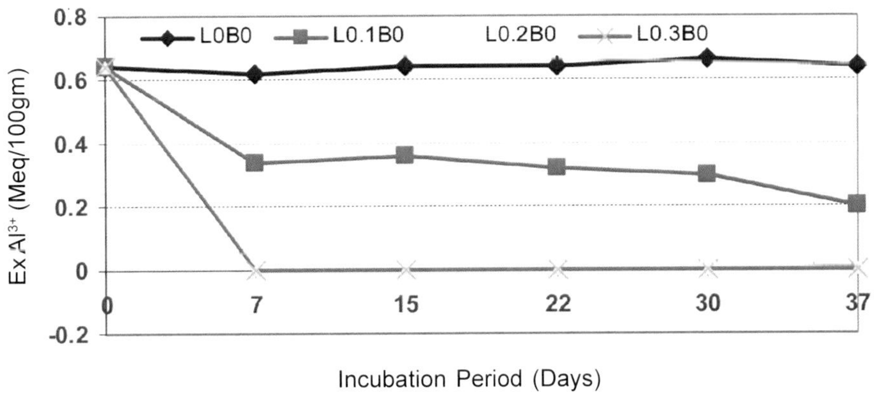

Fig. 5. Effect of lime on Exchangeable Al^{3+}

(iii) Pressmud (PM)

Pressmud is a by-product of sugar industry and can be utilized as an amendment to improve soil physical properties. Efforts were made to characterize pressmud as a liming material for acid soils. Two pressmuds - Nayagarh and Dhenkanal were collected and analyzed for neutralizing value, organic carbon, Ca and Mg content. The pH of the pressmud varied between 6.5 to 6.92 and OC 16-18%. The neutralizing value of 22-24% indicated that it cannot be compared with $CaCO_3$ or PMS as a liming material. However, due to its high organic carbon content, it can be used as an acid soil amendment.

An incubation study was conducted with an acidic upland soil (pH-4.8, exch. acidity-0.6 cmol (P^+) kg^{-1}) to compare the efficiency of PM and PMS. Three different sources of liming materials, such as

PMS, PM-I (pressmud - Nayagarha Sugar factory) and PM-II (pressmud - Dhenkanal Sugar factory) were applied @ 0.1 LR, 0.2 LR, 0.3 LR and 0.4 LR. The LR value was 4.8 t ha^{-1}. The pH increased significantly over control with application of PMS and PM @ 0.2 LR, 0.3 LR and 0.4 LR. The PM-II was better than PM-I. The exchange acidity decreased in all the lime treated soils and attained a value of zero on 7th day of incubation. The study also indicated that application of PMS @ 0.2 LR or PM @ 0.3 LR was adequate for neutralization of exchangeable Al^{+3}.

Field experiments were conducted for three seasons (rabi 2010-2012) to study the combined effect of PMS and PM on growth and yield of groundnut in an upland site of OUAT Central Farm having exchange acidity : 0.96 Cmol (p^+) kg^{-1}; O.C: 3.4 g kg^{-1} and LR: 6.3 t $CaCO_3$ ha^{-1}. PMS and PM was applied alone or in combination of 75:25, 50:50 and 25:75.

Effect of Lime on Soil pH, exch. acidity, exch. Al^{3+} and H^+

Application of PMS, PM and their combinations @ 0.2 LR increased the soil pH over lime control. The peak pH was attained at 25 days of lime application and remained above 5.0 during the entire growth period. Exch. Al^{3+} decreased rapidly within 14 days of lime application and there after slowly, finally, it attained a negligible value between 73-103 days of application. Among the lime sources, the decrease in exchange acidity in PMS (100%) was higher as compared to PM (100%) (Fig. 6).

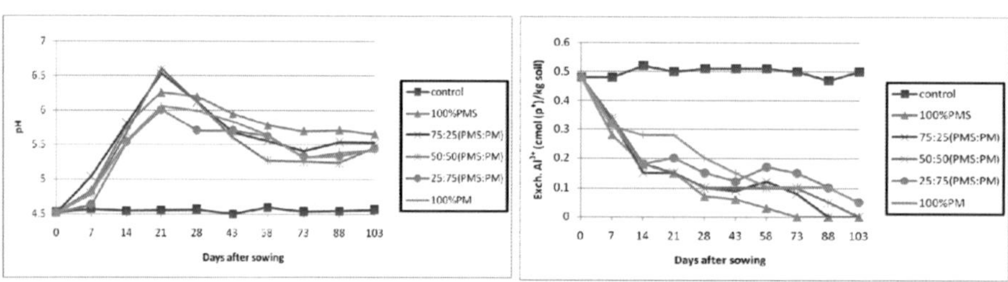

Fig. 6. Effect of liming (PMS, PM) on change in soil pH and Exch. Al^{3+}

Pod yield

The pod yield in lime control varied between 11.63 to 24.21 q ha^{-1} over three seasons. Highest yield response of 95% was obtained when PMS and PM was applied in ratio of 75:25 (Table 4).

Table 4. Effect of combined application of PMS and PM on pod yield (q ha^{-1}) of groundnut

Treatments	Rabi 2010-11	Kharif 2011	Rabi 2011-12	Mean	% response over control
T_1 : Control	24.21	17.0	11.63	17.61	-
T_2 : 100 % PMS	40.36	23.0	35.25	27.53	56.33
T_3: 75 % PMS + 25 % PM	41.78	24.9	36.38	34.35	95.05
T_4 : 50 % PMS : 50 % PM	38.88	21.5	23.88	28.08	59.45
T_5 : 25 % PMS : 75 % PM	36.78	20.5	23.06	26.78	52.07
T_6 : 100 % PM	38.22	20.0	24.31	27.51	56.21
CD (p= 0.05)	9.63	0.65	2.62		

(iv) Stromatolyte

Deposits of about 40 mt of stromatolyte is reported in Koraput districts of Odisha. It is a low grade lime stone commercially not exploited by steel industry. Attempt was therefore made to evaluate its suitability as liming material. The rock was collected from Gupteswar area of Koraput district, ground to pass through different mesh sieves, *viz.* 22, 60, 100, 350. Then, the material was heated at 110°C and at 500°C for 2-4 hours and analyzed for neutralizing value, sulphur, calcium and magnesium contents.

The neutralizing value of rock fragments was 80 % without heating and increased to 81.6% when heated at 110°C for 2h. With increase in temperature to 500°C, the neutralizing value further increased to 93.4%. The fineness of soil particles had no effect on neutralizing value of the rock fragments. On an average, the Calcium content of rock fragment was about 17.3% without heating which increased to 24% as the temperature and duration of heating increased, possibly due to loss of material, other than Calcium.

The amount of Mg present in rock fragments was less than Ca in the rock. It was about 10.5% without heating and the amount increased to 15% due to heating. The sulphur content in the rock increased with increasing particle size, but decreased with heating (Table 5).

Table 5. Ca, Mg and S content (%) of stromatolyte

Mesh No. of sieves	Raw (without heating)			
	N.V. (%)	% Ca	% Mg	% S
22	79.0	16.0	9.6	0.035
60	79.6	17.4	10.5	0.047
100	79.8	17.9	10.8	0.047
350	81.0	17.7	11.1	0.046
Mean	79.9	17.3	10.5	0.044

Stromatolyte as a liming material

Pot experiments were conducted to evaluate the efficiency of stromatolyte and its effect on maize crop. Three doses of stromatolyte *viz.*, 0.1, 0.2 and 0.5 LR were compared with pure $CaCO_3$ @ 0.2 LR and PMS @ 0.2 LR. The test crop, maize, received a common dose of 100-50-50 kg N-P_2O_5-K_2O ha^{-1} through urea and KH_2PO_4. The lime requirement (LR) to raise the pH to 6.5 by Woodruff's Buffer method was 11 tones ha^{-1}. The exchange acidity was 2.08 cmol (P^+) kg^{-1} soil, with exchangeable H and Al of 1.24 and 0.84 cmol (P^+) kg^{-1} soil, respectively. The organic carbon content was 4.34 g kg^{-1} soil.

The soil reaction (pH) initially was 4.47, which increased to 5.33 - 5.85 after 7 days of application of different amendments. Further, soil amelioration with $CaCO_3$ and PMS proved better during early stages of maize crop growth (28th day) as compared to stromatolyte sources. Thereafter, the rock resources maintained higher pH than PMS and $CaCO_3$. The stromatolyte of different size particles had almost similar influence on reducing the exchange acidity, Al^{3+} and H^+.

Results of pot culture study showed that ameliorating acid soil with liming materials significantly increased the biomass yield of maize by complexing and ameliorating the acidity, raising the pH to higher levels. Raw form of stromatolyte of 22 mesh sieve was more effective as compared to standard PMS or $CaCO_3$ and better than the heat treated products of stromatolyte in increasing the biomass production of maize.

(v) LD slag

LD slag, a by-product of steel industry (SAIL, Rourkela) was evaluated for its use as a liming material in acid soils. The pH of

LD slag is 10.49, Ca content 16.82 %, Mg content 2.13 % and neutralizing value 58.8 %. Field trials were conducted at Central Research Station, OUAT, Bhubaneswar, for four seasons (2010-12) with groundnut as test crop. The treatments consisted of 3 dates of applications : (0, 5, 10 day before sowing) and 4 levels of lime (0, 0.1, 0.3 and 0.4 LR). The soil of the experimental site was acidic in reaction (pH 4.53), non-saline, organic carbon 0.33 to 0.55 %, exchange acidity 0.39 cmol (p^+) kg^{-1} and exchangeable Al^{3+} 0.61 cmol (p^+) kg^{-1}. The lime requirement value of the soil was 3.3 t ha^{-1}.

The pH of the soil (Fig. 7) in lime control increased from an initial value of 4.53 to 4.95 during crop growth period. With application of LD slag the soil pH increased to 6.98 and remained above 5.5 till harvest of the crop. There was little change in soil pH when

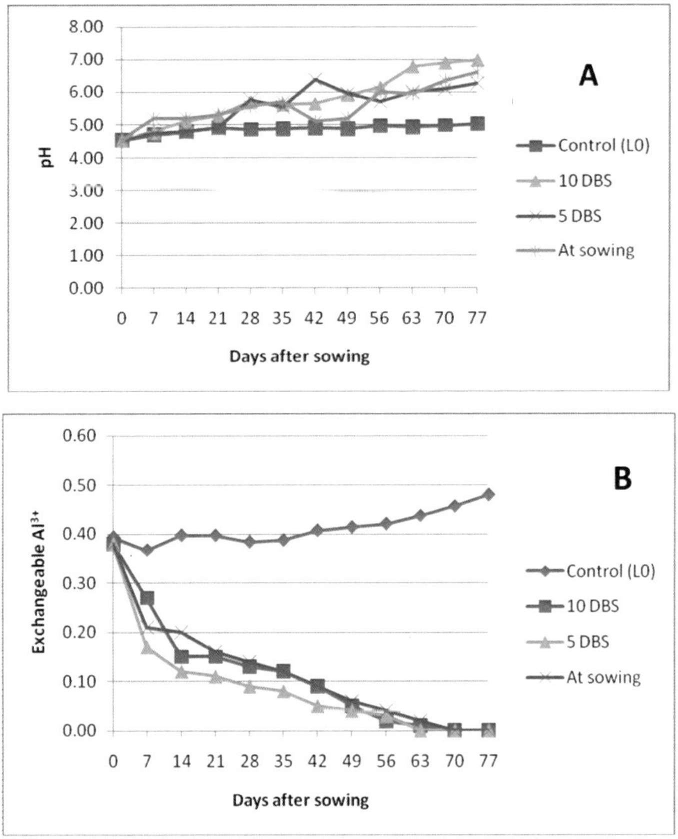

Fig. 7. Change in (A) soil pH and (B) exchangeable Al^{3+} through application of LD slag @ 0.3LR
*DBS: Days before sowing

LD slag was applied at sowing or 5 and 10 days before sowing. Irrespective of the time of application, the pH increased with doses of lime. Application of LD slag @ 0.3 LR recorded higher pH as compared to 0.1 or 0.2 LR throughout the growing season (pH - 4.72 to 6.98).

The exchange acidity and exchangeable Al^{3+} (Fig. 7) in the control plot increased with time, where as in lime treatments, the trend was reversed. About 75% of exch. Al^{3+} was neutralized within 14 days of LD Slag application.

The pod yield of groundnut in lime control varied between 13.9 to 20.6 q ha^{-1}. Application of LD slag @ 0.3 LR at 10 days before sowing resulted in a high pod yield of 31.8 q ha^{-1}.

In another field study, the efficiency of LD slag was compared with PMS as sole or in combinations. The results reveal that sole application of LD slag was inferior to PMS. However, combined application of LD slag and PMS (1:1 ratio) @ 0.3 LR increased the pod yield by 58 to 90 %.

Crop Response to Lime and Fertilizers

Lime requirement of acid soils of India varied from 2.6 to 24.0 t $CaCO_3$ ha^{-1} and it was closely related to soil pH and organic matter content (Sharma and Tripathy 1989). But, liming @ full LR dose is often not economical. For laterite soils of Bhubaneswar, application of $CaCO_3$ @ 0.5 LR was enough for maize (Pradhan 1978), although exchangeable Al was neutralized with 0.25 LR. Application of lime @ 10-20% of LR in furrows increased the yield of green gram, groundnut, black gram, soyabean, lentil, pea and gram over 'no lime' control. Experiments on direct and residual effect of liming in different cropping system conducted in acid sandy loam soil at G. Udayagiri in Odisha showed that, grain yield of mustard, wheat and cowpea increased by 41, 28 and 21%, respectively due to direct effect of liming @ 0.25 LR and maize yield increased by 7-13% under residual condition (Table 6).

Organic amendments also reduce exchangeable Al in soils due to precipitation of Al ions by OH ions released from organic ligands (Hue 1992; Lyamuremye *et al.* 1995). Several workers have suggested for application of organic amendments (FYM) either alone or in combination with lime for controlling the acidity as well as nutrient availability. Liming also stimulates microbial activity leading to mineralization of organic N and fixation of N_2 (Raychaudhuri *et al.* 1998).

Sharma and Sarkar (2005) reported that application of lime @ 2-4 q ha^{-1} increased the yield of rape seed mustard, wheat, green gram, maize, pigeon pea, field pea, black gram, groundnut by 14 to 52% over farmer's practice.

Table 6. Effect of liming on field pea, wheat and mustard and its residual effect on succeeding maize crop

Crop		Direct effect (t/ha)		Residual maize (t/ha)		B:C Ratio
		L_0	L_1	L_0	L_1	
Field pea	F_1	1.02	1.30	2.00	2.21	
	F_2	1.34	1.56	2.16	2.21	
	Mean	1.18	1.43	2.08	2.21	4.25
Wheat	F_1	0.75	0.90	1.00	1.02	
	F_2	0.89	1.15	1.09	1.35	
	Mean	0.82	1.02	1.04	1.18	1.33
Mustard	F_1	0.18	0.24	1.18	1.27	
	F_2	0.29	0.34	1.14	1.31	
	Mean	0.23	0.29	1.16	1.29	1.68

* L_0 : No lime, L_1 : 0.25 LR, F_1 : 50% RDF, F_2 : 100% RDF (recommended dose of fertilizer)
Source : Jena (1996)

Field experiments conducted in several districts of Odisha with varying pH using PMS @ 0.2 LR applied in furrows below the seeds at the time of sowing showed that, liming alone increased groundnut yield by 17-36 % over farmer's practice (FP) in the red soils of Mayurbhanja, Ganjam and Nayagarh. The yield of green gram increased by 5-21 % over FP in laterite soils of Khurda and Dhenkanal. Cabbage and cauliflower responded significantly to lime application in Koraput and Kandhmal districts and response varied between 15-22 % over FP.

On farm trials on liming were conducted by OUAT for many years. The out come of these trials are presented in following sections:

Wheat

Field level demonstrations (Five) were conducted in acidic soils (pH 4.71-5.76) in Mukuna village of Keonjhar with Cv. Sonalika. The soils were acidic with organic carbon medium to high. The yield of wheat in control was 37.3 q ha^{-1}. With application of lime + RDF, the wheat yield increased by 19 per cent over farmers practice (Fig. 8).

Table 7. Response of different crops to liming in acid soil regions of Odisha

Crop	District	pH range	No. of trials	Yield (t ha^{-1})		Yield response (%)	B:C Ratio
				FP	FP + Lime		
Groundnut	Dhenkanal	4.0-6.3	9	0.84	1.14	36	2.34
	Mayurbhanja	4.8-5.2	11	1.00	1.76	17	1.75
	Nayagarh	5.5-5.7	10	1.07	1.27	19	1.20
	Ganjam	5.6-6.1	7	1.91	2.39	25	4.32
Green gram	Khurda	5.5-6.5	12	0.81	0.89	5	1.50
	Dhenkanal	3.8-6.0	10	0.82	0.99	21	1.60
Cabbage	Kandhamal	5.9-6.6	5	11.17	13.00	16	5.40
	Koraput	5.8-7.3	7	22.40	25.69	15	-
Cauliflower	Kandhamal	6.0-6.1	5	9.95	12.10	22	14.05

* Source : Jena (2011)

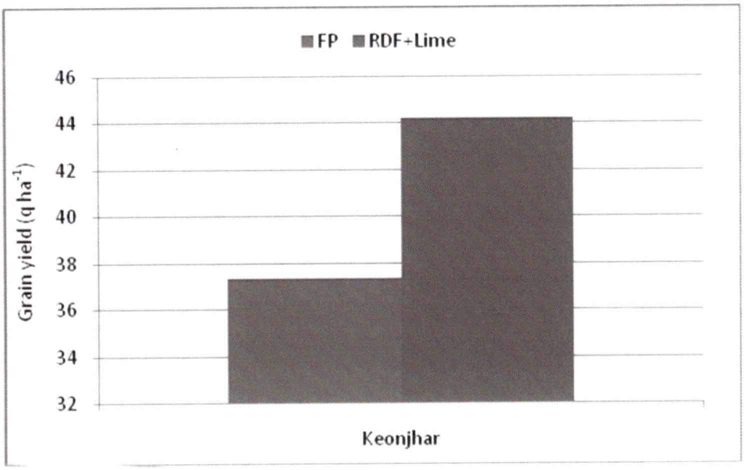

Fig. 8. Effect of lime on grain yield of wheat

Arhar

Ten field trials were conducted in farmer fields in three villages such as Kendumundi of Kalahandi district and Gohirpadar and Chanabeda of Nuapada district in soils with moderately acidic soil reaction (pH-5.21-6.00) and low to medium organic carbon. In Kalahandi and Nuapada district, the arhar yield in FP was 9.84-10.33 q ha^{-1}. Application of PMS @ 0.2 LR increased the pod yield over FP by 90-93 % (Fig. 9).

Response of Arhar to lime

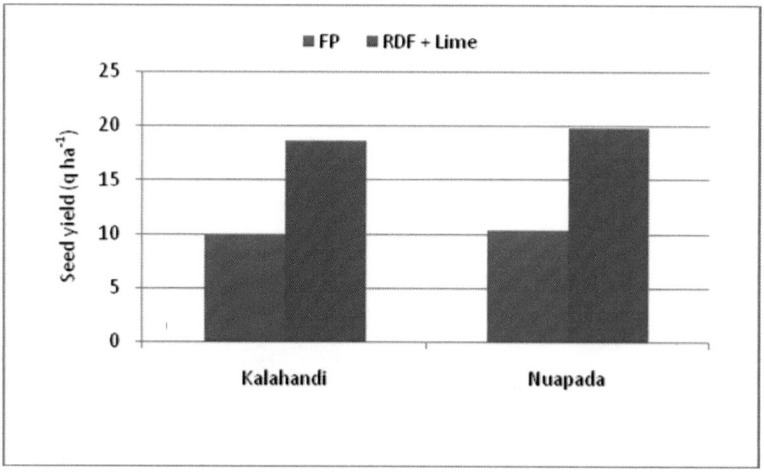

Fig. 9. Effect of lime and fertilizer on yield of arhar

Sunflower

Twenty four field trials on sunflower were conducted in farmers' field of Bramhanipalli village in Sambalpur district and Haja village of Khurda district in soils of low pH and medium organic carbon. The yield of sunflower in FP was 13.5 q ha^{-1} in Sambalpur and 7.1 q ha^{-1} in Khurda district. Application of PMS @ 0.2 LR along with recommended dose of fertilizer increased the sunflower yield over FP by 49 % in Sambalpur and 37 % in Khurda district (Fig. 10).

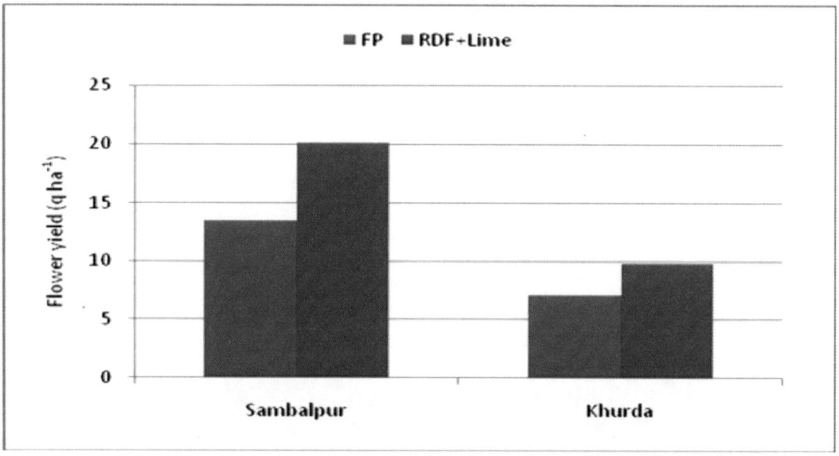

Fig. 10. Effect of lime on yield of sunflower

Response of sunflower to lime and chemical fertilizer

Split Application of Lime

In acid soil regions, leaching of Ca is very common due to heavy precipitation. Leaching of calcium below the root zone in coarse textured soils decreases the liming efficiency. Field studies were undertaken to evaluate the efficiency of split application of lime and its effect on soil pH, exch. acidity, exch. Al^{3+}, exch. H^+ and ground nut yield in acidic upland sites of Central farm of OUAT. The initial soil properties were : pH, 4.88; exch. acidity 0.49 cmol (p^+) kg^{-1} and LR 3.6 t ha^{-1}. The treatments were: main plot-M_1 : lime alone, M_2 : lime + FYM @ 5t/ha, M_3 : lime split (1:1) + FYM @ 5 t/ha and sub plot - T_1 : lime @ 0 LR, T_2 : lime @ 0.1 LR, T_3 : lime @ 0.2 LR, T_4 : lime @ 0.3 LR.

The pod yield increased significantly with dose of lime irrespective of the methods of application. However, the yield in 0.1 LR treatment was optimum and economical as compared to 0.2 and 0.3 LR. Irrespective of the levels of lime, M_2 (lime + FYM) treatment recorded the highest yield as compared to M3 (split application of lime+ FYM) treatment, but both treatments were on par (Fig. 11).

The pH of the soil increased with graded levels of lime and was in the order of 0 LR < 0.1 LR < 0.2 LR < 0.3 LR (Fig. 12). The peak pH value was attained during 7 to 15 days of lime application in M_1 and M_2. But, in case of M_3 (split application), the peak pH value was attained during 30 to 45 days. Comparing different methods of lime application, soil pH was the lowest when lime was applied as basal (M_1), medium when lime was applied along with FYM

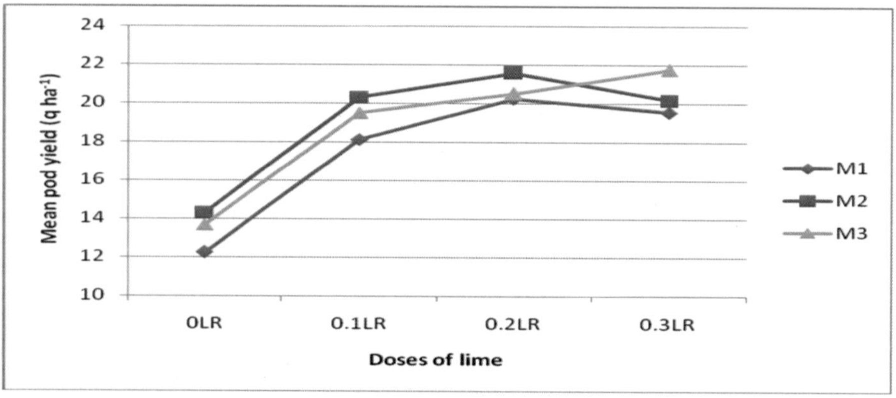

Fig. 11. Effect of split application of lime on ground nut

(M_2) and the highest when lime was applied in split (M_3) at different levels. Lime + FYM treated soils, in general, recorded slightly higher pH than lime treated soils throughout the growth period.

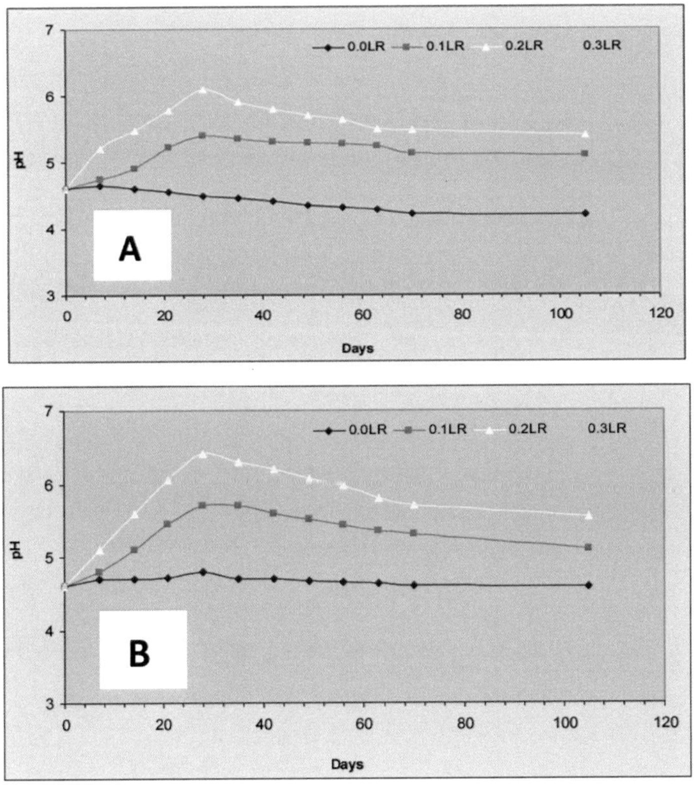

Fig. 12. Effect of (A) Basal application of lime + FYM and (B) split application of lime + FYM on change in soil pH during crop growth period

Exchange acidity was highest in the soils where lime was not applied (T_1). Liming the soil reduced exchange acidity. Since, the exchangeable Al content was zero in lime treated soils (T_2, T_3, T_4), only the exchangeable H^+ contributed to exchange acidity of the soil. The content of exchange acidity varied in the order : 0.1 LR > 0.2 LR > 0.3 LR among the treatments. Comparing different methods of liming, exchange acidity was highest, when lime was applied as basal (M_1), moderate when lime was applied as basal with FYM (M_2) and the lowest when lime was applied in split with FYM (M_3) (Fig. 13).

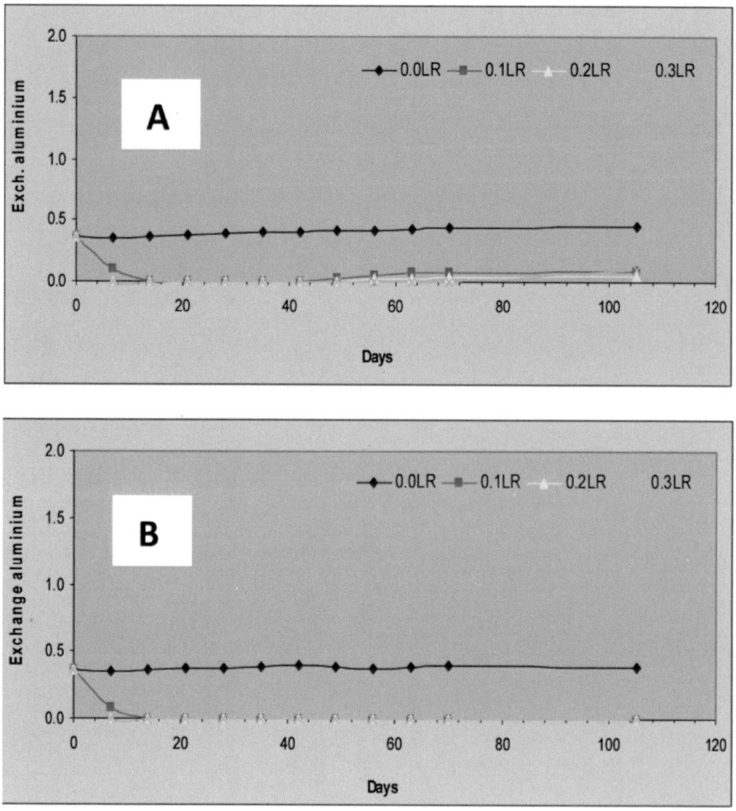

Fig. 13. Effect of (A) Basal application of lime+ FYM and (B) split application of lime + FYM on change in soil exch. Al^{3+} during crop growth

This study showed that split application of 0.1 LR dose of lime was on par with one time application. Basal application of lime @ 0.1 LR along with FYM @ 5 tones ha^{-1} significantly increased the pod yield by reducing the exchange acidity, exchangeable H^+ and exchangeable Al content of the soil. The soil pH was maintained at optimum levels.

Long term effect on soil

Results of long-term experiments conducted in acid soil regions at Ranchi, Bhubaneswar, Palampur and Bangalore over a period of 9-24 years reveal that soil pH decreased by 0.2-0.9 units with application of chemical fertilizer (NPK), but the pH was maintained or increased by 0.3-0.6 units with NPK + lime [Singh and Sarkar (1998), Sahoo *et al.* (1998), Sharma *et al.* (1990), Sudhir *et al.* (1998)]. Average crop yield of soybean, wheat, rice and maize increased by 0.3-0.7 t ha^{-1} due to application of NPK + lime over NPK. The crop response to lime application at Bangalore was negative because the initial pH was higher than 6.0.

Long term effect of liming on soil quality and yield was studied in lateritic soil of Central Research Station, OUAT Bhubaneswar, with groundnut (*cv. Smruti*) as test crop for four seasons. The soil of the experiment site was sandy loam with pH 4.6 and exchange acidity of 1.12 cmol (p^+) kg^{-1}. Lime requirement of the soil determined by Woodruff's buffer method was 7.3 tonne of $CaCO_3$ ha^{-1}. There were 12 treatment combinations as; T_1: Control, T_2 : FYM, T_3 : 0.1 LR, T_4 : 0.1 LR+FYM, T_5 :0.2 LR, T_6 : 0.2LR+FYM, T_7 : 0.3 LR, T_8 : 0.3 LR+FYM, T_9 : 0.4 LR, T_{10}: 0.4 LR+FYM, T_{11} : 0.5 LR, T_{12}: 0.5 LR+FYM. Lime was applied to crops through paper mill sludge having a neutralizing value of 70%. FYM was applied @ 5 t ha^{-1}.

Pod Yield

Generally, the groundnut yield in Odisha during rabi season is higher as compared to kharif. The pod yield in lime control varied between 16.80-24.95 q ha^{-1} during rabi season and 10.35-15.54 q/ha during kharif. With application of lime @0.1 LR, the yield increased to 20.88 q ha^{-1}, which was 23 per cent higher over lime control. The mean yield at higher doses of lime (0.2, 0.3, 0.4 and 0.5 LR) varied between 20.85 to 21.75 q ha^{-1}. The difference in yield between 0.1 and 0.2-0.5 LR was not significant, which might be due to leaching of Ca below the root zone with high precipitation (kharif) or irrigation (rabi). The coarse texture of the sandy loam soil further enhanced leaching. Hence, continuous application of lime @ 0.1 LR is sufficient enough to neutralize soil acidity in groundnut based cropping system. Application of FYM @ 5 t ha^{-1} enhanced the use efficiency of lime and resulted in 18 per cent higher yield over FYM control (Fig. 14).

Effect of Lime and FYM on Soil pH

Continuous cropping of groundnut for four seasons raised the pH of the soil in lime control to 6.0 from the original pH of 4.82. Application of PMS @ 0.1 to 0.5 LR maintained the pH between 6.0 and 7.0 and was higher than that in unlimed soil. Application of FYM also contributed to the rise in pH though the interaction

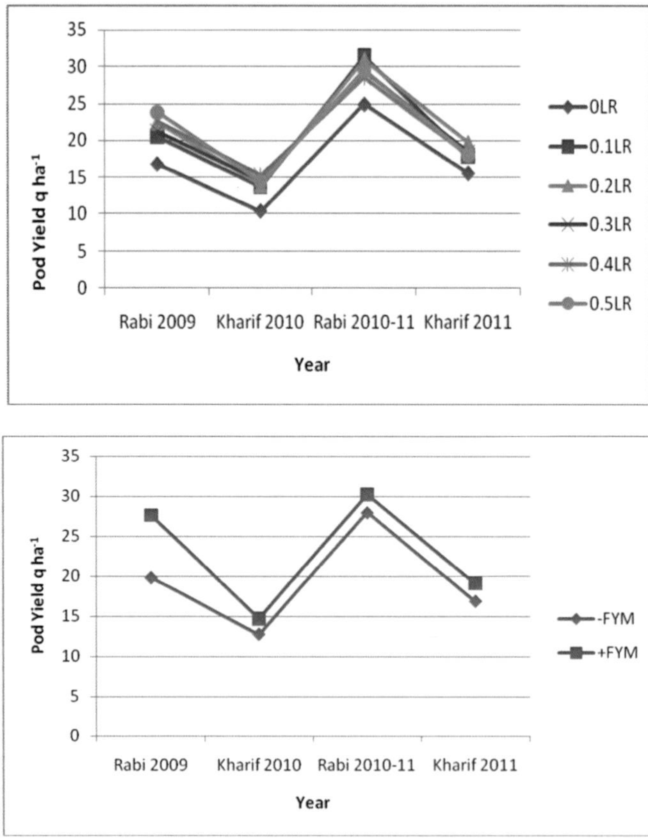

Fig. 14. Effect of continuous liming and application of FYM on pod yield of groundnut

effect was not significant. In general, the pH measured after harvest of rabi crop was higher than that after harvest of the kharif crop. There was seasonal fluctuation of soil pH in different layers due to upward or downward movement of calcium in soil (Fig. 15).

Lime-Boron Interaction in Acid Soils

Deficiency of B can be corrected by applying various boron fertilizers like borax decahydrate ($Na_2B_4O_7.10H_2O$; 10.5% B), boric acid (H_3BO_3, 17% B), granubor II ($Na_2B_4O_7.5H_2O$; 14.6% B). Boron is generally applied through broadcast and mixed properly prior to sowing or transplanting of crops @ 0.75 to 1.5 kg ha^{-1}. Band placement 1-2 kg B ha^{-1} was superior to broadcasting in increasing cud yield of cauliflower (Singh 2007). Productivity of cereals, chickpea, pigeon-pea, groundnut, sunflower, sesame, linseed and mustard increased significantly with application of 1- 2.5 kg B ha^{-1}. Application of granubor significantly increased the yield of rice-groundnut in laterite soil of Bhubaneswar (Jena *et al.* 2006).

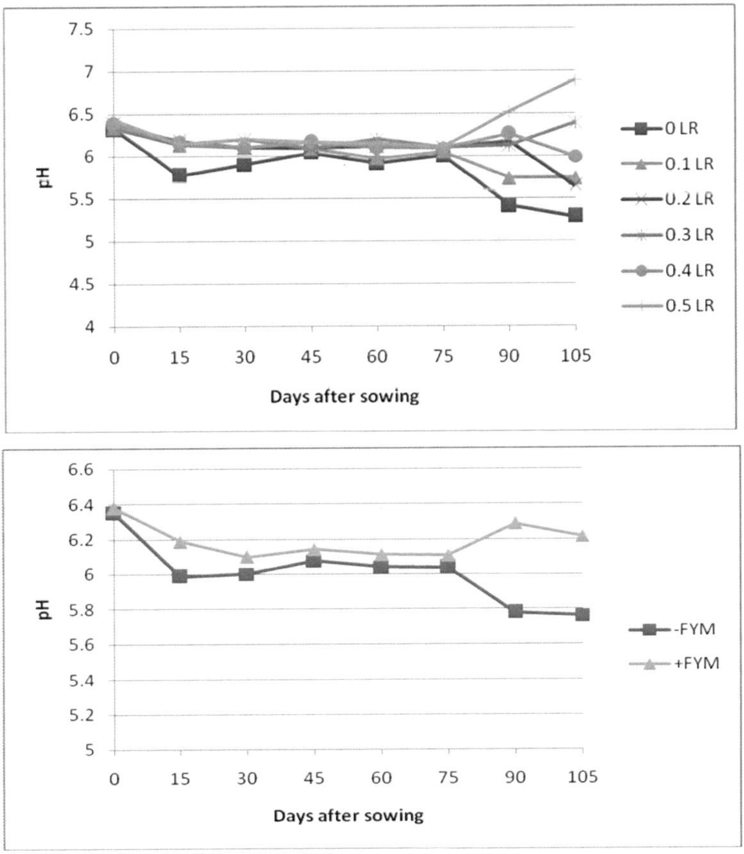

Fig. 15. Effect of lime and FYM on soil pH over four seasons of cropping

Recovery of applied B in acid soils is about 36-54% due to higher precipitation with hydrous oxide of Al and Fe (Mandal & Mandal 1992). Mandal (1995) and Jena (2008) found that combined application of B with lime together had significant effect on increasing the yield of wheat and cauliflower, respectively.

Effect of Lime and Boron On Cabbage-Okra-Cowpea Cropping System in Acid Laterite Soils

Field experiments were conducted in boron deficient lateritic soils of Bhubaneswar during November, 2007 (first crop); March, 2008 (second crop); July, 2008 (third crop) and November, 2008 (fourth crop) to study the direct, residual and cumulative effect of lime and boron on yield and nutrient availability in cabbage - okra - cowpea cropping system. The experiment was laid out with 12 treatments (four levels of lime- 0, 0.1, 0.2 and 0.3 LR and three levels of B - 0, 1 and 1, 2 kg B ha^{-1}). Uniform recommended fertilizer dose of 125 : 40 : 60, 120 : 80 : 80 and 25 : 50 : 25 kg N: P_2O_5 : K_2O per hectare was applied to cabbage, okra and cowpea, respectively. Lime through PMS was applied to the first crop where as B through borax was applied to first and third crop (cabbage).

The soil of the experimental site was acidic in reaction and LR value was 35 q ha^{-1}. The exchange acidity was 0.2 [cmol (p$^+$) kg^{-1}]. The soil was rich in organic carbon (7.8 g kg^{-1}), medium in N and K. The exch. Ca^{2+} content was 2.7 [cmol (p$^+$) kg^{-1}].

Effect of Lime and Boron on Yield

Lime and B application had significant effect on cabbage yield. Highest yield of 621 q ha^{-1} was obtained with lime @0.2 LR + B @ 2 kg ha^{-1} .The yield of cabbage increased over control (L_0B_0) by 21 to 31% at 0.1 LR to 0.2 LR. There was reduction of yield by 17% (over

Table 8. Direct, residual and cumulative effect of lime and boron on crop yield (q ha^{-1})

Treatment		1st Crop (Cabbage) Direct(L & B)	2nd Crop (Okra - Fruit) Residual	3rd Crop (Cowpea-Pod) Residual	4th Crop (Cabbage) Residual (L) Direct (B)
LimeLevels	L_0	436.0	86.50	41.74	440.0
	$L_{0.1}$	529.0 (21.3)*	90.71 (4.87)	49.42 (18.38)	466.0 (5.70)
	$L_{0.2}$	573.0 (31.4)	90.94 (5.14)	48.68 (16.63)	537.0 (21.85)
	$L_{0.3}$	509.0 (17.0)	95.97 (10.94)	53.70 (28.65)	562.0 (27.52)
BoronLevels	B_0	455.0	88.37	47.26	473.0
	B_1	526.0(16)	90.49 (2.40)	48.71 (3.08)	510 .0(7.76)
	B_2	554.0(22)	94.22 (6.62)	49.20 (4.11)	521.0 (10.10)
C.D.(0.05)	L	7.38	NS	1.79	20.80
	B	6.39	NS	NS	NS
	L x B	12.79	NS	3.10	36.03

* Values in parenthesis represent the per cent increase or decrease over control

control) at 0.3 LR indicating that the interaction between lime and B at higher level of lime (0.3 LR and 2 kg B ha^{-1}) is antagonistic. With application of B @ 1 and 2 kg B ha^{-1}, there was increase in yield by 16 and 22 % over control (B_0), respectively although the differences were not significant (Table 8).

The mean fruit yield of okra, which was grown under residual lime and B ranged from 80.26 to 99.19 q ha^{-1} as against 80.33 q ha^{-1} in control. There was difference in the fruit yield due to residual effect of lime and boron. The fruit yield increased by 5-10% with lime and 2-6% with B treatments.

The pod yield of cowpea (third crop) was 39.75q ha^{-1} in control, which increased by 16-28% with lime and 3-4 % with B. Significant residual effect was observed at higher dose (0.3 LR) of lime application.

After harvest of cowpea, cabbage was grown (fourth crop) under residual lime but received two sprayings of borax. Significant residual effect of lime was observed in the fourth crop and the yield with lime was 6-28% higher over lime control (342.38 q ha^{-1}). Sprayings of B was not as effective as soil application. However, the yield increased by 8-10% with B sprays.

Change in Soil pH

The pH of the soil in lime control after harvest of first cabbage crop increased from 5.77 (initial) to 6.69. With application of lime @ 0.2 LR and 0.3 LR, this increased to 6.77 and 6.85, respectively whereas in B treatments, the pH values decreased compared to B control.

The pH of the soil at harvest of second crop was lower as compared to the first crop (cabbage). The pH in lime treatments was 2-6% higher over lime control whereas in B treatments it was 3-5% higher over B control.

Similarly, the pH of the soil after harvest of 3rd crop (cowpea) was lower as compared to second crop, okra and it remained between 5.85 to 6.98 as against 5.85 in control. There was increase in pH by 2-11% over control due to residual effect of liming. (Fig. 16).

Effect on Exchange Acidity

The exchange acidity of initial soil was 0.2 cmol (p^+) kg^{-1} which decreased to 0.17 cmol (p^+) kg^{-1} in lime control after harvest of the

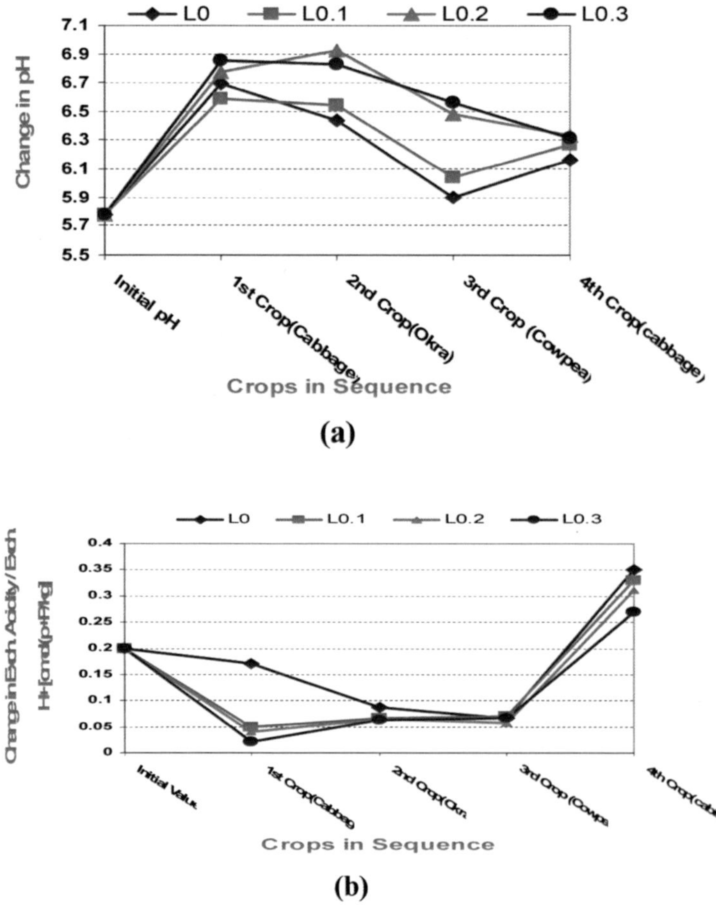

Fig. 16: Change in (a) pH and (b) exchange acidity, Al^{3+} and H^+ after harvest due to application of different levels of lime in Cropping Sequence

first cabbage crop (Fig. 16). With application of lime, there was significant decrease in exchange acidity by 71 to 90%. Boron had no effect on exchange acidity.

At the end of second crop, the exchange acidity increased as compared to first crop (Cabbage) due to decrease in bases under residual lime and boron. The values varied between 0.02 to 0.13 cmol (p^+) kg^{-1} as against 0.06 in control. However, the exch. acidity in lime treatments was 23-27% lower as compared to lime control. Residual effect of lime had no effect on exchange acidity during 3rd and 4th season. At the end of third crop, the exchange acidity ranged from 0.04 to 0.10 cmol (p^+) kg^{-1} in lime treatments as against 0.07 cmol (p^+) kg^{-1} in control.

There was significant increase in exchange acidity during 4th cropping season which ranged from 0.15 to 0.52 cmol (p^+) kg^{-1} in lime treatments. In lime control, it increased from 0.20 (at beginning) to 0.40 cmol (p^+) kg^{-1}. However, the residual effect of lime helped to reduce the exchange acidity by 5-23% compared to lime control.

Interaction of lime and B on Ca content of crops

Accumulation of Ca in lime control treatment was 0.99% in cabbage, 2.16% in okra, 0.16% in cowpea and 1.57% in cabbage (4th crop). Accumulation of Ca due to liming increased by 31% in cabbage, 12-20% in okra and 2-37% in cowpea. Application of lower dose of B (1 kg ha^{-1}) increased the Ca content by 17% over B control in cabbage. But, there was reduction in Ca concentration at higher rates of Boron application.

Interaction of lime and Boron on B content of crops

Boron content of cabbage increased by 20.7% at 0.1 LR. There was reduction in B content by 6-15% when the crop received higher dose of lime (0.2 LR, 0.3 LR). Application of B @1 2 kg ha^{-1} increased the rate of B accumulation by 31.5 to 36.8 %.

Content of B in okra fruit varied from 8.02 to 16.29mg/kg. The interaction of lime and B was significant and highest B content (16.29mg/kg) was recorded in 0.2 LR + B @ 2 kg ha^{-1}. Content of B in cowpea pod varied from 26.16 to 51.21 mg/kg, which was higher than that in okra. The B content increased by 56 to 58% with lime and by 2 to 12 % with B. Accumulation of B in cabbage (4th crop) increased by 41 to 55% in residual lime treatments. Two sprayings of B significantly increased the accumulation of B by 24-29% over B control.

Front line demonstrations in farmer's fields

Field demonstrations (Thirteen) were conducted with cabbage as test crop in village Barhaguda of Bargarh district and in village Sataguda of Kandhamal district. The pH of the soil in Bargarh district varied between 4.87-5.98 and in Kandhmal 4.50-5.26. The soils were medium to high in available B. The cabbage yield in FP treatment was 317 q/ha in Bargarh and 78 q ha^{-1} in Kandhmal. Integrated use of B with lime and RDF increased the cabbage yield over FP by 27 % in Bargarh and 91 % in Kandhamal indicating a significant response of cabbage to B application (Fig. 17).

Response to lime and B on yield of cabbage

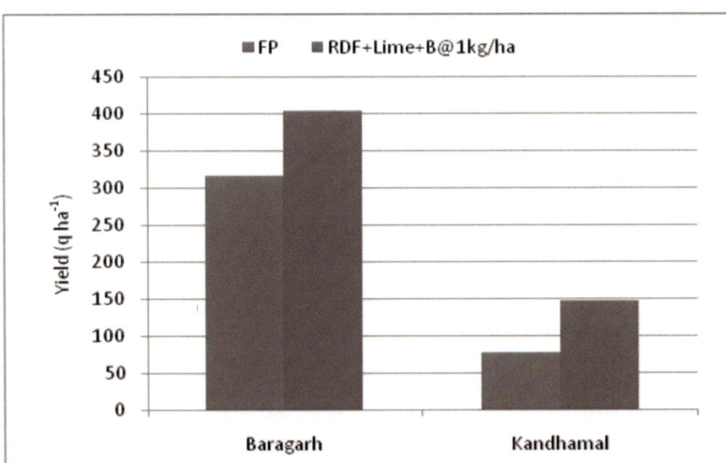

Fig. 17. Effect of lime and B on yield of cabbage

Twenty FLDs were conducted in red soils of Kandhamal, Gajapati and Deogarh districts of Odisha to study the effect of lime and B on cauliflower . Soils of the experimental site were acidic with pH ranging from 5.24-5.34, low to medium in OC (3.7-5.6 g kg^{-1}). Available B in soils of Gajapati and Deogarh was low to medium, whereas it was high in Kandhmal. Cauliflower responded significantly to lime and B in all sites. By application of lime and recommended dose of fertilizer, the yield increased over FP (34.1 q ha^{-1}) by 102% in Gajapati district. With application of B + Lime + RDF, the yield increased by 106% in Kandhmal (FP-95 q ha^{-1}) and 32% in Deogarh (FP - 250 q ha^{-1}) district.

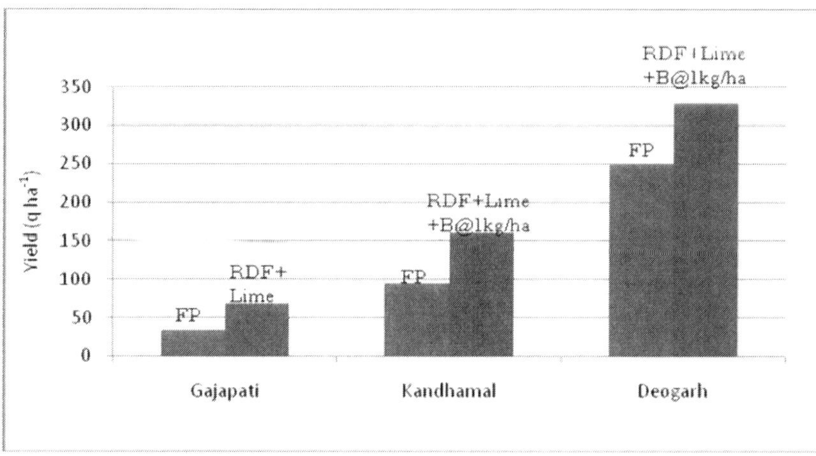

Fig. 18 Effect of lime and B on yield of cauliflower

Secondary Nutrients in Acid Soils

Most acid soils are deficient in Ca, Mg and S except acid sulphate soils, which contain high amount of S. Soils having Ca saturation less than 25% of the total cation exchange capacity require Ca application to most of the crops.

Coarse textured acidic upland soils low in organic matter show S-deficiency. Sulphur deficiency of 17-87% was recorded in acid soil regions of the country (Singh 2007) and 28 % in Odisha (Jena 2008). Continuous depletion of S lead to significant decline in crop yields compared to the yields achieved in NPK + S treatment. Deficiency of S can be corrected by use of various S sources such as gypsum (15-18% S), SSP (12%S), ammonium sulphate (24% S), ammonium phosphate sulphate (15% S), sulphate of potash (18%), gromor sulphur bentonite (90% S) in acid soils. The efficiency of gromor bentonite S was compared with SSP in hybrid rice-potato sequence in acid soils of Bhubaneswar (Jena *et al.* 2006). Results reveal that application of S @40 kg ha^{-1} increased the yield of several crops by 16.6-68.6 % (Table 9).

Table 9. Crop response to Phosphogypsum application in Odisha

Crop	Soil	Yield (q ha^{-1}) NPK	NPK + S	% Response
Groundnut	laterite	5.08	6.95	36.8
Kharif				
Rice	laterite	25.50	29.73	16.6
Hybrid rice	laterite	43.13	52.33	21.3
Rice	alluvial	52.30	63.30	21.0
Potato	laterite	178.0	212.7	19.5
Green gram	laterite	6.30	8.53	35.4
Groundnut *(Rabi)*	alluvial	17.50	29.50	68.6

**Source* : Jena (2008)

Effect of Lime and Sulphur in Groundnut Based Cropping Systems

Field experiments were conducted in sulphur deficient acidic red soils of Central Farm of OUAT, Bhubaneswar to study the direct and residual effect of lime and sulphur on yield, nutrient content, oil content of groundnut (Cv. *Smruti*) and changes in soil properties. There were 12 treatments consists of 4 levels of lime (0, 0.1, 0.2 and 0.3 LR) and three levels of S (0, 20 and 40 kg ha^{-1}). Sulphur through gypsum and lime through PMS was applied only to rabi crop (first and third groundnut crop).

The soil was acidic (pH 4.82) and non saline. The LR value determined by Woodruff's Buffer method was 7.3 t ha^{-1}. The exchange acidity and exchangeable Ca of the soil was 0.96 and 1.26 cmol (p^{+}) kg^{-1}. The soil was low in available N and K, but high in available P. The soil was medium in available SO_4-S (29 kg ha^{-1}).

Pod yield of groundnut

There was significant increase in pod yield with graded doses of lime and sulphur. The pod yield over lime control (11.5 q ha^{-1}) rose by 45, 58 and 72 % at 0.1, 0.2 and 0.3 LR, respectively. With application of S @ 20 and 40 kg ha^{-1}, the pod yield increased by 16-18% over control. Highest pod yield of 25.4 q ha^{-1} was noted in treatment that received lime @ 0.3 LR + S @ 40 kg ha^{-1}. The interaction of lime and S was significant.

Soil pH

The pH of soil after harvest of groundnut ranged between 4.16 to 5.36 compared to the initial value of 4.82. The pH of the soil in 0.1 LR treatment was lower than the initial value (pH 4.82). This showed that, application of lime @ 0.1 LR was not sufficient to maintain the soil pH at the end of the cropping season. However, with application of higher dose of lime (@ 0.2 LR and 0.3 LR, soil pH increased by 26 - 29% over unlimed plots, suggesting the benefits of liming.

There was further decrease in pH which ranged 4.48-4.61 at the end of second crop and 4.41-4.61 at the end of third crop in lime treatment. During same period the pH in lime control plot decreased from 4.26 to 4.20. The interaction effect of sulphur and lime on pH was significant (Table 10).

Table 10. Direct, residual and cumulative effect of lime and sulphur on soil pH

Treatment		1st Crop (Rabi 07-08) Direct	2nd Crop (Kharif 08) Residual	3rd Crop (Rabi 08-09) Cumulative
Lime Levels	L_0	4.16	4.26	4.20
	$L_{0.1}$	4.57(9.85)	4.51(5.79)	4.41
	$L_{0.2}$	5.25(26.20)	4.61(8.14)	4.59
	$L_{0.3}$	5.36(28.85)	4.48(5.16)	4.64
Sulphur Levels	S_0	5.10	4.38	4.45
	S_1	4.76(-6.67)	4.42(0.86)	4.43
	S_2	4.66(-8.63)	4.60(5.03)	4.50
C.D.(0.05)	L	0.05	0.032	1.01
	S	0.04	0.028	0.61
	LxS	0.09	0.056	1.62

Exchange Acidity, Exchangeable Al^{3+} and Exchangeable H^+

The exchange acidity of the soil before sowing of groundnut was 0.96 cmol (p^+) kg^{-1}, which reduced by 95% (0.04 cmol (p^+) kg^{-1}), when the crop received lime @ 0.3LR. Residual effect of lime (applied to first crop) had little effect on exch. acidity. In 0.3 LR lime treatment, the exch. acidity increased from 0.04 (1st crop) to 0.65 cmol (p^+) kg^{-1} after the harvest of second crop and 0.39 cmol (p^+) kg^{-1} at the harvest of third crop.

Exchangeable Ca^{2+}

Application of lime @ 0.2 and 0.3 LR maintained the exch. Ca content of soil above the initial value (1.26 cmol (p^+) kg^{-1}) at the end of 1st and 2nd crop. There was a gradual decline in exch. Ca content of soil in lime control over years. Again, when lime and S was applied to 3rd crop, the status of exch. Ca increased. (Fig. 19).

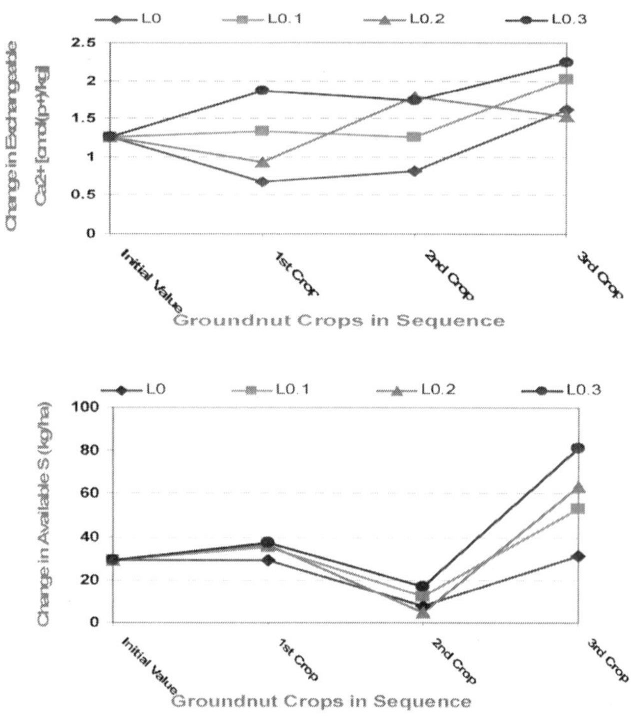

Fig. 19. Change in Exchangeable (a) Ca^{2+} and (b) S (kg ha^{-1}) after harvest due to application of different levels of lime in groundnut cropping system

Available Sulphur

After the harvest of first rabi groundnut crop, available S content of soil in control (L_0S_0) was 21.75 kg ha^{-1} (Fig. 19), which increased by 22-29% with lime application. Combined application of lime and sulphur maintained the available S status of 47 kg ha^{-1}.

S, Ca, Mg and oil content of groundnut kernel

Application of lime enhanced the accumulation of S, Ca and Mg in groundnut. Liming @ 0.3 LR, increased the S content of groundnut kernel by 70%.

Calcium in groundnut kernels increased by 20-25% with liming. Application of S increased the S content by 18-35% over S control (0.45%). The interaction of lime and sulphur on Ca content of groundnut was significant.

The oil content of groundnut varied between 43.2 to 45.9 %. The oil content in lime control was 44.4%, which increased by 0.6% in lime treatments. Addition of S increased the oil content by 2.8 to 3.8 % over S control (43.53%).

Effect of Lime and S on microbial population

Bacterial, *Azotobacter*, PSM and *Rhizobium* population of the soil was 24 x 10^5, 19 x 10^5, 23 x 10^5, and 18 x 10^4 CFU g^{-1}, respectively. Application of lime @ 0.2 LR and S @ 40 kg ha^{-1} recorded highest bacterial population (150 x 10^5 CFU g^{-1}) in the root zone of groundnut. The effect of sulphur on PSM population was not significant. Similar observations were recorded for *Rhizobium* and Azotobacter population.

Table 11. Effect of Lime on soil micro-organisms

Levels of lime	Total bacterial population x 10^6	Azotobacter x 10^6	PSM x 10^6	Rhizobium x 10^6
		CFU g^{-1}		
Control	4.6	2.7	2.4	0.25
0.1 LR	11.0	3.3	4.3	0.37
0.2 LR	15.0	8.6	8.3	0.53
0.3 LR	3.8	2.8	4.2	0.37

Farmer's Field Demonstrations

Groundnut

Sixty three farmer's field demonstrations on groundnut were conducted in different sites of Sambalpur, Sunergarh, Dhenkanal, Khurda, Baragarh and Kandhmal districts. The pH of the soils varied from 4.25-6.50. The pod yield of groundnut in FP in Sambalpur, Kandhmal and Sundergarh districts ranged between 5.6-16.8 q ha^{-1}. With application of lime, the yield increased over FP by 45-104 %. On the other hand, integrated use of S @ 40 kg ha^{-1} with lime increased the pod yield over FP by 47- 122 % in laterite soils of Dhenkanal and Khurda and alluvial soil of Bargarh district (Table 12).

Table 12. Response to lime and chemical fertilizers on pod yield of groundnut

Sl. No.	District	No. of demonstrations	Soil pH	Pod yield (q ha^{-1}) FP	RDF + lime	% response over FP
1.	Sambalpur	7	4.80- 6.16	16.8	24.3	45
2.	Kandhamal	15	4.55- 5.37	5.6	9.8	75
3.	Sundergarh	11	5.10- 5.37	7.4	15.1	104
				FP	RDF + lime + S @ 40kgha^{-1}	
4.	Dhenkanal	10	4.25- 5.43	8.2	17.1	109
5.	Khurda	10	5.10- 6.50	9.5	21.1	122
6.	Bargarh	10	4.87-5.98	16.2	23.8	47

Black Gram

Black gram occupies about 23 lakh ha during both kharif and rabi seasons. The average yield of black gram in Odisha is about 4.4 q ha^{-1}, which is low. The low crop yield is attributed to soil acidity and non-application of adequate amount of calcium and phosphorus to the crop. Field trials were conducted in 6 districts of Odisha to evaluate the effect of lime and sulphur on yield. The soils had a pH range of 4.42-6.88. The yield of black gram in FP treatment varied between 2.7-4.2 q ha^{-1}. Application of lime with RDF increased the black gram yield by 52 per cent in Nuapada and 69 per cent in Kalahandi. Application of lime + RDF + S enhanced the yield of blackgram by 130 per cent in Nuapada, 71 per cent in Dhenkanal, 95 per cent in Koraput and 84 per cent in Nawarangapur. This showed that application of lime (PMS @ 0.2 LR) and gypsum significantly increased the black gram yield (Table 13).

Table 13. Effect of lime and sulphur on seed yield of black gram

Sl. No.	District	No. of demonstrations	Soil pH	Seed yield (q ha^{-1}) FP	RDF + lime + S @ 40 kg ha^{-1}	% response over FP
1.	Nuapada	10	4.99-6.69	4.2	9.8	133
2.	Dhenkanal	7	4.42-6.14	3.8	6.4	71
3.	Koraput	7	4.44-6.88	2.9	5.7	95
4.	Nawarangpur	7	4.66-4.79	3.3	6.0	84
				FP	RDF + lime	
5.	Nuapada	9	5.62-6.28	3.0	4.8	52
6.	Kalahandi	9	5.21-5.36	2.7	5.7	69

Maize

Field trials were conducted in red soils of Keonjhar, Koraput and Sambalpur district to study the effect of lime and S on maize. The soils were acidic (pH 4.44-5.57) with low to medium organic carbon status.

Table 14. Effect of lime and S on maize

Sl. No.	District	No. of demonstrations	Soil pH	Green cob yield (q ha^{-1}) FP	RDF + lime	% response over FP
1.	Keonjhar	20	4.71-5.57	111.3	160.5	44
2.	Koraput	20	4.44-4.79	147.3	203.2	38
3.	Sambalpur	5	5.2-5.34	108.8	157.4	45

Optimum and maximum yield of cob were obtained, when the soil was treated with lime @ 0.2 LR along with RDF and S @ 40 kg ha^{-1} (Table 14).

Response of lime and S on yield of green cob

Integrated Nutrient Management

In a long term experiment conducted over 10 years in an acidic red soil of DLAP, Phulbani, Odisha with rice-horse gram cropping system, Jena *et al* (2000 b) observed that, water stable aggregates, porosity, available water capacity, organic C, available P and K content of the soil were higher with gliricidia green leaf manure or FYM application either alone or in combination with 50% RDF over control or 100% RDF. While full dose of NPK gave higher yield in years of favourable distribution of rainfall, INM practice recorded higher yield in the years of uneven distribution of rainfall, suggesting the sustainability of crop production with INM in rainfed acid soil regions (Table 15). Combined application of chemical fertilizers, FYM @ 5 t ha^{-1} and lime @ 0.25 LR to maize crop in acid red sandy loam soils of Phulbani district gave significantly higher crop yield and recorded better soil water utilization than that with application of chemical fertilizers alone (Jena *et al.* 2000).

Table 15. Long term effect of fertilizers and manure on physical properties of an Alfisol

Treatments	BD ($Mg\ m^{-3}$)	Water stable aggregates (%)	Porosity	Water retention ($kg\ kg^{-1}$)		Saturated moisture content ($kg\ kg^{-1}$)
				33 k Pa	1500 k Pa	
T_1 : 100% RD (60-40-40)	1.69	15.8	36.2	0.096	0.033	0.290
T_2 : 50% RD	1.67	25.3	36.9	0.095	0.033	0.285
T_3: 50% N through *Gliricida* leaves + 50% K	1.58	32.1	40.3	0.145	0.073	0.298
T_4: 50% N through FYM + 50% PK	1.57	33.5	40.7	0.121	0.046	0.351
T_5: T_2 + T_3	1.60	32.8	39.6	0.111	0.022	0.218
T_6: T_2 + T_4	1.60	31.5	39.6	0.117	0.026	0.290
T_7: Control	1.62	26.4	38.8	0.085	0.018	0.215

Effect Of Amendments On Maize, Mustard and Groundnut Biomass Yield

The efficiency of three different amendments namely PG, pure $CaCO_3$, PMS @ 0.2 LR with or without FYM @ 10 t ha^{-1} was evaluated in the root zone of three different crops- maize, mustard and groundnut in pots.

The soil used for the study has clay texture, pH-4.95, low in OC and available N, P, K but high in available S. Exchange acidity was

1.16 cmol (p^+) kg^{-1} and acidity is mostly due to exch. Al [1.04 cmol (p^+) kg^{-1}]. Exch. H^+ was 0.12 cmol (p+) kg^{-1}.

Biomass yield (average for two years) of maize in control was 84.0 g pot 1, which increased by 33.3 % with Phosphogypsum. With application of $CaCO_3$ or PMS, the biomass yield was further increased by 40.5 to 47.6 %. Integrated application of FYM and $CaCO_3$/PMS increased biomass yield over control by 48-93%.

Mustard produced biomass yield of 39.24 g pot^{-1} at harvest in unlimed pot. By application of PG, the biomass yield increased over control by 33 %. Application of $CaCO_3$ or PMS increased the biomass yield over control by 38 to 49 %. Integrated application of FYM with $CaCO_3$/PMS further increased the biomass yield over control by 85-93 %.

The biomass yield of groundnut increased over control by 7-30 % with addition of PG, $CaCO_3$ or PMS. Inclusion of FYM with $CaCO_3$ or PMS increased the efficiency of liming materials and increased the biomass yield by 31 % over lime (Fig. 20).

Effect on pH

The pH of initial soil was 4.95. In lime control, it varied between 4.76 to 4.89 in maize, 5.04 to 5.21 in mustard and 5.00 to 5.08 in groundnut. The pH in maize root zone was lower as compared to mustard and groundnut. The pH declined by 0.06-0.16 units with PG. Application of $CaCO_3$ or PMS with or without FYM maintained the soil pH near to neutral (Table 21).

From this study, it may be concluded that integrated use of lime and FYM was beneficial to correct soil acidity. The pH increased significantly from 4.95 to 7.04 and availability of phosphorus increased from 4.26 to 24.6 kg ha^{-1}. Availability of Ca and S was higher in lime treated soil than control. Integrated use of FYM and lime increased the biomass yield over control by 93% in maize and mustard and 31% in groundnut. Liming enhanced the concentration of P, S and Ca in crops.

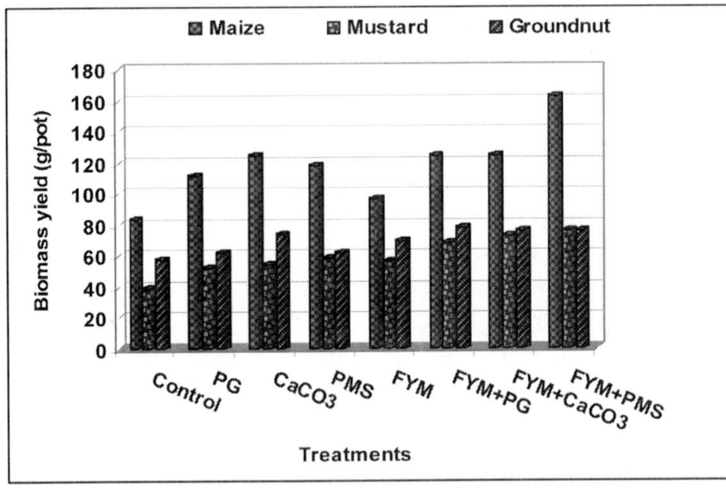

Fig. 20. Effect of amendments with or without FYM on biomass yield of crops

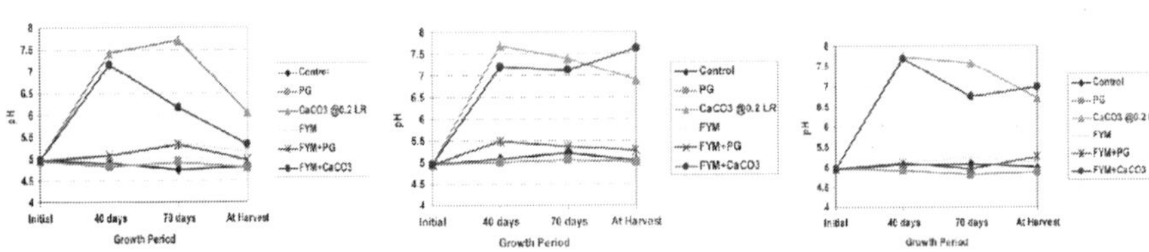

Fig. 21. Effect of Phospho gypsum, lime and FYM on soil pH in the root zone of (a) Maize, (b) Mustard, (c) Groundnut

Field demonstrations on Tomato

Thirteen farmer's field demonstrations were conducted in acidic (pH 5.1-5.5), low organic carbon status soils of Gajapati and Deogarh districts on tomato. Integrated application of RDF + lime was beneficial for tomato in red soils. In Deogarh district, the yield response of the crop was 28% over farmers' practice (294 q ha^{-1}) and maximum response was obtained, when the soil was treated with lime + RDF + FYM. In Gajapati, the yield in FP treatment was lower (195 q ha^{-1}) as compared to Deogarh. However, about 37% higher yield was recorded, when the crop received lime @ 0.2 LR along with RDF and FYM @ 5 t ha^{-1} (Fig. 22).

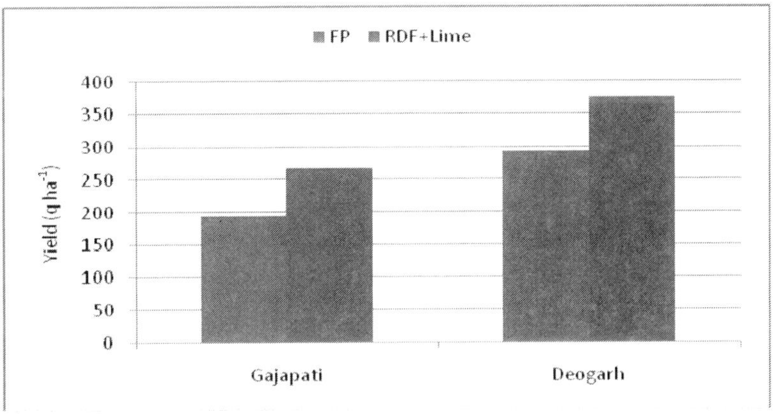

Fig. 22. Effect of lime on yield of tomato

Okra

Field trials were conducted in five villages of Kandhmal district, to study the effect of lime and FYM on okra. The yields significantly increased by 88 % due to application of PMS @ 0.2 LR over farmers' practice (8.5 q ha^{-1}) or RDF (10.2 q ha^{-1}). The optimum and maximum yield (15.9 q ha^{-1}) was obtained, when the soil was limed with PMS @ 0.2 LR along with application of RDF and FYM @ 5 t ha^{-1} (Fig.23).

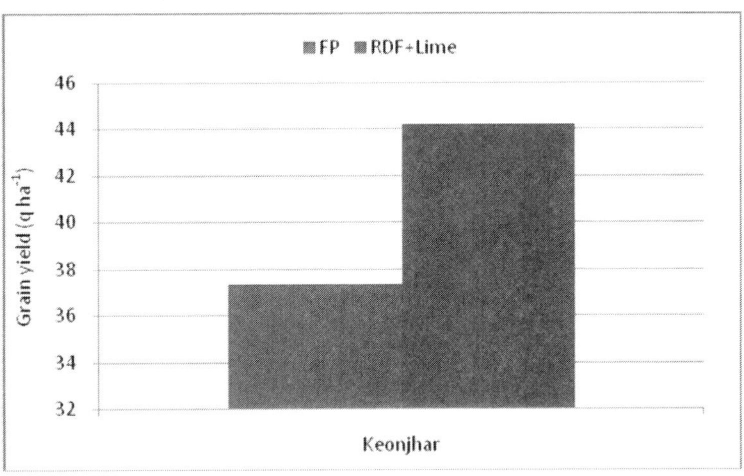

Fig. 23. Effect of lime on yield of okra

Lime, B, S interaction

Radish

Field experiments were conducted to study the interaction effect of lime, boron and sulphur on yield of radish at Chiplima on acidic (pH-5.3) sandy loam soil, deficient in B, S and available N and medium in status with respect to P and K. The test crop radish (Cv. Pusa Chetki) received four doses of B (0, 0.5, 1.0 and 1.5 kg B ha^{-1}) super imposed over four doses of S (0, 15, 30 and 45 kg S ha^{-1}) and two doses of lime (0, 0.2 LR) as PMS applied before sowing of seeds. The fertilizer dose of 80-75-100 kg N-P_2O_5-K_2O and 5 t FYM ha^{-1} was common to all the treatments.

Lime application @0.2 LR increased the yield (Table 16) of radish by 11 per cent compared to no lime application (13.7 t ha^{-1}). The crop responded positively to graded doses of B application, resulting in increase in yield over B control (19 t ha^{-1}) by 15, 24 and 33 per cent with sub-optimal, optimal and super-optimal doses of B, respectively. Similar to B, S application resulted in 15, 41 and 59 per cent increase due to suboptimal, optimal and super-optimal doses of sulphur compared to no sulphur application (17.5 t ha^{-1}). Application of lime @ 0.2 LR along with application of S @ 45 kg ha^{-1} and B applied @ 1.5 kg ha^{-1} increased the yield of radish in B and S deficient acid soil.

Table 16. Interactive effects of boron, sulphur and lime on yield of radish (t ha^{-1})

Treatments	B_0		$B_{0.5}$		$B_{0.1}$		$B_{1.5}$		Mean
	L_0	$L_{0.2}$	L_0	$L_{0.2}$	L_0	$L_{0.2}$	L_0	$L_{0.2}$	
S_0	13.7	14.9	16.7	18.3	17.0	19.8	18.9	20.9	17.5
S_{15}	15.5	17.7	18.0	20.0	20.0	23.1	22.0	24.3	20.1(15)
S_{30}	19.9	22.2	22.2	25.0	24.2	27.3	27.1	29.4	24.6(41)
S_{45}	22.8	25.4	26.8	29.0	27.7	31.0	28.9	31.4	27.9(59)
Mean	17.9	20.0	20.9	23.1	22.2	25.3	24.2	26.5	—
Boron	19.04	—	21.98	—	23.8	—	25.3	—	
CD	S = 0.36	B=0.36	L=0.26	S xB	S x L	B x L	S x B x L		
(0.05)				=0.72	=25.3	=0.51	=1.02		

Potato

Twenty eight trials on potato were conducted in acidic red soil of Kandhamala district with pH ranging from 4.90-5.37. Integrated application of lime with RDF and B @ 1 kg ha^{-1} increased the potato yield by 58 per cent over control (241.9 q ha^{-1}). Application of lime and S brought about 13-15% additional yields (Fig. 24).

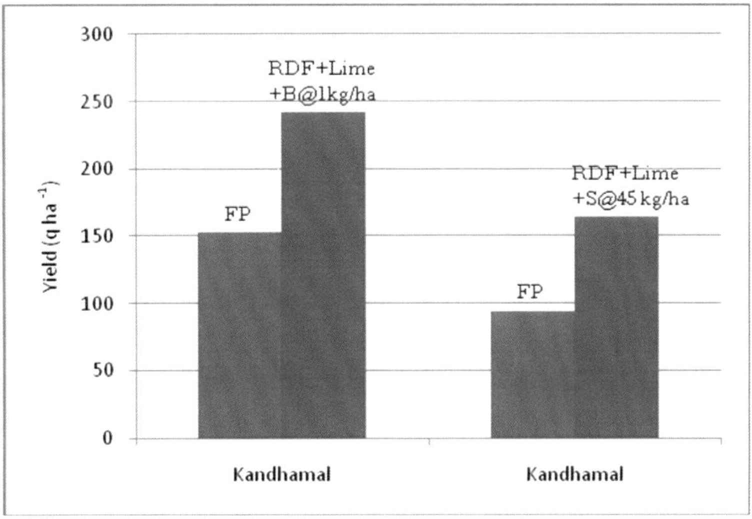

Fig. 24. Effect of lime and S or B on yield of potato

Seed cotton

Forty five trials on cotton were conducted in intensive cotton growing areas of Rayagada, Kalahandi and Nuapada districts. The soils were low to high in organic carbon and high in available B. The results showed that cotton significantly responded to lime, B and S. Application of lime @ 0.2 LR + RDF increased the cotton yield by 71 per cent in Kalahandi and 77 per cent in Nuapada. In absence of lime, application of B with RDF increased the yield by 36 percent over FP in Rayagada. In Nuapada, the cotton yield was increased by 44% with application of RDF+ lime + S (Table 17).

Table 17. Effect of lime and S or B on yield of seed cotton

Sl. No.	District	No. of demonstrations	Soil pH	Yield (q ha^{-1}) FP	RDF + lime + S	% response over FP
1.	Nuapada	5	5.0-5.44	12.5	18.1	44
				FP	RDF + B	
2.	Rayagada	28	7.47-8.19	5.1	6.9	36
				FP	RDF + lime	
3.	Kalahandi	6	5.21-5.36	12.5	21.2	71
4.	Nuapada	6	5.62-6.13	12.1	22.9	77

Management of Iron Toxic Soils

Iron toxicity occurs in hill bottom red and lateritic soils (Alfisol, oxisol, ultisol) under undulating topography and impeded drainage condition. These soils are characterized by poor base saturation and poor supply of available nutrients like K, Ca, Mg, P, Zn and Cu. Iron toxicity symptoms in rice is seen as bronzing, when Fe^{2+} concentration in soil solution goes up to 250-500 mg kg^{-1} due to reduced conditions under prolonged submergence (Jena *et al.* 2008). The concentration of Fe^{2+} further increases due to lateral flow of Fe from an adjacent upland to low land rice fields during rainy season (Sureshkumar, 1999). In most of the rice soils, the concentration of Fe^{2+} increases upon flooding and attains peaks after 2-5 weeks (Mandal 1961).

In iron toxic soils, substantial amount of iron is oxidized and get deposited on the active roots making physical barrier for absorption of plant nutrients from soil solution. Under extreme conditions, more number of roots become blackish in colour and only a few remain whitish to brownish in colour. Plants get stressed and are forced to produce more new roots at the expense of shoot growth. Production and decay of roots continue throughout the plant growth period. Iron toxicity in plants promotes sterility and the grain yield is reduced by 20-80% depending upon the situation.

Iron toxicity can be corrected by providing intermittent drainage. It reduces uptake of iron and increases availability of other plant nutrients. Liming of soils @ 1-2 t ha^{-1} along with K application had beneficial effect on alleviating Fe toxicity in rice by 27% over lime control in soils of Odisha (Mitra and Sahu 1992). Application of fresh cowdung @ 5 t ha^{-1} increased grain yield by 9.31 q ha^{-1} over control (19.04 q ha^{-1}). Application of lime @ 0.5 LR or potash @ 40 kg ha^{-1} or Zn @ 10 kg ha^{-1} increased grain yield by 8.25, 11.41 and 11.63 q ha^{-1}, respectively over control (Jena 2007). Foliar application of $MnSO_4$ @ 0.6% had no beneficial effect on rice in iron toxic soil. Balanced fertilization helps in alleviating iron toxicity in rice. Singh *et al* (1993) reported that application of 90 kg P_2O_5 ha^{-1} to iron toxic soil at Barapani farm of Meghalaya resulted in reduction of Fe^{2+} from 3.60 to 1.63 mg kg^{-1}. Dipping rice seedling in boronated SSP and FYM slurry before transplanting, helped to increase rice yield by reducing Fe toxicity.

Jena (2007) evaluated several rice genotypes in iron toxic laterite soils of Bhubaneswar for three consecutive years and reported that rice genotypes like Kalinga III, Udayagiri, Panidhan and Tulasi are tolerant to iron toxicity while, IR 36, Konark, Birupa, Gajapati, Samalai and Indrabati are moderately tolerant to iron toxicity. Savithri & Sree Ramulu (1980) evaluated rice varieties in iron toxic acid soils (pH 4.7, DTPA Fe- 31 ppm, Aquic Hapludalf) in Kanyakumari district. Genotypic difference in the degree of susceptibility to excess Fe was confirmed by changes in the content of metabolically active Fe^{2+}, chlorophyll and enzymatic activity in plant parts. Metabolically-active Fe^{2+} contents in the leaves, stems and roots of tolerant genotype (ASD-16) was 212, 392 and 4674 ppm as compared to highly susceptible rice genotype (ADT-36) having 281, 442 and 5933 ppm Fe content, respectively. Genotypic differences in degree of susceptibility to excess Fe were attributed to leaf tissue tolerance of high level of Fe, reduced translocation from root to shoot and ability of roots to resist its entry inside the plant

In Odisha, about 1 lakh hectare of medium-low land rice suffers due to iron toxicity resulting in yield loss to the extent of 60 to 80 %. A field experiment was conducted in Central Farm, OUAT, Bhubaneswar to evaluate the dose and time of lime application to rice grown in an iron toxic soil. The treatments were : Main plot - M_1 : Basal application of lime, M_2 : Split application of lime, M_3: Basal application of lime with Zn @5kg ha^{-1}, Sub Plot treatments - S_1 : PMS applied@ 0 LR, S_2: PMS applied@ 0.3 LR, S_3 : PMS applied@ 0.5 LR, S_4: PMS applied@ 0.7 LR. Paper mill sludge (PMS) was used as lime source. In M_2, lime was applied at transplanting and 20 days after transplanting in 1:1 ratio. In M_3, Zn was applied in the form of $ZnSO_4$ @ 25 kg $ZnSO_4$ ha^{-1} at planting.

The soil of the experimental site had high iron content of 484 ppm, pH-4.51, medium organic carbon- 4.3 g kg^{-1}, low available potassium 106.8 kg ha^{-1} and low soil available Zn 0.85 ppm.

Soil pH And Available Potassium

In control, the pH of the soil slightly increased up to 14th days after transplanting and thereafter it decreased (Fig. 25). With application of lime, the soil pH increased up to 21 days of lime application and thereafter decreased. But, when lime was applied in splits, the increase in pH extended beyond 28 days.

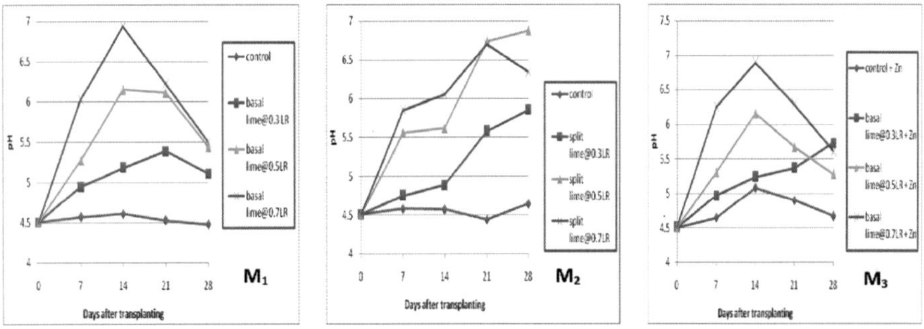

Fig. 25. Changes in soil pH as influenced by method and dose of lime application

The availability of K increased with higher doses of lime. The split application of lime induced release of K for a longer period as compared to single application (Fig. 26).

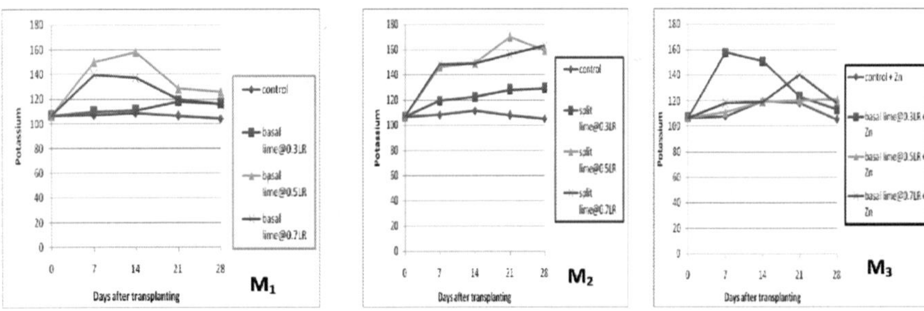

Fig. 26. Changes in available K as influenced by method and dose of lime application

Available Zinc

The concentration of Zn in soil increased upto 7 days of planting and there after remained constant. The concentration of Zn was minimum in lime control and increased with levels of lime. Availability of Zn further increased with addition of Zn. Split application of lime resulted in an increase in the availability of Zn as compared to basal application (Fig. 27).

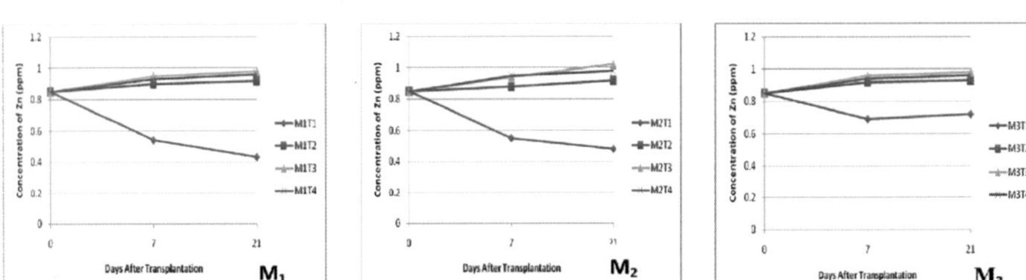

Fig. 27. Influence of liming on concentration of Zn

Available Manganese and Iron

With increase in duration of submergence, the content of Mn decreased up to 7 days and, thereafter it increased. Further, unlike Zn, the concentration of Mn decreased with dose of lime. Split application of lime had little influence on the concentration of Mn (Fig. 28). The maximum content of Fe was recorded in control, which reduced gradually in lime treated plots (Fig. 29).

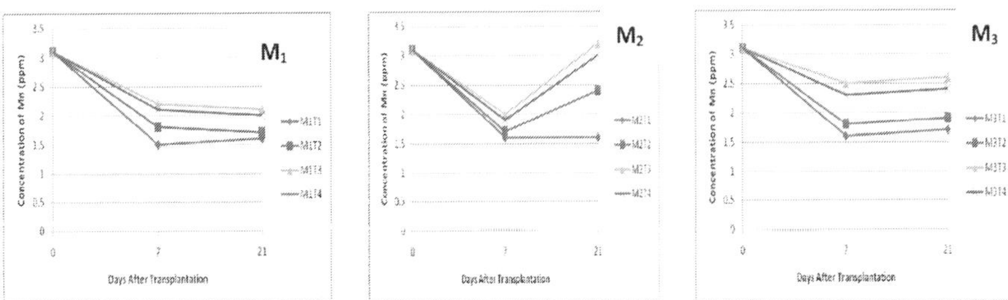

Fig. 28. Influence of liming on concentration of Mn

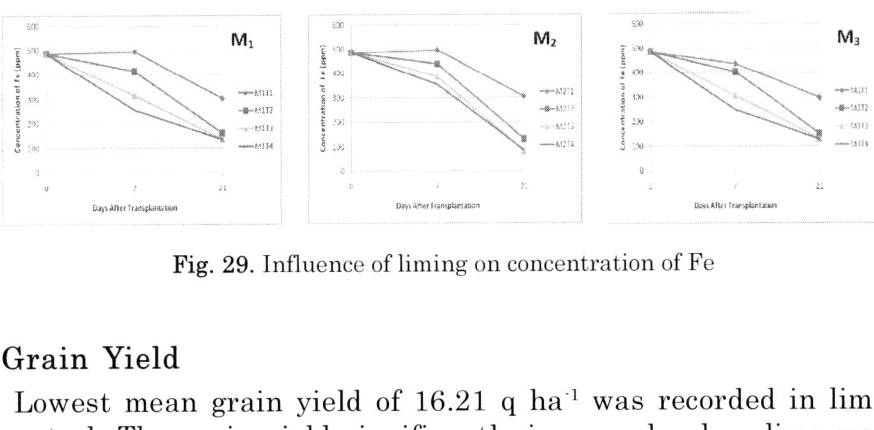

Fig. 29. Influence of liming on concentration of Fe

Grain Yield

Lowest mean grain yield of 16.21 q ha^{-1} was recorded in lime control. The grain yield significantly increased, when lime was applied @ 0.5 LR. Split application of lime was more effective than basal application. Split application of lime @ 0.5 LR reduced the concentration of Fe and also increased the concentration of Mn, Zn and K.

Table 18. Yield of rice (q ha^{-1}) as influenced by treatments

Treatments	Doses of lime				Mean
	0.0 LR	0.3 LR	0.5 LR	0.7 LR	
M1Basal lime	14.03	16.11	17.13	17.55	16.21
M2Split lime	14.09	16.80	21.21	19.09	17.80
M3Basal lime + Zn	15.38	16.33	20.20	18.42	17.58
Mean	14.50	16.42	19.52	18.35	
$CD_{(0.05)}$	Main (M) -0.056 Sub (S) -0.033 M x S -0.057				

Zinc Management in Acid Soils

Zinc deficiency is fairly widespread in Assam and north-eastern hill regions due to high organic matter content. In terai soils, the deficiency was about 49-51% as compared to 20-46% in red and laterite soils. Zinc deficiency in soils of Odisha has been reported to be 19% (Jena 2008). Basal application of ZnSO4 is the common practice to correct Zn deficiency. Application of $ZnSO_4$ @ 25 kg ha^{-1} in coarse textured soils and 37.5 kg ha^{-1} in fine textured soils was optimum for rice, wheat, maize and other field crops. If soil application is missed at sowing/planting, top-dressing up to the pre-flowering stage in rice, wheat and other crops is recommended. Organic manures raise the use efficiency of synthetic materials due to higher chelation and subsequently slow release (Singh 2007, Jena 2006). In a study, Zinc @ 2.5 kg ha^{-1} was thoroughly mixed with 200-300 kg of fresh cowdung or well decomposed FYM and incubated for 20-25 days under optimum moisture condition to facilitate chelation of Zn. At the end, the material was dried under shade, finely ground for uniform distribution on the field. The results revealed that Zn-enriched organic manure @ 2.5 kg Zn ha^{-1} recorded similar yield of groundnut, mustard, sesame, bengalgram, pea as compared to 5 kg Zn ha^{-1} (Jena 2006).

Several field demonstrations with Zinc were conducted in medium acidic alluvial soils of Odisha on rice. The soils were acidic with pH range of 4.8-6.16 .The soils were medium to high in organic carbon content.

Table 19. Effect of Zn spraying on rice yield

Sl. No.	District	No. of demonstrations	Soil pH	Grain yield (q ha^{-1}) FP	Grain yield (q ha^{-1}) RDF + Zn spray (twice) @ 0.5% at tillering and PI stage	% response over FP
1	Sambalpur	7	4.8-6.16	42.4	54.6	29
2	Baragarh	8		39.6	48.8	23
3	Nayagarh	15	4.79-6.56	39.6	48.8	23

The DTPA Zinc content of the soil ranged from 0.40-0.72 mg kg^{-1}. Medium duration rice responded to Zn spraying. The yield in FP was 39.6-42.9 q ha^{-1}, which increased by 23-29% with two sprayings of Zn, one at tillering, and the other at the flowering stage.

Thirty five FLDs on rice were conducted in laterite soils of Khurda district and alluvial soils of Jajpur district to study the effect of Zn (soil application) on rice yield. The pH of the soil varied between 5.13-6.64. The DTPA zinc content of the soil ranged from 0.45 to 0.81 mg kg^{-1}. Zinc sulphate was applied @ 25 kg ha^{-1} at transplanting. Application of Zn with recommended dose of fertilizer increased rice yield by 46% in Zn deficient soil of Khurda and by 40% in Jajpur district (Fig. 30).

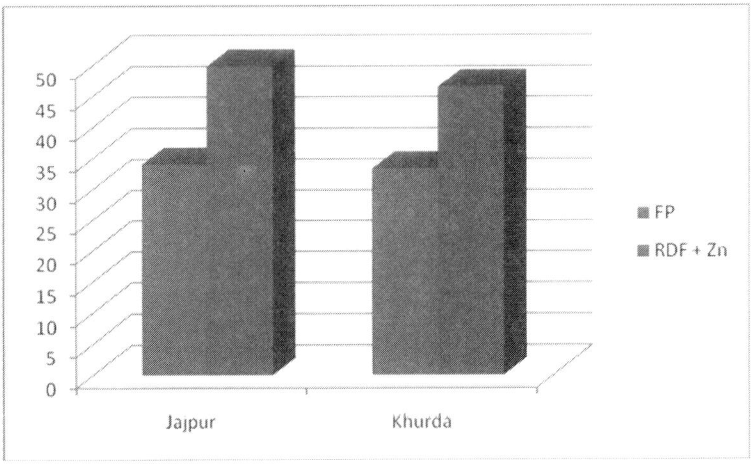

Fig. 30. Effect of Zn soil application on grain yield of rice

Phosphorus Management on Acid Soils

Panda (2007) reported that North-carolina rock phosphate of 35 mesh was superior to 100 mesh size of Indian phosphate rocks such as Udaipur, Musoorie, Hirapur, Kasipatnam, Maton and Purulia. Udaipur rock phosphate containing dolomite and calcite was found economical for maize-mustard cropping system in Alfisol. (Jena *et al* 2004). The effectiveness of low reactivity phosphate rock can be increased by applying it to green manure crop preceding the main crop or by inoculation of the field with either phosphate solubilizing micro-organisms or mycorrhiza (Misra 2004). Rice grown in iron toxic soils is benefited from application of Udaipur rock phosphate (Misra 2004).

Several workers have successfully attempted to reduce the cost of P fertilization in acid soils by direct use of rock phosphates (Misra and Pattanayak, 1998) to soil (pH < 5.5) or use of rock phosphates and single super phosphate mixture in 1:1 ratio to soil (pH 5.6-6.5) or to apply rock phosphates to green manure crops prior to rice crops taken in sequence or use of compacted products of Jhamarkotra rock phosphates (JPR).

Further, it has been recommended to apply the entire P requirement of the cropping sequence, particularly for groundnut-rice cropping system in form of rock phosphates to the groundnut crop grown during *rabi* season and the residual effect be realized in rice crop during *kharif* season. This is because the rock phosphate applied to dry season groundnut gets solubilized to greater extent and the portion that gets fixed during dry season becomes available to rice crop due to waterlogged conditions (Misra and Pattanayak 1998).

Leaching loss of Ca

A laboratory study was conducted on the leaching behaviour of calcium in an acid soil. A soil column of 15 cm length and 8.5cm diameter was packed with soil (pH 4.8) of Bhubaneswar to a bulk density of 1.5 gm cm^{-3}.The soil was saturated with 1N NH_4Cl solution. 3 g Calcite was spread on soil surface of the column and leached with 1 N NH_4Cl (Fig 31) at an infiltration rate of $21 cm^3 hr^{-1}$. The leachate was collected periodically for about 200 hr and analyzed for calcium content.

The concentration of Ca in the leachate over time is presented in Fig 32. The amount of calcium leaching down the column was computed by integrating the area under break through curve of calcium. It was found that, about 40% of the calcium applied leached down the column by displacing 100 cm of 1N NH_4Cl solution over a period of 200 hr.

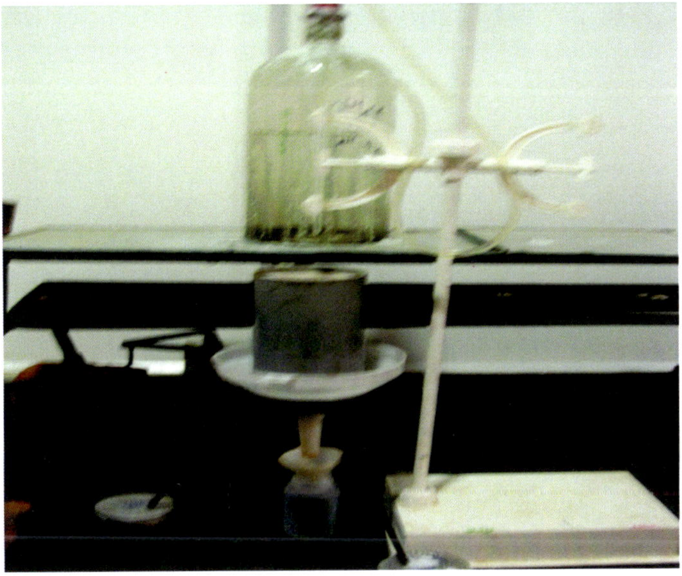

Fig. 31. Experimental setup on leaching study

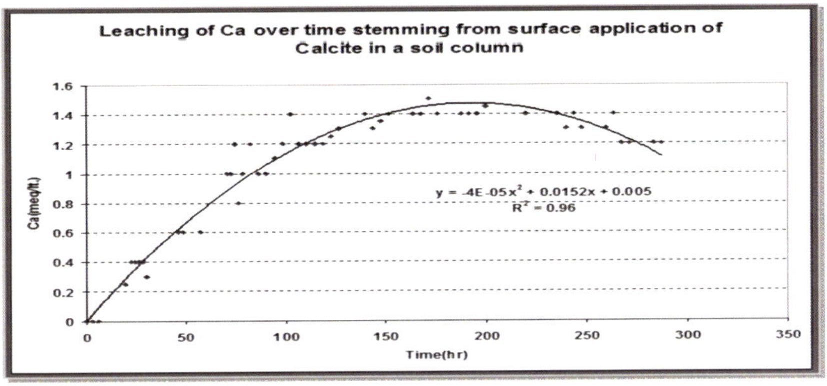

Fig. 32. Leaching of calcium over time in a soil column

Epilogue

The physical constraints of acid soils are high percolation rate, loss of water and nutrients, low water holding capacity and soil crusting. Use of organic manure increases water holding capacity and organic carbon status of soil. Utilization of sub surface water by growing deep rooted crops like groundnut, blackgram, green gram, arhar etc. will benefit the farmers. Use of lime for amending crusting soils for different crops like cotton soybean, cowpea, finger millet etc. will be helpful.

Naturally occurring sources like calcite, dolomite, industrial wastes like paper mill sludge (paper industry), pressmud (sugar industry), basic slag (steel industry), spent wash etc. can be used as liming material.

A small dose of lime @10-20 % of LR to each crop instead of one time application of full dose give better results. Furrow application of lime is economical. For crops sown in line, method of application is placement. How to do this, when the farmers follow the broadcast method of sowing crops?

Vegetable crops respond to lime. However, method of lime application in vegetables has not been standardized. Dose, method and time of lime application for vegetable crops as well as spices especially turmeric and ginger, tuber crops need to be studied. In hill slope and hill bottom soils, agroforestry and horticultural crops can be grown successfully with lime application. However, the dose, time and method of lime application for these systems need detailed study.

Efficiency of liming material can be increased by the use of organics like compost, vermicompost, biofertilizers and secondary and micronutrients like sulphur, boron or zinc as per soil tests. For this, laboratory incubation studies need to be conducted. Long term effect of doses of lime with or without organic manure on major crop sequences should be studied. Lime induced immobilization of heavy metals in soils, bio-solids and mine tailings reduce their availability to crops. Hence, research is needed to develop methods to quantify lime-enriched mobilization of deficient plant nutrients and lime induced immobilization of heavy metals in soils.

In acid soils of Odisha, the deficiency of micro- and secondary nutrients is emerging as yield limiting factors and is threatening the sustainability of major crop production systems. Hence, regular monitoring of changes in the nutrient status of soil is essential to formulate and adopt suitable remedial measures.

Industrial wastes, which have been identified as lime source should be utilized locally as far as possible. Economic distances of transport of such material may be worked out. Steel mill slag and blast furnace slag being slow acting and its production cost being high due to grinding, needs further research for promotion.

Utility of Phosphogypsum, a byproduct of phosphate industry though not used to ameliorate surface soil acidity, may counter the aluminum toxicity of sub-surface soil acidity. This requires study.

Crop species tolerant to soil acidity need be identified beyond what is already known. Screening of acid tolerant genotypes of oilseeds, vegetables grown in rainfed/irrigated areas should find a place in acid soil management. The yield of pulses grown in large acid soil areas after kharif rice is very low. Research is needed to find suitable acid tolerant varieties of pulses through varietal screening and management practices for growing them under rice-fallow area. Iron toxic soil can be effectively managed through varietal screening and amelioration measures.

Use of Udaipur rock phosphate as a source of Ca and P in acid soil has been worked out and found suitable. It has been used extensively in Kerela because the Kerela Govt has subsidized the transport cost. Similar initiatives should be taken by other Govts. to use it in acid soil regions of Odisha, Assam, Chhatisgarh, Jharkhand and West Bengal.

To reduce fixation and increase availability of applied phosphate instead of water soluble sources of P, insoluble P sources like ground rock phosphate in soils of pH less than 5.5 and 1:1 mixture (on P Basis) of rock phosphate and single super phosphate in soils of pH 5.6 to 6.5 can be preferably used.

Management of biological properties of soil can be done through identification of efficient strains of indigenous Rhizobium for biological nitrogen fixation. Studies on effect of lime, P and Mo on the activity of above organisms, rhizospere studies, organic acid release pattern for effective nutrient acquisition and synchronization, soil biotechnology, microbial diversity and, lime v/s biofertilizer interaction are necessary.

The dose of well decomposed organic residues to be used for slightly acidic soils should be worked out through multilocational trials, since such organic amendments help to minimize the leaching loss of basic cations by improving physical conditions. Residue incorporation, minimum tillage, conservation agricultural practices need to be promoted in acid soil regions for higher crop productivity.

Farmer - Scientist participatory field demonstrations should be carried out in farmers' fields in larger areas to popularize the liming technology. Farmers should be trained on the use of lime, source, dose, method of lime application, selection of crop etc. through training. Govt. Extension functionaries, KVKs, ATMAs and NGOs should work in a integrated manner to create awareness among farmers for management of acid soils for sustainable crop production.

References

Biswas, T.D. and Khosla, T.D. (1971) International Symposium On Soil Fertility Evaluation. Indian Society of Soil Science, New Delhi.

Hue, N.V. (1992) Correcting soil acidity of a highly weathered ultisol with chicken manure and sewage sludge. *Communucation in Soil Science* 152: 241-264

Jena, D. (1996) Annual Report, RRTTS, G. Udayagiri, OUAT, Bhubaneswar

Jena D., Pani, B.K. and Subudhi, C.R. (2000) Effect of FYM and lime on yield and water use by maize. *Indian Journal of Soil Conservation* 28: 88-90.

Jena D., Pradhan, N.K., Pani B.K. and Senapati, P.C.(2000b) Long – term effect of integrated nutrient management on properties of an Alfisol and rice yield under dry land agriculture. *Indian Journal of Dry land Agriculture Research and Development* 15: 37-41.

Jena, D; Jena, M.K; Pattnayak, S.K; Sahu, D. and Mitra, G.N. (2004) Evaluation of compacted phosphate rocks for maize-mustard cropping sequence in a Typic Haplaquept of Northeastern ghat zone. *Journal Indian Society of Soil Science* 52: 80-84.

Jena, D; Sahoo, R; Sarangi, D.R. and Singh, M.V. (2006) Effect of different sources and levels of sulphur on yield and nutrient uptake by groundnut-rice cropping system in an Inceptisol of Orissa. *Journal of the Indian Society of Soil Science*, 51: 126-129.

Jena, D. (2006) Annual Progress Report. All India Co-coordinated research project of micro and secondary nutrients and pollutant elements in soils and plant. OUAT, Bhubaneswar.

Jena, D. (2007) Annual Progress Report. All India Co-coordinated research project of micro and secondary nutrients and pollutant elements in soils and plant. OUAT, Bhubaneswar.

Jena, D. (2008) Management of acid soils for sustainable crop production. NAE-Acid soil Bulletin. No. 4, OUAT, Bhubaneswar.

Jena, D; Singh, M.V; Pattnaik, M.R. and Nayak, S.C. (2008) Scenario of micro and secondary nutrient deficiencies in soils of Orissa and management. Technical Bulletin 1. OUAT, Bhubaneswar.

Jena, D. (2011) Acid Soil Management in India – Challenges and Opportunities. In Strategies for arresting Land Degradation in South Asian Countries" published by SAARC Agricultural Centre, BARC Complex.

Lyamuremye, F ; Dick, R.P. and Baham, J. (1995) Organic amendment and P dynamics:1 Phosphorus chemistry and sorption. *Soil Science* 16:426-434

Mandal, L.N. (1961) Transformation of iron and manganese in waterlogged rice soils. *Soil Science*: 91, 121-126.

Mandal, B. and Mandal, L.N. (1992) Chemistry of submerged soils. Proceedings of the workshop on Micronutrients held at Bhubaneswar. pp. 240.

Mandal, B. (1995) Final Report of the ICAR Ad-hoc Project, Studies on the chemistry of Boron and Molybdenum availability in Acid Alluvial soils. BCKV, Kalyani, West Bengal: pp 1-138.

Misra,U.K.; Satapathy, S. and Panda,N.(1989) Charectetrisation of some acid soils of India- 1.Nature of Soil acidity. *Journal of the Indian Society of Soil Science* 37:22-28

Misra, U.K. and Pattanayak, S.K. (1998) Characterisation of rock phosphates for direct use in different cropping sequences. Technical Report of the U.S. India Fund, *Project No. In: AES*-708, 1991-1995.

Misra, M. (2002) Response of crops to graded doses of lime amended with or without FYM. M.Sc. (Ag) Thesis, OUAT, Bhubaneswar.

Mishra, U.K. (2004) Acid soil and its management. *Journal of the Indian Society of Soil Science*, 52: 332-343.

Mitra, G.N. and Sahu, S.K. (1992) Effect of potassium application on alleviating iron toxicity and nutrient uptake by rice in laterite soils. Proceeding of the workshop on Micronutrients, 22-23 January, 1992, pp. 269-272.

Raychaudhuri, M; Kumar, K. and Raychaudhury, S. (1998), Response of soyabean to lime and phosphorous on an ultisol of Manipur. *Journal of the Indian Society of Soil Science* 46: 628-632.

Panda, N. (2007) Management of acid soils. *Bulletin of the Indian Society of Soil Science* 25: 1-9

Panda, D; Jena, D; Nanda, S. and Mahapatra, P.K. (2007) Cropping systems in acid soils: Enhancing their productivity and sustainability. *Bulletin of the Indian Society of Soil Science*, 25: 59-87.

Pradhan, N.K. (1978) Ph.D. Thesis, Orissa University of Agriculture & Technology, Bhubaneswar.

Sahu, G.C. and Patnaik, S.K. (1990) Influence of lime or organic matter on pH and ion exchange phenomenon of some alfisol. *Journal of the Indian Society of Soil Science* 38: 154-157.

Sahoo, D; Rout, K.K. and Misra, V.V. (1998) Effect of twenty five years of fertilizer application on productivity of rice-rice system. In : Long-term Soil fertility management through Integrated Plant Nutrient supply (A. Swarup, D.D. Reddy and R.N. Prasad, Eds), Indian Institute of Soil Science, Bhopal, pp. 206-214.

Savithri, P. and Ramulu, S. (1980) Effect of continuous application of Zn, Cu, and Fe on yield and availability of nutrients in cropping systems. *Journal of the Indian Society of Soil Science* 28: 371-374.

Sharma, S.P. and Tripathi, B.R. (1989) Studies on lime requirement of acid soils of India. *Journal of the Indian Society of Soil Science* 37: 363-370

Sharma, S.P; Sharma, P.K. and Tripathy, B.R. (1990) Forms of acidity in some acid soils of India. *Journal of the Indian Society of Soil Science* 38: 189-195.

Sharma, P.D. and Sarkar, A.K. (2005) Managing acid soils for enhancing productivity. Technical Bulletin. 23. ICAR, NRM, Division, KAB II, New Delhi, pp. 1-24.

Singh, K.P. and Sarkar, A.K. (1998) Effect of continuous cropping and fertilizer use on crop yields and fertility of red and laterite soil. In : Long term Soil Fertility Management through Integrated Plant Nutrient Supply (A. Swarup, D.D. Reddy and R.N. Prasad, Eds.) Indian Institute of Soil Science, Bhopal, pp. 146-153.

Singh, B.P; Das, Madhumita, Dwivedi, B.S; Ram; M. and Prasad, R.N. (1993) Response of rice to phosphorous and potassium grown in iron toxic Haplaquent under mid altitude of Meghalaya. *Annals of Agricultural Research* 14: 100-102.

Singh, M.V. (2007) Problems of micro and secondary nutrients in acid soils of India on their management. In. R.K. Rattan Eds. Nutrient management in Acid soils. *Bulletin of the Indian Society of Soil Science* 25: 1-88

Sudhir, K., Mariswamy Gowda, S.M. and Srikanth, K. (1998) Long term effects of continuous cropping and fertilizer use on sustenance of finger millet-hybrid maize system. In : Long term fertility Management through Integrated Plant Nutrient Supply (A. Swarup, D.D. Reddy and R.N. Prasad, Eds.) IISS, Bhopal, pp. 139-145.

Sureshkumar, P. (1999) Variability in iron and zinc availability in laterite and lateritic soils of central Keral with reference to rice nutrition. Final Report of ICAR Adhoc- Project, Kerala Agricultural University, Vellanikara, Thrissur. pp. 1-24.

9

Secondary and Micronutrients in Acid Soils

Surendra Singh and A.K. Ghosh

Acid soils occupy one fourth of the land in India. The formation and distribution of acid soils depend upon several factors like temperature, vegetation, parent material and hydrological conditions etc. For this reason, these soils have low pH, poor base saturation, high sexquioxide content, which affects the transformation and availability of plant nutrients. Secondary and micronutrient deficiency as well as toxicity are major problems in cultivation of crops in acidic soils. These soils are generally deficient in Ca, Mg and S and adequate in Fe, Mn and Cu. Apart from the secondary nutrients, Zn, B and Mo deficiencies affect crop production. In lowlands, rice crop suffers from toxicity of Fe. Aluminum toxicity, common in highly acidic soils, also pose major problems in crop production. Management of secondary and micronutrients in acid soils for agriculture is a national priority to achieve food sufficiency and sustainability.

Importance

Calcium

In acid soils of high rainfall areas, Ca^{2+} and Al^{3+} dominate the CEC of the soil. Exchangeable and solution Ca^{2+} are in dynamic equilibrium and provide the needed Ca^{2+} for plants. Ca deficiency is not common, but can occur in highly leached and unlimed acid soils. Acid soils of very low CEC (sandy soils, soils with poor organic matter) may contain low calcium level for most crops. In addition, low soil pH (high Al^{3+} and H^{+}) impedes Ca^{2+} uptake by plants. As pH decreases, higher Ca^{2+} concentration is required to maintain root growth. Most crops respond to Ca application, when Ca concentration is below 25% of CEC. Type of clay also influences Ca^{2+} availability. 2:1 clays require higher % Ca saturation than 1:1 clays. Montmorillonite clays require > 70% Ca saturation of the cation exchange capacity (CEC) for adequate Ca availability, while kaolinite clays are able to supply sufficient Ca^{2+} even at 40-50% Ca saturation of the CEC.

Magnesium

Mg deficiency is not a major factor in most acid soils. Conditions conducive to Mg deficiency in acid soils are low CEC, liming with Mg free materials, higher rate of NH_4^+ fertilizer application, higher rate of K fertilizer application, excess leaching and growing crops that require high levels of Mg. The crops susceptible to Mg deficiency in acid soils are tobacco, citrus, potato, cotton and soybean. Mg saturation for optimum plant growth is about 7 to 10% of the CEC. Reduced Mg^{2+} uptake in many strongly acid soils is caused by high level of exchangeable Al^{3+}. Al saturation of 65 to 70% of the CEC is often associated with Mg deficiency. Mg deficiency can also occur in soils with a high Ca:Mg ratio of 10:1 to 15:1. Further, recommended K:Mg ratio is <5:1 for field crops, 3:1 for vegetables and sugarbeets, and 2:1 for fruits. Mg deficiency observed in Potato grown in some acid soils is mainly due to combined effect of low soil Mg, high rates of NH_4+ and K fertilizer and an excess solution Al.

Sulphur

Sulphur concentration in plants range from 0.1 to 0.5%. S content increases in the order Gramineae < Leguminosae < Cruciferae. Soils low in organic matter, non-irrigated, leached, coarse textured soils are prone to sulphur deficiency. The capacity of a soil to retain SO_4^{2-} in a solid phase mineral form is highly pH dependent, particularly in soils high in Al and Fe oxides. As soil pH decreases below pH 6.0, the capacity of the soil to remove SO_4^{2-} from soil solution either by surface adsorption or precipitation increases considerably. Sulphur mineralization in soil is a slow process of supply of SO_4^{2-} to plants and is ideal at near neutral soil pH to encourage microbial activity. There is a close relationship among organic C, total N and total S in soils, which is about 120:10:1.4 for well drained soils. Normally C:S ratio of <200:1 in crop residues encourage S mineralization while > 400:1 lead to sulphur immobilization.

Micronutrients

Among the micronutrients, in strongly acid soils, Mn, Zn, and Cu are often in high concentration. Boron is leached from soil due to high rainfall and sloppy terrain in which many acid soils have developed. Molybdenum is often deficient in acid soils. The pH of soil solution has a major influence on the solubility of iron. The solubility of Fe^{3+} decreases about 1000 fold per unit rise in soil pH. The amount of Zn in solution is also regulated by soil pH. Solubility of Zn, Mn, Cu and Fe^{2+} increases 100 fold for each unit lowering of soil pH. Among the cationic micronutrients (Zn, Fe, Cu and Mn), toxic concentrations of Mn are more likely. This is partly because some soils have high total Mn and partly because strongly acidic soils can dissolve toxic levels of Mn. Molybdenum deficiency is common in acidic coarse textured soils due to leaching losses and strong molybdate adsorption. Solubility of Mo increases about 10 fold per unit rise in soil pH above 7.0.

Role and Deficiency Symptoms

Calcium and magnesium are sufficient in normal soils, but their importance is felt in acid as well as alkali soils. Sulphur deficiency is widespread not only in acid soils, but also in other soils possibly due to the use of sulphur free fertilizers, low sulphur di oxide emissions etc. Secondary nutrients are becoming increasingly limiting in acid soil regions, because of intense leaching, runoff losses and crop uptake. Among the micronutrients, Zn, Fe, Cu and Mn deficiencies are not likely to occur frequently in acid soils; their toxicities may be important under some situations. B and Mo are likely to be deficient in acid soils. Important role of these micronutrients in plant nutrition is briefly outlined along with their symptoms of deficiency.

Role in plant	Symptoms of deficiency
Calcium	
• Constituent of cell wall and plays important role in cell elongation and division, structure and permeability of cell membrane • Promotes early root development and plant growth • Carbohydrate translocation	• Failure of terminal buds to develop • Immobile inside the plant; hence symptoms start with tips of shoot and root • Blossom end rot of tomato; black heart of peanut; internal browning of Brussels sprout; tip burn of lettuce and cauliflower
Magnesium	
• Constituent of chlorophyll and imparts green color to leaves • Structural component of ribosome • Acts as carrier of phosphorus • Activates the formation of polypeptide chains from amino acids • Cofactor of certain enzymes	• Induced chlorosis followed by marginal scorching • Mobile inside the plant; hence symptoms start with older plant parts • Initially older leaves become chlorotic and chlorosis spreads interveinally towards the leaf base and mid rib resulting in a herringbone pattern • Buckwheat, spinach, cauliflower, pea, oats, corn have high Mg requirement
Sulphur	
• Forms disulphide bonds between polypeptide chains and important constituent of sulphur containing amino acids: cystine, cysteine and methionine • Needed in synthesis of Coenzyme A, biotin, thiamin and glutathione • Component of S-adenosyl methionine, liplic acid and sulpho lipid • Required for chlorophyll synthesis	• General yellowing of young leaves • Immobile inside plant and symptoms first appear on younger leaves • Chlorosis and failure to synthesize proteins resulting in accumulation of amino acids • In older leaves both interveinal chlorosis and necrotic areas may be found
Zinc	
• Act as a metal component of enzymes and as an enzyme cofactor • As a factor of RNA polymerase enzyme, Zn influences protein synthesis	• Zn deficiency include interveinal chlorosis· Leaf malformation resembling virus disease possibly because it is involved in synthesis of tryptophan a precursor of I.A.A. • The purplish colour of lower leaf surface of Zn deficiency trees is called bronzing
Iron	
• Helps in chlorophyll formation • Structural component of porphyrin	• Chlorosis between veins while viens remain dark green

Contd.

Role in plant	Symptoms of deficiency
and non heme molecules (ferredoxin) • Component of enzyme system	• Immobile inside plant and symptoms first appear on younger leaves • Die back of shoots
Copper • Is a constituent of certain enzymes including ascorbic acid oxidase and tyrosinase • Has some indirect effects on nodule formation • Helps in the utilization of iron during chlorophyll synthesis • Act as electron carrier in enzymes which bring about oxidation-reduction reactions in plants	• Various degree of stem deformity • Interveinal chlorosis, leaf mottling, defoliation and dieback of terminal shoots • A deficiency of Cu occasionally occurs in some forest trees but not often in fruit trees
Manganese • Essential for chlorophyll synthesis • Activation of enzyme systems • Role in the oxygen evolution step	• Reduced contents of chlorophyll • Leaves malformed and development of chlorotic areas between major viens • Chlorosis does not occur in young leaves
Nickel • Cofactor of urease • Disrupts N metabolism • Abnormal accumulation of glycine and oxalic and lactic acids	• Growing tips and margins become chlorotic
Boron • Helps in uptake and utilization of calcium • Helps in protein synthesis	• Scorched leaf margins, dead shoot tops and weak lateral shoots giving a bushy appearance • Leaves thick, dark green and margins roll upwards • Younger leaves are dwarf eg. tap rot of tobacco
Molybdenum • Helps in fixation of atmospheric N by nodule bacteria • Helps in protein synthesis	• Chlorosis of young leaves, burning of tips and margins of older leaves and leaf abscission
Chlorine • Involved in water splitting step of photosynthesis	

Mobility in plant

The relative mobility of a nutrient in plant is helpful in understanding the reason for the differential appearance of nutrient deficiency. Deficiency symptoms in older leaves indicate that the nutrient is mobile nutrient, because the plants can transport some quantities of these nutrients from old to new leaves. Symptoms of deficiency on younger leaves indicate the shortage of an immobile nutrient. Deficiency symptoms can serve as a guide for diagnosing limiting nutrients and the need for corrective measures. However, chlorotic and necrotic leaves might also result from the toxic effects of nutrients, pollution and also from disease and pest incidence. Therefore, confirmation regarding the cause of symptoms is required, before corrective measures are taken up. Some of the diagnostic criteria used for secondary and micronutrients are:

- Ca deficiency is first seen on growing tips and on the youngest leaf. This is the case with all nutrients that are not very mobile in the plant. Typical symptoms of deficiency are cup-shaped, crinkled and dark green leaves.
- As Mg is readily translocated from older to younger plant parts, its deficiency symptoms first appear in the older leaves. Typical symptom of Mg deficiency is interveinal chlorosis in older leaves, in which the veins remain green but the area between them turn yellow.
- Under low S conditions, mobility of S is low as S is a structural component and can not be translocated. Deficiency of S starts with the appearance of pale yellow or light green leaves. S deficiency symptoms in most cases appear first on the younger leaves, and are present even after N application. Plants deficient in S are small and spindly with short and slender stalks.
- Boron is relatively immobile in plants and frequently B content increases from the lower to the upper parts of plants. B deficiency usually appears on the growing points of roots, shoots and youngest leaves.
- Cu is not readily mobile in the plant and its movement is strongly dependent on the Cu status of the plant. Cu deficiency symptoms are first visible in the form of narrow, twisted leaves and pale white shoot tips.
- Absorbed Fe is immobile in the phloem. Fe deficiency first appears on younger leaves. Yellowing of the interveinal areas of leaves (commonly referred to as iron chlorosis) occurs. In severe deficiency, leaves become pale white due to the loss of chlorophyll.

- Mn deficiency symptoms resemble Fe and Mg deficiency and interveinal chlorosis occurs in the leaves. However, Mn deficiency symptoms are first visible on the younger leaves, whereas in Mg deficiency, the older leaves are affected first.
- Common symptoms of Zn deficiency are: stunted plant growth, chlorotic bands on either side of the midrib in monocots (particularly maize), brown rusty spots on leaves in some crops.
- Mo appears to be moderately mobile in plants and deficiency symptoms appear in the middle and older leaves. Mo deficiency in legumes resembles N deficiency because of its role in nitrogen fixation. Mo deficiency can cause marginal scorching and rolling or cupping of leaves and yellowing and stunting in plants.

Reasons of Deficiency

Deficiency of secondary and micronutrients in acid soils emanate from soil and nutrient losses arising from runoff and leaching owing to high rainfall. High porosity and sandy nature of soils accentuate the problem. Besides, these soils not only contain a comparatively low amount of total Ca, Mg, Zn, Cu, B and Mo, but also much of it, particularly Zn, Cu, B and Mo remain strongly bound by oxides of Fe and Al. Their application in acid soil leads to rapid fixation by these oxides, which make these unavailable to plants. This is particularly true for Zn and Cu.

Reasons for secondary nutrient deficiencies in acid soils are (i) coarse texture of soils formed from rocks low in Ca^{2+} mineral (ii) removal of Ca^{2+} and Mg^{2}+ ions from surface layers due to weathering, erosion, leaching and uptake by plants (iii) low availability of Ca^{+2} and Mg^{2}+ to crops in acid sandy soils due to low CEC (iv) high levels of exchangeable Al^{3+} in strongly acid soils reducing Ca^{2+} and Mg^{2}+ uptake by crops (v) red and lateritic soils, which are low in organic matter (vi) use of S-free concentrated fertilisers, and (vii) low or no addition of organic manures and residues to crops grown in acid soils.

Micronutrients in acid soils vary due to management factors, such as: (i) B and Mo existing in soil in anionic from, are fairly deficient in soils of humid regions as these leach out of the soil profile because of high rainfall (ii) intensification of agriculture due to raising of more crops per unit time and per unit area involving

Magnesium deficiency in potato and tomato
Interveinal of older leaves with veins remaining green, pinkish/brown colour of old leaves

Tomato plants with marginal chlorosis & necrosis and forward curled leaf margins due to Ca deficiency

Magnesium deficiency in potato and tomato

Sulphur deficiency in groundnut

Boron deficiency in cauliflower

Molybdenum deficiency in mustard

A visual guide to deficiency symptoms of some secondary and trace elements

use of high analysis fertilisers (iii) continuous application of nitrogenous fertilizer (urea) alone common with farmers of acid soil regions. For example, red soils of Ranchi under maize-wheat cropping system continuously fertilized with 110 kg N ha^{-1} $crop^{-1}$ through ammonium sulphate for 34 years turned highly acidic (pH 4.0) releasing higher concentrations of micronutrients in soil solution, which leached to lower horizons due to high rainfall and low CEC of soil. (iv) soil erosion in a slopy terrain results in depletion of soil organic matter and micronutrient cations in different groups of acid soils .

Status in acid soils

Analysis of 338 surface samples from the acidic soils of Assam, Himachal Pradesh, Jharkhand and Meghalaya indicated wide variation in the status of exchangeable Ca^{+2} and Mg^{+2} ranging from 0.20 to 12.15 and 0.58 to 2.91 cmol (P+) kg^{1} (Table 1).

Table 1: Exchangeable Ca^{2+} and Mg^{2+} in acidic soils of different states

States	No. of samples	pH	Exchangeable Ca^{2+} Cmol (P^{+}) kg^{-1}		Exchangeable Mg^{2+} Cmol (P^{+}) kg^{-1}	
			Range	Mean	Range	Mean
Assam	50	4.0.5.1	1.92-2.97	2.32	1.48-2.00	1.75
HimachalPradesh	100	5.0-5.6	1.08-4.70	3.10	1.0-2.8	1.80
Jharkhand	125	4.4-6.5	2.28-12.15	5.57	0.58-2.91	1.59
Meghalaya	63	4.3-5.6	0.20-5.80	1.10	0.60-1.50	0.94
Over all	338		0.20-12.15		0.58-2.91	

(Sarkar and Singh, 2003)

Continuous use of mineral fertilizers considerably reduced the soil pH from the initial value of 5.5 in permanent manurial trial of Ranchi (Table 2). The acidity continuously increased over the 40 years, which can be mainly ascribed to the cumulative effect of acid forming fertilizers. Contrary to this, FYM+NPK or Lime+NPK application resulted in much less soil acidity. Results also point out that exchangeable Ca and Mg in soil was lowest in N (ammonium sulphate treated plots).

Table 2. Effect of long term fertilization of mineral fertilizers, FYM and lime on pH and exchangeable Ca and Mg.

Fertilizers Treatments	pH	Exchangeable cations Cmol (P^+) kg^{-1}	
		Ca	Mg
Unfertilized	5.4	3.05	2.21
N	3.9	1.10	1.75
NP	4.0	2.36	2.15
NPK	4.2	2.33	2.00
Lime+NPK	6.5	8.10	2.05
FYM	6.3	8.25	4.16

Sarkar *et al.* (2000)

Sulphur deficiency is reported to the extent of 20 per cent in acidic soils of the Assam (Table 3). Acid soils of Jharkhand, Odisha and hill soils of Himachal Pradesh are prone to S deficiency (21.0-75 per cent) . Mean available S content in acidic soils from different states varied from 7.4 mg kg^{-1} in Himachal Pradesh to 16.4 mg kg^{-1} in Odisha, with deficiency to the extent of 43.9 per cent.

Table 3: Available S content and magnitude of S deficiency in acidic soils

States	Soil Goups	Available S (mg kg^{-1})		Percent sample deficient in S
		Range	Mean	
Assam	Entisol	9.5-72.6	12.2	20.0
Himachal Pradesh	Oxisol	0.5-25.0	7.4	74.0
Jharkhand	Alfisol/Ultisol	0.33-94.7	10.9	50.5
Orissa	Alfisol	Tr-64.2	16.4	21.0
	Alfisol	Tr-28.1	13.1	43.0
	Entisol	Tr-25.7	8.0	75.0
Over all		Tr-94.7	11.16	43.9

Tr = traces
(Sarkar and Singh, 2003)

Data collected for different states representing mostly acid soils, indicate the deficiency of Zn, Cu, Fe, Mn, B and Mo in 31, 3, 4, 3, 45 and 46% of the samples analysed, respectively (Table 4). The deficiency of Zn was minimum (8%) in soils of Pondicherry and maximum (53%) in the acid soils of Chhattisgarh due to low organic matter content (Singh 2000).

Table 4. Extent of micronutrient deficiencies in acidic soils of India

State	No of samples	Per cent samples deficient				B*		Mo*	
		Zn	Cu	Fe	Mn	No of samples	% samples	No of samples	% samples
Andhra Pradesh	489	15	1	2	1	380	52	380	53
Tamil Nadu	6428	37	7	16	7	1031	21	-	-
Karnataka	5573	41	0.1	4	2.6	8513	32	-	-
Kerala	650	34	3	0.9				50	78
Pondicherry	4108	8	4	2	3				
Assam	12165	34	1	2	2	122	17		
Jharkhand	6621	42	3	1	1	455	57	100	55
Chhattisgarh	2837	53	1	1	1	544	49	550	30
Orissa	16041	19	2	1	1	10142	69	100	87
West Bengal	6547	36	0	0	3	3177	68		
Meghalaya	95	57	2	0	23	-	-	-	-
Maharashtra	120	30	1	2.4	1	-	-	-	-
Jammu & Kashmir	93	12	-	-	-	-	-	-	-
Himachal Pradesh	155	42	1	27	5	155	1	-	-
Uttarakhand	6758	28	7	14	9	100	14	-	-
Total	68680	31	3	4	3	24619	45	1180	46

*Data presented represent only certain districts instead of whole state
Singh (2007)

Systematic studies were undertaken to assess the micronutrient status in acid soils of Assam and North east Hill (NEH) region. Taking DTPA extractable micronutrient levels as 0.6 ppm Zn, 0.2 ppm Cu, 4.5 ppm Fe, 2 ppm Mn, 0.5 ppm hot water soluble B and 0.15 ppm ammonium oxalate extractable Mo as the critical levels, to categorize deficient / sufficient soils, 45 and 12% of the soil samples were found to be deficient in B and Mo, respectively. Widespread deficiency of Zn ranging from 23-54% was recorded in soils of Assam, Jharkhand, Chhattisgarh, Uttarakhand, Odisha and West Bengal. High organic matter content in Assam soils induced Zn deficiency in brown hill and forest soils. To a large extent, cold and wet soils of Assam are found to be suffering from Zn deficiency. Zinc deficiency has also been recorded in tea soils of the NEH regions. Singh (2001a,b) reported wide variation (ranging from 12 to 52%) in Zn deficiency in soils across the agro-ecological zones of India. Data reveal that most of the acid soils are adequate in Fe, Cu and Mn, whereas perceptible deficiencies of Zn and B occur in the old alluvial soils of Assam, red and yellow soils of Odisha, red and lateritic soils and alluvial soils of Terai and Teesta in West Bengal (Singh 2001a) and 18.2% soils of the red and yellow soils of the Santhal Parganas in Jharkhand.

In West Bengal, out of 6457 soil samples assessed for micronutrient disorders, occurrence of Fe and Cu deficiency was noticed only in few locations (Ghosh *et al.* 1992). Manganese deficiency was observed in 4 to 8% soils in Terai and Teesta alluvium. On the contrary, Zn deficiency was widespread and of a higher order (49-51%) compared to that in red and laterite soils (20-46%). The deficiency of zinc and boron in Terai and Teesta alluviual soils of acidic reaction, may be attributed to higher sesquioxide content, which are largely responsible for their fixation.

Deficiency of B to the order of 82-96 % has been reported from Terai and Teesta alluvial soils (Singh 2001a) while in the red and lateritic soils of Bankura, Purulia and Midnapur districts, the deficiency ranged for 40 to 90%. Soils of the Chhotanagpur plateau are also deficient in boron due to continuous depletion of B as a result of intensive vegetable cultivation. In Mayurbhanj district of Odisha, 69% soil samples analysed were deficient in boron.

Meager information is available on Mo deficiency in acidic soils of India and often Mo deficiency has been judged by response to Mo application. Beneficial effect of Mo use was observed in soybean grown on acidic Oxisols (Subba Rao and Adinarayana 1995), which might be partly due to enhanced nitrogen fixation. Available Mo content in acidic red and lateritic soils at high altitudes of Srikakulam and Vijaynagar districts of Andhra Pradesh ranged from 0.03 to 0.39 mg kg^{-1}, with mean value of 0.17 mg kg^{-1}. Nearly, 49% of the soils are reported to be deficient in Mo and these would need Mo fertilization for higher crop yields (Bhupal Raj *et al.* 2005).

Critical limits

Critical limit of a nutrient varies from one soil to another, between crops and varieties and also between methods of its estimation. On the basis of existing information on soil and plant analysis and response obtained in pot culture and field experiments conducted, the following limits of secondary and micronutrients in acidic soils and plants, have been reported (Table 5).

Table 5: Critical levels of secondary and micronutrients in soils as determined by different analytical methods for different crops

Secondary /Micronutrients	Extractant	Crop	Deficiency	Excess
Ca	NH_4OAc	Rice	<1.0 c mol kg^{-1} or <8.0% saturation of CEC	>20% saturation of CEC (optimum)
Mg	NH_4OAc	Rice	<1.0 c mol kg^{-1}	> 3.0 c mol kg^{-1} (optimum)
S	$Ca(H_2PO_4)_2$	Rice	<9.0 mg kg^{-1}	>50 mg kg^{-1}
Zn	DTPA	Rice and wheat	0.86 mg kg^{-1}	10.0-20.0 mg kg^{-1}
Cu	DTPA	Rice, Corn	0.20 -0.30 mg kg^{-1}	>17.0-25.0 mg kg^{-1}
Fe	DTPA	Rice	2.5-4.5 mg kg^{-1}	>300.0 mg kg^{-1}
Mn	DTPA	Rice	1.0 mg kg^{-1}	140.0-200.0 mg kg^{-1}
B	Hot-$CaCl_2$	Wheat	0.50 mg/kg mg kg^{-1}	>3.0-5.0 mg kg^{-1}
Mo	Ammonium oxalate	Common crops	0.05 mg kg^{-1}	>0.5-1.0 mg kg^{-1}

Bhattacharyya *et al.* (1998)

Table 6: Critical limit of S in soils and crops (45/50 days) in acidic soils

States	Crops	Extractants	Critical limit Soil (mg kg^{-1})		Plant (%)	
			S	B	S	B
Jharkhand	Mustard	0.15% $CaCl_2$	7.3	-	0.24	-
		$Ca(H_2PO_4)_2$	17.0	-	-	-
	Niger	0.15% $CaCl_2$	9.0	-	0.24	-
		$Ca(H_2PO_4)_2$	15.5	-	-	-
	Blackgram	0.15% $CaCl_2$	8.7	-	0.21	-
		$Ca(H_2PO_4)_2$	18.6	-	-	-
	Lentil	0.15% $CaCl_2$	9.9	-	0.21	-
	Greengram	0.15% $CaCl_2$	9.5	-	-	-
		$Ca(H_2PO_4)_2$	33.8	-	-	-
		$Ca(H_2PO_4)_2$	14.3	-	-	-
	Soybean	Hot water	-	0.46	-	21.5 (mg kg^{-1})
		0.1 M $CaCl_2$	-	0.65	-	-
Meghalaya	Pea	Hot water	-	0.70	-	23.8 (mg kg^{-1})
Manipur	Rapseed	Morgans' reagent	40.0	-	0.35	-

Sarkar and Singh (2003)

0.15% $CaCl_2$ extractable S of 7.3 to 9.9 mg kg^{-1} in acidic soils of Jharkhand was found as critical limit for mustard, niger, blackgram, lentil and green gram. The critical limit of S establish by $Ca(H_2PO_4)_2$ extraction was higher ranging from 15.5 to 33.8 mg kg^{-1}.Crtical limit of S in rapseed with Morgans' extractant was 40 mg kg^{-1} for acidic soils of Manipur. The critical limit of S in plant ranged from 0.21 to 0.35 % for these crops. The critical limit of hot water extractable B was 0.46 and 0.70 mg kg^{-1} in soils of Jharkhand and Meghalaya, respectively (Table 6).

Nutrient ratio

A Ca / total cation ratio of 0.10 to 0.15 is desirable for adequate Ca^{2+} supply to most crops. Mg concentration of 0.2% represents its critical level for many crops. Sulphur content in young fully developed leaves of 40-50 days old oilseed and pulse crops categorizes them as severely S deficient (<0.15%), deficient (0.15 to 0.30%), moderately sufficient (0.30 to 0.45%) and sufficient (>0.45%) (Tandon, 2011). Apart from critical concentration, the N:S ratio of 15-16:1 for alfalfa, soybean, maize and groundnut, 3:1 for mustard and 7:1 for wheat plants can serve as a guide to S deficiency. Approximate concentration of secondary and micronutrients in plants is presented in table 7 to diagnose these as deficient, sufficient or toxic.

Table 7: Approximate concentration of secondary and micronutrients in mature leaf tissue of various crop plants

Nutrient	Deficient	Sufficient or normal	Toxic
Ca(%)	-	0.2-1.0	-
Mg(%)	-	0.1-0.4	-
S(%)	<0.15	0.2-0.4	>500
Fe (mg kg^{-1})	<50.0	100-500	300-500
Mn (mg kg^{-1})	15-25	20-300	100-400
Zn (mg kg^{-1})	10-20	27-150	100-200
Cu (mg kg^{-1})	2-5	5-30	50-20
B (mg kg^{-1})	5-30	10-20	>100
Mo (mg kg^{-1})	0.03-0.15	0.1-2.0	

Rattan and Goswami (2002)

Response of crops

Deterioration in soil fertility is often observed in crops/cropping systems, even with adequate use of NPK fertilizers. This is due to the imbalance of certain secondary and micronutrients in soils and plants. Acid soils most often are adequate in cationic micronutrients (Zn, Cu, Fe and Mn), but deficient in anionic micronutrients (B and Mo) which limit crop productivity. Poor soil physical conditions that favour low retention of water and nutrients can significantly reduce plant growth in these coarse textured or highly leached acidic soils with low CEC. Trend of response of Ca, Mg, S, B and Mo to crops grown in acidic soils is summarized below:

Secondary nutrients

Lime mainly provides calcium to soil, where it is changed into a form available to plants. A substantial amount of calcium is removed by crop from the soil like N, P and K. Calcium as a nutrient for soybean and groundnut, in acidic soil of Jharkhand was studied (Table 8). Furrow application of 2-4 ha^{-1} calcium carbonate and gypsum increased grain yield of soybean and pod yield of groundnut to the tune of 26.7 and 47.6 per cent, respectively. The improvement in crop yield due to gypsum was mainly due to Ca rather than increase its effect on soil pH.

Table 8: Mean crop yield due to different sources of calcium in acidic soils of Jharkhand

Treatments	Soybean yield (q ha^{-1})	Response (%)	Groundnut Yield(q ha^{-1})	Response (%)
No lime	23.1	-	17.5	-
1 LR (28.5)* Broadcast in the beginning	37.2	30.2	24.7	41.0
$CaCO_3$ as furrow application (2.0)	27.5	18.7	25.8	47.0
$CaCO_3$ as furrow application (4.0)	30.5	31.6	27.2	21.0
Gypsum as furrow application (2.0)	27.7	19.4	24.2	38.2
Gypsum as furrow application (4.0)	28.2	21.6	22.2	26.7
CD (P=0.05)	1.36	-	1.85	

*(Lime q ha^{-1}) Mathur *et al*: (1983)

Magnesium is often deficient in acid soils. Field experiments were conducted in acidic soils of Ranchi on groundnut, wheat, soybean and potato, to study the response of crops to added magnesium as magnesium sulphate (Table 9). Response of Mg in these crops was higher with application of 30 kg Mg ha^{-1}. Groundnut and soybean responded well to Mg application followed by potato and wheat. Per cent response to Mg application ranged from 21 to 41.6, 16 to 25.6, 12.6 to 27.8 and 5.9 to 22.7 in groundnut, soybean, potato and wheat, respectively.

Table 9: Effect of Mg application on crop yield in acidic soils of Jharkhand (average of 3 years)

	Crop yield (q ha^{-1})							
Treatment *Mg (kg ha^{-1})	Groundnut	Response (%)	Wheat	Response (%)	Soybean	Response (%)	Potato	Response (%)
Control	13.8	-	34.64	-	19.3	-	192.9	-
10	16.71	21.0	36.67	5.9	22.40	16.0	217.2	12.6
20	17.90	29.7	39.24	13.2	23.19	20.2	225.2	16.7
30	19.54	41.6	42.50	22.7	24.25	25.6	245.7	27.8
CD (P= 0.05)	2.82	-	4.06	-	4.35	-	37.9	

*Mg as $MgSO_4$ $7H_2O$
Lal (1999)

Response of crops to S application can be judged based on nature and severity of deficiency on plant foliage, plant analysis, soil analysis and /or responses of crops to the applied S. The magnitude of response differs widely among the crops and their cultivars, soil types and degree of S deficiency. Beneficial effect of S application on increasing yield of several cereal, oilseed, pulse and cash crops has been reported in S deficient acidic soils of eastern India. The response of sulphur application to sources of S differed among crops grown in these soils (Table 10).

Table 10: Response of crops to S application in acidic soils of eastern India.

States	Crop	S rate and sources (kg ha^{-1})	Yield without S	Yield with S	Response (%)
Assam	Tea	20Gy	21.2	25.5	20.2
	Toria	20Gy	13.9	15.0	6.67
Jharkhand	Niger	45PG	3.3	4.6	40.0
	Groundnut	40Gy	23.5	30.0	27.9
	Mustard	60PG	12.0	15.0	26.3
	Soybean	60PG	12.1	17.2	40.8
	Balckgram	24PG	18.6	23.4	26.1
	Lentil	30PG	14.9	17.5	17.0
	Frenchbean	40 sulf-95	47.4	61.5	29.7
	Cowpea	40 sulf-95	68.5	117.7	63.1
	Onion	60 sulf-95	88.5	123.5	39.2
	Garlic	60 sulf-95	34.9	57.8	55.5
Odisha	Rice	30SSP	37.0	42.3	14.2
	Groundnut	30GY	6.7	10.3	53.6
	Sunflower	20SSP	4.4	7.8	77.6
	Mustard	20SSP	9.2	11.2	17.0
West Bengal	Rice	30GY	47.6	52.3	9.8
	Mustard	40GY	12.4	13.5	8.4
	Seasmum	30GY	13.3	15.7	18.3
	Jute	90 AS+SSP	19.3	23.4	21.0

PG= Phosphogypsum SSP= Single superphosphate AS= Ammonium sulphate
Gy= Gypsum Sulf-95= Sulfur 95
Sarkar and Singh (2003)

S application had beneficial effect on tea in Assam. The highest response of Tea to S application (20.2 per cent) was at 20kg S ha^{-1}. Response of toria was of lower magnitude (6.67 per cent) at 20 kg S ha^{-1}. In acidic soils of Jharkhand, S application from 24-60 kg ha^{-1} increased crop yield appreciably. The response due to S application ranged from 26.3 to 40.0, 17 to 26 and 29.7 to 63 per cent in oilseed, pulse and vegetable crops. Response of S to rice, mustard, groundnut and safflower in red and lateritic soils of Odisha was studied at 20-30 kg ha$^{-1,}$ where, the yield response varied from 14.2 (rice) to 77.6 per cent (safflower). The extent of response of groundnut to S was higher, when it was applied after liming. In red and lateritic soils of West Bengal, yield response in rice, mustard and sesamum was 9.8, 8.9 and 18.3 per cent, respectively.

There is an active involvement of S in protein synthesis and oil production. Sulphur is a constituent of amino acids like methionine, cystine and cysteine, which are building blocks of protein. The quality of tea and oil content in toria in Assam improved with 20 kg S ha^{-1} added as elemental sulphur (Table 11).

Table 11: Effect of S on the quality of crop produce in acidic soils of eastern India

State	Crop	S rate and sources (kg ha^{-1})	Content	Without S (%)	With S (%)	Response (%)
Assam	Tea	20ES	Thealflavimms	2.6	-	0.10
		20ES	Thearubigins	13.0	-	0.90
		20ES	Brightness	37.7	-	0.80
		20ES	Colour	7.6	-	0.30
		20ES	Tester score	61.3	-	12.7
	Toria	20GY	Oil	23.4	25.8	2.4
Jharkhand	Groundnut	40GY	Oil	36.3	41.8	5.5
	Mustard	60PG	Oil	36.8	43.6	6.8
	Niger	45PG	Oil	36.1	39.3	3.2
	Soybean	60PG	Oil	15.5	19.5	5.8
			Protein	35.0	37.3	6.5
	Blackgram	24PG	Protein	15.8	19.6	3.8
	Lentil	30PG	Protein	16.6	19.1	2.5
Odisha	Groundnut	30GY	Oil	30.3	34.2	3.9

ES= Elemental Sulphur Gy= Gypsum PG= Phosphogypsum
(Sarkar and Singh, 2003)

In acidic upland soils of Jharkhand, sulphur application from 40 to 60 kg ha^{-1} in groundnut, mustard and niger increased the oil content from 3.2 to 6.8 per cent. Protein content in blackgram and lentil increased from 2.5 to 5.8 per cent with 24 to 30 kg ha^{-1} of S application. Groundnut is Odisha produced maximum oil content (39 per cent) at 30 kg S ha^{-1} added as gypsum.

Sulphur application, in general, benefits more than one crop grown in sequence and produces a significant residual response. Some recent results on sulphur management in cropping systems in acidic soils are summarized in Table 12. Depending upon the system, the direct effect of S contributed to an increase in crop yield from 13.1 to 18.8 per cent, while residual S, contributed to yield by19.2 to 30.0 per cent. In groundnut-wheat crop sequence, direct and residual responses of S were almost similar. Cropping systems involving oil seed or pulse crops benefited more from the residual S.

Table 12: Direct and residual effect of S in cropping systems

State	Croping Systems	Rate & source(kg ha^{-1})	Yield (q ha^{-1})					
			Direct crop			Residual crop		
			(+)S	(-)S	Response%	(+)S	(-)S	Response%
Jharkhand	*Mustard - **Blackgram	60PG	15.2	12.8	18.8	10.8	8.8	22.7
	Groundnut- Wheat	30SSP	19.0	16.0	18.8	26.7	22.4	19.2
	Groundnut- Mustard	45SSP	15.1	12.9	17.1	10.4	8.0	30.0
	Rice-Pea	30PG	27.9	23.8	17.2	15.1	12.4	21.8
Odisha	Rice-groundnut	60 GY	-	-	13.1	-	—	19.3

Direct crop* Residual crop** SSP= Single superphosphate PG= Phosphogypsum
(Sarkar and Singh, 2003)

Micronutrients

Response of crops to boron fertilization was recorded in large number of field trials in rice and wheat with average yield response of 15%. Gram, groundnut and oilseeds responded more to boron fertilization (Table 13).

Meager information is available on Cu fertilization and response of crops to its application in acid soils. Responses to Cu application was recorded in organic matter rich acid red loam soils of Kerala (Sreedharan and George (1969); boro rice in acid leached old alluvial soils of northern part of West Bengal (Samui and Bhattacharya 1973); rice grown in Pattambi (Singh *et al.* 2004).

Beneficial effect of Mo application has been reported in rice, wheat, soybean, cauliflower, mustard and lentil grown in acidic soils. Molybdenum application increased the rice yields due to better nitrogen utilization in acid soils of West Bengal (Das Gupta and Bera 1972), but the results were not equivocal in rice grown in old alluvial soils of Odisha (Nayar *et al.* 1980). However, yield increase to the tune of 38% to foliar spray of 0.5 kg sodium molybdate per hectare was reported in wheat in light textured alluvial soil of West Bengal (Bhattarcharya 1992).

Response of crops to zinc application has been convincing in Zn deficient soils. Response of zinc use in Wheat was higher than rice in 148 experiments conducted on farmers' fields in Himachal Pradesh. In long-term field experiment in hill soils of Palampur, zinc application significantly increased maize and wheat yields in the initial years, but not in the subsequent years (Sharma *et al.* 2005). Response of potato to zinc application was 33-45 q per hectare in north Indian hills and 15-40 q per hectare in red and lateritic soils of South India (Grewal and Trehan 1990).

Table 13. Response of crops to micronutrients in acid soils of India

Element	Crop	No of experiments	Yield (q/ha)		Increase Response	
			NPK	NPK + Micronutrients	(q/ha)	%
B	Maize	4	42.29	49.67	7.38	17.45
	Rice	58	32	36.60	4.6	14.38
	Wheat	77	19.66	22.59	2.93	14.9
	All cereals mean	139	26.03	30.15	3.85	11.74
	Gram	1	7.72	13.63	5.91	76.55
	Groundnut	6	14.64	18.32	3.68	25.14
	Raya	5	9.73	13.99	3.46	35.56
	Rapeseed	7	8.90	10.43	1.53	17.19
	All oilseed crops mean	18	11.04	13.83	2.79	25.27
	Cabbage/ Cauliflower	55	366.48	412.58	46.1	12.58
	Potato	16	146.30	161.5	15.2	10.3
	All vegetables mean	71	316.86	356.00	39.14	12.07
Cu	Wheat	2	33.7	39.26	5.56	16.5
Mo	Maize	2	50.1	56.80	6.7	13.37
	Rice	1	32.06	39.36	7.3	22.77
	Wheat	63	18.27	19.78	1.51	8.26
	Rapeseed	8	8.18	8.91	0.73	8.92
	All oilseed mean	8	8.18	8.91	0.73	8.92
Zn	Rice	156	39.82	49.75	9.93	24.94
	Wheat	80	18.24	20.05	1.81	9.92
	All cereals mean	237	32.37	39.51	7.15	19.76
	Groundnut	2	14.65	22.44	7.79	53.17
	Mustard	14	11.54	14.59	3.05	26.43
	All oilseed crops mean	18	10.60	13.84	3.24	26.43
	Green gram	5	20.84	27.60	6.75	29.96
	Cauliflower/cabbage	24	401.53	451.8	50.27	12.52
	Sugarcane	1	7.48	8.62	1.14	15.24

Singh (2007)

Table 14: Response of crops (other than cereals) to soil application of B in acidic soils of eastern India.

States	Rate of B and source (kg ha^{-1})	Crop	Yield (q ha^{-1})		Response (%)
			With B	Without B	
Assam	1.0 kg B	Mustard	7.6	6.5	16.9
Jharkhand	3.0 kg B	Groundnut	15.3	11.4	34.0
	2.0 kg B	Pigeonpea	16.7	10.4	60.5
	2.0 kg B	Gram	9.3	6.8	36.7
	2.0 kg B	Cauliflower	220	123.1	19.1
	6.0 kg B	Lucerne*	733.0	636.0	15.3
	2.0 kg B	Soybean	16.2	19.9	16.5
Odisha	2.0 kg B	Groundnut	14.1	10.6	29.9
Sikkim	1.0 KgB	Frenchbean	91.7	84.3	8.8
Nagaland	1.0 Kg B	Soybean	28.8	25.5	12.9

*Green forage yield (Sarkar and Singh, 2003)

Significant responses of oilseed, pulse, vegetable crops (1 to 2 Kg/ha) and forage crops (6 Kg/ha) to B application have been observed on B deficient acidic soils of Assam, Jharkhand, Odisha and Sikkim (Table 14). Response of B to mustard and french bean was 16.9 and 8.8 per cent in Assam and Sikkim at 1 Kg B ha^{-1}, respectively. In red and lateritic soils of Jharkhand and Odisha, cauliflower and groundnut responded to 2 kg B ha^{-1} with response of 19.1 and 29.9 per cent, respectively. Application of 2.0 kg B ha^{-1}in acidic soils of Jharkhand resulted in higher response in pigeonpea (60.5%) followed by gram (36.7%). Based on the pattern of response of crops to Boron application in acidic soils, the recommendations have been provided for use by extension agencies (Table 15).

Table 15: Boron recommendation for crops grown in B deficient acid soils

States	Crops	Recommendation
Assam	Coconut	25 g borax $palm^{-1}$ in the soil
	Cabbage and cauliflower	8 kg borax ha^{-1} as basal or 0.2-0.3 % spray of borax
	Pea	10 kg borax ha^{-1} in North Bank Plain Zone
	Rapseed and mustard	10 kg borax ha^{-1} in North Bank Plain Zone and 5-10 kg ha^{-1} in Upper Brahmputra Valley
Jharkhand	Cauliflower, maize, groundnut, lentil, greengram, gram and wheat	Boron deficient acidic soils of Jharkhand should be treated with 1 - 1.5 kg B ha^{-1}(10-15 kg borax ha^{-1}) as basal application in intensively growing areas of vegetable, oilseed, pulse/legumes and cereal crops Foliar application of 0.2% borax or boric acid have been found effective in controlling brown rot of cauliflower. Among the three B sources tested (borax, boric acid and granubor II), Granubor II is least soluble. Both sources of Boron borax and Granubor II are equally effective for ameliorating B deficient acid soils in Jharkhand.
Meghalaya	Citrus	Spray of B as part of multi –micronutrients spray
	Radish	5 kg borax ha^{-1} to soil
Odisha	Cabbage and cauliflower	0.3 % boric acid spray 2 weeks after planting and 2 weeks before curd/head formation
West Bengal	Cauliflower	10 kg borax ha^{-1} in terai region
	Citrus	Foliar spray with 2 g borax per litre water 4 & 8 weeks after planting and 10-12 kg borax ha^{-1} during soil preparation
	Mustard	10 kg borax ha^{-1} in terai region
	Wheat	10 kg borax ha^{-1} at land preparation

Tandon (1995)

Responses of oilseed, pulse and forage crops to Mo application were observed in acidic soils of Jharkhand and Odisha (Table16). Soil application of 0.6 to 1.5 kg. Mo ha^{-1} as ammonium molybdate to groundnut, soybean, cauliflower and lucerne grown in a red sandy loam soils of Jharkhand was beneficial with a response of 19.5, 25.8, 32.5 and 9.3 per cent, respectively. Greengram also responded to Mo application as seed treatment in acidic red loam soils of Odisha with a yield response was 39.9 per cent.

Table 16: Response of crops to molybdenum application in acidic soils

States	Rate of Mo and source (kg ha^{-1})	Crop	Yield (q ha^{-1})		
			With Mo	Without Mo	Response (%)
Jharkhand	*0.6 kg Mo	Groundnut	18.4	15.4	19.5
	*1.0 kg Mo	Lucerne***	709.0	649.0	9.3
	*0.8 kg Mo	Soybean	22.9	18.2	25.8
	*1.5 kg Mo	Cauliflower	200	151.0	32.5
Odisha	**8g Mo/ 25 kg seed	Green gram	5.96	4.26	39.9

*Soil application **Seed treatment ***Green forage yield

Sarkar and Singh (2003)

Boron blended macronutrient fertilizers

The gap between toxicity and deficiency for B in soils and crops is very narrow and hence fertilization with boron need to be done carefully. Undesired and excessive use of B fertilizers leads to its high build up in soil and may cause deleterious effect on crop productivity. Boronated single super phosphate with 0.2% B was approved under FCO, but it was not accepted by farmers as the amount of boron was insufficient to meet crop needs. Now, as a modification, blending of 0.3% B with NPK 14-35-15 on weight basis has been advocated. Results of several field experiments suggest that efficiency of boronated NPK was high in crops like mustard, wheat, lentil, potato and coriander in acid soils, with sequence of response in the order: potato>wheat>coriander > mustard> lentil (Mandal and Martin, 2006).

Recovery of B applied in acid soil is about 36-54% due to its higher precipitation with hydroxides and enhanced adsorption in the hydrous oxides of Al and Fe etc. Liming acid soils often reduces the recovery of native and applied B by 15-40% (Saha and Singh 1997). Mandal (1995) found that the combined application of B with lime had the greatest effect on increasing the dry matter yield of wheat variety Sonalika by 39.3 and 55.8% at half and full doses of lime over no B and no-lime treatments. Boron left significant residual effect on yield of the following cowpea crop by 55.5 and 35.8% at half and full doses of lime as compared to the 13.3% increase in yield recorded with B alone.

Micronutrient foliar sprays

Basal application of micronutrients to soil through broadcast is commonly recommended to correct their deficiencies in most of the crops. It has the advantage of meeting micronutrient needs quite effectively besides the residual value. Foliar application of Fe and Mn has been found to be more efficient compared to soil application, because of the transformations of added $FeSO_4$ and $MnSO_4$ in soil to relatively unavailable forms. Foliar spray of B and Mo is also preferred, as it is found to be either at par or better than soil application. The efficiency of 0.5-1.0% ferrous sulphate solution sprays improves, when solution is sprayed un-neutralized along with 0.1% citric acid. Bhattacharya (1992) suggested sprays of 0.5%, zinc sulphate solution neutralized with 0.25% lime solution. If required, 2-3% urea can also be sprayed along with micronutrients. Use of suitable additives or surfactants like teepol or gum is required to promote or ensure higher sticking of solution on to the foliage. Studies showed that foliar spray of 0.25% borax + 0.5% zinc sulphate + 0.05% sodium molybdate solution at three weeks' interval twice or thrice in clear weather significantly increased the productivity of various crops over that with NPK alone.

Deficiencies of micronutrient in horticultural and plantation crops

Oxisols of Konkan region are characterized by low pH and high Fe, Mn and Cu content. Deficiencies of B and Zn limit the productivity of field as well as plantation crops. Mango and cashew are the prominent plantation crops and about 1.47 lakh hectares are planted with Alphanso mango alone which on an average yields 3 t ha^{-1}. But, physiological disorders like spongy tissue severely reduces mango yields and economy. Chavan and Gupta (1987) reported that spongy tissue had lower concentration of iron and copper but 4-5 times higher concentration of Zn as compared to healthy pulp. They found that when the iron concentration in the healthy pulp of spongy fruit approached a level of 60 ppm, zinc was not detectable but, when Fe concentration was 17 ppm, the Zn concentration increased to significantly higher level of 36.9 ppm in the spongy tissue. Thus, the Fe:Zn ratio in fruits decreased appreciably from 5 in the healthy fruits to only 0.77 in spongy tissues. This is attributed to the emergence of deficiencies of Zn and B in the bench terraces of soils of mango orchards to the extent of 87 and 90%, respectively (Mahajan 2001).

Banana is an important plantation crop of Tamil Nadu and application of lime to the extent of 25% of lime requirement to bananas grown on acid soils increased the bunch yield to 26.3% over control. Soil application of NPK along with foliar spray of N, P, K, Zn, Cu and B was the best treatment that gave a yield of 39.4 t ha^{-1} as compared to 22.6 t ha^{-1} in control (Suresh 1976).

Manganese toxicity seriously affected apple productivity in acid soils of Himachal Pradesh. Singh *et al.* (1977) found that 50% of the orchards grown in acid soils aggravated "apple measles", which is a nutritional imbalance induced by Mn toxicity . Application of 10 kg Zn ha^{-1} and 0.5% $ZnSO_4$ foliar spray twice at first pink bud stage in March and early flowering stage in May resulted in new flush formation that helped in increasing the fruit yield of apple by 20-50 kg fruits per tree per year.

Liming is recommended only for matured tea fields that are under regular plucking. It is recommended to lime such fields 4-6 weeks before application of NPK. The lime requirement is made taking into account soil pH, rainfall, yield level and length of pruning. Since, the liming material does not move laterally in the soil, 2-4 t lime per hectare should be uniformly broadcast over the soil and if required added in splits. The native Mg level of tea soils is generally low to medium and is not adequate to maintain high productivity of tea. Hence, application of dolomitic limestone (12.6% $MgCO_3$) once or twice in a pruning cycle is recommended as a regular practice to correct soil acidity and prevent emergence of Ca and Mg deficiencies in the soil. Mg deficiency is seen in the tea fields as yellowing of the old leaves and formation of mosaic of brown yellow tint between the veins in an inverted 'V' shape. This may be corrected by applying 200-400 kg $MgSO_4$ per hectare in two spits depending upon the severity of the deficiency.

Sulphur plays a vital role in mineral nutrition of tea. Deficiency of S is described as "tea yellow" in younger leaves, where the leaflet takes a striking yellow colour in the vein down to the finest branching standout prominently dark green. Sulphur deficiency is, however not common problem of tea soils because of regular use of ammonium sulphate in the manuring schedule. Requirement of S works out to 13.22 to 17.6 kg for every 1000 kg of made tea. This could be adequately met by soil application of 20% of the annual nitrogen requirement in the form of ammonium sulphate.

The tea soils are adequate in Fe, Mg, and Al. However, Zn deficiency is widespread in tea gardens and its correction is a sustainable practice for high tea productivity. Zn deficiency is characterized by very short internodes, chlorotic and small sickle shaped leaf and standard auxiliary shoots (rosettes). Affected bushes often become dormant, adversely affecting the yield. Foliar spray of zinc sulphate is highly efficient in correcting Zn deficiency. Foliar sprays of 0.5% zinc sulphate solution along with 500 ppm Mg, 100 ppm B and 10 ppm NAA help in better absorption and mobilization of micronutrients into the pluckable shoots. Some tea gardens also require application of 0.5 kg sodium molybdate per hectare.

Other trace elements of importance

Delineation of trace metals in soils around steel city of Bokaro (Jharkhand) and Durgapur (West Bengal) reveal wide variation among soils. Among the different chemical forms, exchangeable form of these metals were found to be higher than that of organically bound, carbonate or sulphide forms. Clay content and soil pH controlled bulk of the variation in trace metal content of these soils. Field experiments were conducted on groundnut, soybean and wheat with graded levels of Ni and Pb application. Results reveal wide variations in these crops to tolerate high trace metal levels in acidic soils. A drastic reduction in groundnut yield beyond 10 kg Ni /ha was noted in acidic soil of Jharkhand (Table 17).

Table 17 Effect of graded levels of Ni and FYM on groundnut yield and Ni concentration

Ni level (kg/ha)	Pod yield (q/ha)		Concentration (mg/kg)	
	+FYM	-FYM	+FYM	-FYM
0	13.19	10.89	0.20	0.22
5	10.88	10.72	2.11	2.41
10	10.88	9.31	2.42	2.56
20	4.35	3.65	2.49	2.63
40	2.93	0.97	2.58	2.61

Singh *et al.* (2007)

Amelioration of iron toxicity

Fe^{3+} commonly present in $Fe(OH)_3$ is reduced to Fe^{2+} and the levels of Fe^{2+} in submerged soils increases and a concentration ranging between 300-500 mg Fe^{2+} L^{-1} commonly occurs at pH below 5. This may be toxic to rice plants. In large areas particularly of Odisha and Meghalaya, toxicity of Fe and Mn to rice plants is a serious problem. Such high concentration of Fe^{2+} in soil solution not only per se causes toxicity but also induces Zn deficiency.

Iron toxicity can be alleviated by intermittent drainage, which favours conversion of Fe^{2+} to Fe^{3+} and reduces iron availability to the plant. Other management techniques include:

- Provision of deep drains around the field
- Checking the lateral seepage carrying reduced Fe by construction of check embankments across the slope and devising the soluble iron through diversion weirs.
- Liming soils as per lime requirement
- Application of high dose of K helps in increasing oxidizing zone around rice rhizosphere grown under Fe- toxic condition, thereby preventing reduction of Fe^{3+} to Fe^{2+}. This ultimately restricts the Fe uptake by rice and yield is improved.
- Balanced fertilization helps to alleviate iron toxicity.
- Application of P and FYM has been reported to be beneficial. Dipping rice seedlings in boronated SSP and FYM slurry before transplanting helps reduction of Fe toxicity.
- Utilization of varietal tolerance to iron toxicity is also an important strategy.

Important recommendations

1. Because of heavy leaching and runoff, the red and lateritic soils suffer from acute deficiencies of Ca and Mg. To prevent Ca and Mg deficiencies, application of farm yard manures and dolomite is recommended. Use of single superphosphate (13-20%Ca) or basic slag (19-24% Ca) may also be useful for arresting Ca deficiency

2. Use of gypsum, phosphogypsum and pyrites as sulphur source is advocated to combat the problem of S deficiency in red and lateritic acidic upland soils for increasing the productivity oilseeds and pulses. Application of S @ 20-30 and 30-40 kg S ha^{-1} particularly for pulses and oilseeds is highly remunerative in such soils.

3. Because of B and Mo deficiencies in acidic soils, yields of crops such as oilseeds, pulses and vegetables are badly affected. Application of these two elements, both as soil and foliar spray can prevent the deficiencies. Boron applied a Borax to soil @10-20 kg B ha^{-1} can prevent B deficiency. Application of 0.5 to 1.0 kg ammonium molybdate ha^{-1} is enough for arresting the deficiency of Mo. When the symptoms are visible in standing crops, application of these nutrients as foliar spray is recommended.

4. Zn deficiency can be prevented by selection of Zn efficient crop varieties, dipping seedlings or pre –soaking seeds in ZnO (2-4%) and application of Zn fertilizer (Zinc sulphate 10 Kg ha^{-1}) or foliar application of 0.5% Zinc sulphate solution.

5. To prevent Fe-toxicity in red and lateritic soils particularly for rice, selection of variety tolerant to Fe-toxicity, delayed transplanting of rice seedlings and balanced fertilization, particularly K may be followed.

Policy issues

For management of secondary and micronutrients in acidic soils (pH<5.5), the following issues need to be addressed:

- Soil testing needs to be popularized among farmers especially for diagnosis of secondary and micronutrient deficiencies in states. GIS-based soil nutrient maps should be used for planning and implementation of the programme.

- Organic manure and crop residues need to be added to soils in addition to fertilisers to promote Integrated Nutrient Management in crops.
- Secondary and micronutrient fertilizers should be used judiciously for higher benefits. Water saving technologies such as drip, sprinkler, fertigation etc. should be used especially for close spaced crops for improved efficiency of inputs.
- In low pH soils, every effort should be made to ameliorate the soil acidity. This should be followed by balanced use of NPK fertilisers and secondary and micronutrients based on soil tests.
- Efficient varieties of crops need to be identified for their tolerance to secondary and micronutrient deficiencies. These need to be promoted in regions known for such problems.
- Toxicities of nutrients in acid soils need attention especially in soils having pH below 5.0. Liming will help in reduction of this problem.
- Farmers in most cases are unaware about the problems of secondary and micronutrients in acid soil regions. Training and empowerment of farmers in this direction will help to create awareness.

References

Bhattacharya, B. K., Mukherjee, D., Ali, A.H., Trivedi, N.Bhatterjee, P.K. and Mandal, B (1998) Micronutrients in soils and plants of West Bengal, Directorate of Agriculture, Govt. of West Bengal .Calcutta pp 28.

Bhattacharya, S (1992) Micronutrient response of crops in soils of West Bengal.Proceedings of the workshop on micronutrients. Indo-British Fertilizer Educational Project. –Hindustan Fertilizer Corporation Ltd held at Bhubaneswar. pp. 183- 186.

Bhupalraj, G., Patnaik, M.C. and Khadke, K.M. (2005) Molybdenum status in soils of Andhra Pradesh. Progress Report. All India coordinated Research project on Micro and Secondary Nutrients and Pollutant Elements in soils and Plants of Andhra Pradesh. ANGRAU, Hyderabad. 25, 1-123.

Chavan A.S. and Gupta, D.N. (1987) Relationship of micronutrients with incidences of spongy tissues in Alphango Mango. Proceedings of the National Symposium on Micronutrients stress in crop plants. MPKV, Rahuri.

Das Gupta, D.K. and Bera, G.S. (1972) Effect of molybdenum on nitrogen utilizaton in rice. *Fertilizer News* 17(6), 89-90.

Ghosh, A.K., Sinha, T.K. and Chaudhary, S.R. (1992) Micronutrient status in soils of Teesta in West Bengal. Proceedings of the Workshop on Micronutrients. Indo-British Fertilizer Educational Project. –Hindustan Fertilizer Corporation LTD held at Bhubaneswar. pp. 118-124.

Grewal, J.S. and Trehan, S.P. (1990) Micronutrient deficiency and amelioration in potato. Central Potato research Institute, *Technical Bulletin* 23, 1-26.

Lal, Suresh (1999) Status and forms of magnesium in some major soil series of Chotanagpur and Santhal Parganas region and response of crops to its application. Ph.D. thesis submitted to BAU, Ranchi (Jharkhand).

Mahajan (2001) Status and distribution of micronutrients in relation to soil properties in lateritic soils under mango orchards in south Konkan. Ph.D. Thesis, BSKKV, Dapoli.

Mandal, B. (1995) Final Report of the ICAR Ad-hoc Project, Studies on the Chemistry of Boron and Molybdenum availability in Acid Alluvial Soils. BCKV, Kalyani, W.B., pp 1-138.

Mandal, B. and Martin, P. (2006) Final Report. Contractual Project on Response of Granubor to crops. BCKV, Kalyani, W.B., pp. 1-138.

Mathur, B.S., Sinha, H., Rana, N.K. and Lal, S. (1983) Use of calcium as a nutrient for legume crops in acidic red loam soils of Ranchi. *Journal of the Indian Society of Soil Science* 31:334-336.

Nayar, P.K., Misra, A.K., Rao, A.V.S. and Patnaik, S. (1980) Effect of nitrogen rates with and without foliar application of molybdenum on the yield, micronutrient content and uptake by rice plant. *IL Riso* 29: 35-41.

Rattan, R.K. and Goswami, N.N (2002) Essential nutrients and their uptake by plants, pp 309-332. In Fundamentals of Soil Science, G.S. Sekhon *et al.*, (eds) *Indian Society of Soil Science*, New Delhi.

Saha, J.K. and Singh, M.V. (1997) Effect of CaCO3 on the adsorption and desorption of boron in swell-shrink soils in Madhya Pradesh. *Journal of the Indian Society of Soil Science* 45: 57-61.

Samui, R.C.and Bhattacharya, A.K. (1973) Grain, straw and dry matter yields of the late sown wheat as affected by micronutrients and varieties. *Indian Journal of Agronomy* 24: 432-434.

Sarkar, A. K. and Singh, Surendra (2003) Crop response of secondary and micronutrients in acidic soils of India, *Fertiliser News* 48 (4) : 47-54.

Sarkar, A.K., Singh K.P., Singh, B.P. and Singh, R.P.(2000) Long term effect of fertilizer, manure and amendment on crop productivity and soil fertility. SSAC (BAU) *Tech. Bull.* No.2/2000.

Sharma, S.P., Singh, M.V., Subehia, S.K., Jain, P.K., Kaushal, V.and Verma, T.S.(2005) Research Bulletin No. 2, IISS, Bhopal and CCSHPKV, Palampur, H.P. pp. 1-88.

Singh, A.L., Basu, M.S. and Singh, N.B. (2003) Potentials of groundnut in north eastern states of India: Mineral disorders of groundnut. Technical Bulletin NRCG, Junagarh.pp 80-82.

Singh, D., Leelawathi, Krishnan, C.R. and Sarup, S. (1979) Responses of crops to micronutrients fertilization. Indian Agricultural Statistical Research Institute (ICAR), New Delhi. *Research Bulletin* 1:1-16.

Singh, M.V. (2001a) Evaluation of micronutrient status in different agro-ecological zones of India. *Fertiliser News* 46(2): 25-42.

Singh, M.V.(2001b) Responses of micronutrients to crops in different agro- ecological regions – Experiences of AICRP on micronutrients. *Fertiliser News* 46(10): 43-49.

Singh, M.V. (2007) Problems of micro and secondary nutrients in acid soils of India and their management. *Bulletin of the Indian Society of Soil Science* .25 pp31.

Singh, M.V.(2000) Micro- and Secondary- Nutrients and Pollutant Element Research in India. Progress Report of the AICRP of Micro-and Secondary- *Nutrients and Pollutant Elements in soils and Plants.* 29. 1-122.

Singh, M.V., Wanjari, R.H. and Adhikari, Tapan (2004) Nutrient dynamics, crop productivity and sustainability of long –term fertilizer experiments in India. AICRP Long –term Fertiliser Experiments, *Indian Institute of Soil Sciences*, Bhopal. pp. 1-120.

Singh, R., Singh, M.V and Pant, P. C. (1977). Manganese status in the hill soil of Pithoragarh. *Indian Journal of Agricultural Chemistry*. 10: 35-41.

Singh, R.P., Sarkar, A.K.,Mishra, B.and Rusia. D.K. (2007) Technologies for efficient management of soil, water and plant nutrients for crop production in Jharkhand. SSAC (BAU) *Tech. Bull.* 3:33.

Sreedharan, C. and George, C.M (1969). Effect of micronutrients on paddy in a red loam soil of Kerala. *Agricultural Research Journal of Kerala* 7 (2): 61-66.

Subba Rao, A. and Adinarayana (1995) Molybdenum Research and Agricultural Production. In: Micronutrient Research and Agricultural Production (H. L. S. Tandon, Ed). FDCO, New Delhi. 1: 60-77.

Suresh, S. (1976) Nutrition of rice and banana in soils prone to iron toxicity in high rainfall zone of Tamil Nadu. PhD Thesis, Tamil Nadu Agricultural University, Coimbatore.

Tandon, H.L.S. (Ed) (1995) Micronutrient research and Agricultural production. Fertlizer Development and Consultation Organization, New, Delhi. pp 164+ix.

Tandon, H.L.S. (2011). Sulphur in soils, crops and fertilizers. Fertlizer Development and Consultation Organization, New, Delhi. pp 14-15.

10

Paper Mill Sludge as an Acid Soil Ameliorant

Sushanta Kumar Pattanayak

Soils become acidic due to natural and artificial (anthropogenic) reasons. The natural reasons include their origin from acidic parent materials namely granite, rhyolite, pegmatite and dacite etc. and areas under high rainfall (rainfall exceeding evapotranspiration) removing basic cations through leaching and erosion particularly in coarse textured soils with undulated, hilly and sloping topography. Under natural condition, accumulation of organic matter and its decomposition also releases organic acids into the soil system, turning soils acidic. The anthropogenic factors responsible for turning soils acidic, are (i) removal of basic cations through crop cultivation, (ii) continuous use of acid forming fertilizers and (iii) acid rain in industrial areas.

The acidic soils develop physical, chemical, nutritional and biological constraints for crop production in terms of soil crusting (affecting seed germination), high infiltration rate, low water holding capacity, high permeability (all affecting water availability for crop production), low pH, low cation exchange capacity, (due to dominance of 1:1 type of clay) low base saturation (16-67 %), high Al, Fe and Mn saturation percentage, high P fixing capacity

(92 %) (Pattanayak and Misra, 1989), poor availability of essential plant nutrients like Ca, Mg, P, Mo, B and Si, poor microbial activity and biologically mediated nutrient transformation processes, poor N_2 fixation due to poor *Rhizobial* activity etc. All these lead to poor root growth, root injury, root discoloration, restricted root branching, damaged root membrane causing poor uptake of plant nutrients.

Restoration of lost basic cations,amelioration of acidity, supplementation of deficient nutrients as per crop requirement, judicious use of chemical fertilizers, proportionate use of organics can help managing acid soils. Liming of acid soil is the way to raise pH, base status, cation exchange capacity, inactivate Al, Fe and Mn in soil solution and reduce P fixation (Panda and Koshy, 1982, Misra *et al.*, 1989, Sahu and Patnaik, 1990, Mishra and Pattanayak, 2002).

Liming materials

The materials commonly used for liming acid soils are oxides, hydroxides, carbonates and silicates of calcium and magnesium. The presence of Ca and Mg in liming materials is important,but in addition ,the accompanying anion must be one that will reduce the activity of hydrogen and hence aluminium in the soil solution. A number of liming materials are available for agricultural use(Table 1).

Reaction of Lime in Soil

The reaction mechanisms of liming materials in acid soils are complex. Lime reactions in soils depend upon the nature and fineness of the liming materials. Liming materials are classified as calcitic ($CaCO_3$), dolomitic [$CaMg(CO_3)_2$] or mixture of the both. These are sparingly soluble in pure water, but become soluble in water containing CO_2. Greater the partial pressure of CO_2 in the system, the more soluble is the liming material. **The reaction of $CaCO_3$ in soil can be written as:**

$$CaCO_3+H_2O+CO_2 \rightarrow Ca(HCO_3)_2 \rightarrow Ca^{2+}[\text{Cation exch. site}]+ HCO_3^- + OH^-$$

$$H^+ \text{ (soil solution)} + OH^- \rightarrow H_2O$$

$$H^+ \text{ (soil solution)} + HCO_3^- \rightarrow H_2O + CO_2 \uparrow$$

The hydrogen ions (H^+) in the soil solution react to form weakly dissociated water, and the calcium (Ca^{2+}) ion from liming material

Table 1: Liming materials, their composition and reactivity

Sl.No.	Local name Natural deposit	Chemical formula	CCE (%)	Ca	Mg (%)	Na	K
1.	Calcite	$CaCO_3$	100	40	-	-	-
2.	Dolomite	$Ca\text{-}Mg(CO_3)_2$	192	55	3.0	0.22	0.12
3.	Calcitic limestone		75-100				
4.	Dolomitic limestone		75-108				
	Alternative agricultural liming materials						
5.	Quick lime/ burntlime/ Oxide lime	CaO	178		71.4		
6.	Hydrated lime	$Ca(OH)_2$	135		54		
	Marl	$CaCO_3$+ Aluminium silicate clay	80				
7.	Acetylene lime	$CaC_2+H_2O \rightarrow Ca(OH)_2 + C_2H_2$	125				
8.	Egg Shells	$CaCO_3$ + others	60				
	Industrial wastes						
	Highgrade						
9.	Basic slag	Phosphate and Silicate of lime	50-70				
10.	Boiler Wood Ash		38-70				
11.	Coment kiln dust		60-80		0.6-2.8	1-28	4-10 (s)
12.	Papermill sludge (Emami)		37.4	12.2	1.4	0.02	0.03
	Papermill sludge (Chaudwar)		61-84	20.2-29.0	2.3-14.8	0.04-0.34	0.04 (k)
	Papermill sludge (J.K.Rayadaga)		55-92.3	20.8-45.6	2.2-2.4	0.23-1.4	0.02
	Papermill sludge(Brajarajnagar)		66.6-84.1	27-50	2.1-13.2	0.39-2.5	0.1
	Papermill sludge(Koraput)		84.8-90.1	23.5-46.6	-	0.8-1.2	0.02-0.2
	Papermill sludge(Jeypur)		52-86.6	20.6-33.0	-	0.3-0.9	0.04
13.	Press mud (Aska-Sugar Factory)		28	4.5	16.5	0.01	0.2
	Low Grade						
14.	Poultry:Broiler litter		10	3.6	0.8	0.12	-
15	Layer litter		20	7.0	1.7	0.22	-
16.	Municipal: Biosolid		08	1.8	0.32	0.24	0.3

undergo cation exchange reactions. As a result, acidity in soil is ameliorated and the base saturation per cent of the colloidal material is increased.

In acid soils, concentration of the H^+ ions in solution is related to the hydrolysis of Al^{3+} or hydroxyaluminium or hydroxy–Fe^{3+} ions. Their hydrolysis in turn, is influenced by the amount of clay and organic matter in the system. Continuous removal of H^+ from the soil solution ultimately results in the precipitation of Al^{3+} and Fe^{3+} ions and their replacement on the adsorption sites with Ca^{2+}/Mg^{2+} and other basic cations. The exchangeable Al^{3+} and the hydroxyluminium and hydroxyl – Fe^{3+} are precipitated as $Al(OH)_3$ and $Fe(OH)_3$.

Reaction with exchangeable Al

$$\underset{\square}{2Al} + 3\,CaCO_3 + 6\,H_2O \rightarrow \underset{\square}{3Ca} + 2Al\,(OH)_3\downarrow + 3H_2CO_3$$

$$H_2CO_3 \rightarrow H_2O + CO_2\uparrow$$

Reaction of $CaSiO_3$

Calcium, the cation constituent of $CaSiO_3$ function simply as the calcium cation found in the liming materials. The silicate anion function in different way as compared to carbonate anion. The silicate anion does not produce CO_2 like carbonate, that release to atmosphere as greenhouse gas or remain in soil potentially contributing to acidic conditions. The reaction of calcium silicate in acid soil is as follows:

$CaSiO_3 + 2H^+ + H_2O$ [Soil solution]$\rightarrow Ca^{2+} + H_4SiO_4$ [monosilicic acid]

The H_4SiO_4 acid complexes with toxic Al^{3+} and Al-hydroxy species form nontoxic aluminosilicate and hydroxyalumino silicate compounds, which precipitate at the root zone.

Increase in pH after liming:

During the process of amelioration of acidity, the pH of acid soil is increased, which can be illustrated as below:

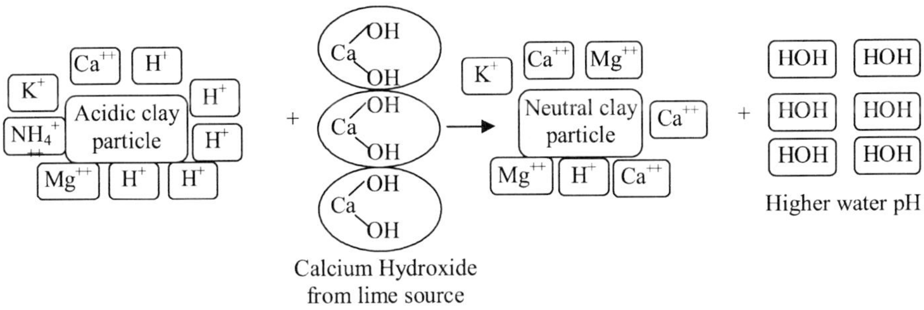

Comparative performance of liming materials

Liming acid soils is an established and efficient management practice. Due to industrial use of $CaCO_3$, its cost is increasing over the years, which is increasing the cost of cultivation and discouraging the farmers for its use. Moreover, amelioration of acidity once in soil is not a permanent proposition. After each harvest of the crop, soils turn acidic. Applied lime also get lost to deeper layers after each rain or irrigation. The industrial wastes provide better options for liming of acid soils (Table 1). Among the sources, paper mill sludge and basic slag are important.

Lime neutralize and reduce soil acidity (Pattanayak and Pandey, 2012, Fig.1) there by raising the pH of acid soils, improve base saturation and cation exchange capacity (Table 2), inactive Al, Fe and Mn in soil solution by precipitating (Table 3), reduce phosphate fixation, influence N mineralization keep K in available mode (Table 4).

Table 2. Change in CEC [cmol (p^+) kg^{-1} soil] and base saturation per cent after liming and soil*

Lime source (LR) and dose	Days of crop growth					
	0		30		70	
	CEC	BS (%)	CEC	BS (%)	CEC	BS (%)
No lime	13.4	39.6	13.3	26.6	13.2	23.4
$CaCO_3$ @ 0.2 LR	13.4	39.6	14.6	45.1	14.0	33.2
Paper mill sludge @ 0.2 LR	13.4	39.6	15.1	46.1	14.8	31.2
Stromatolyte @ 0.2 LR	13.4	39.6	14.9	44.7	14.2	30.7

*Sandy loam soil, pH 4.47, LR to raise pH to 6.5:2.3 t CaCO3 ha^{-1} (Pattanayak and Pandey, 2012)

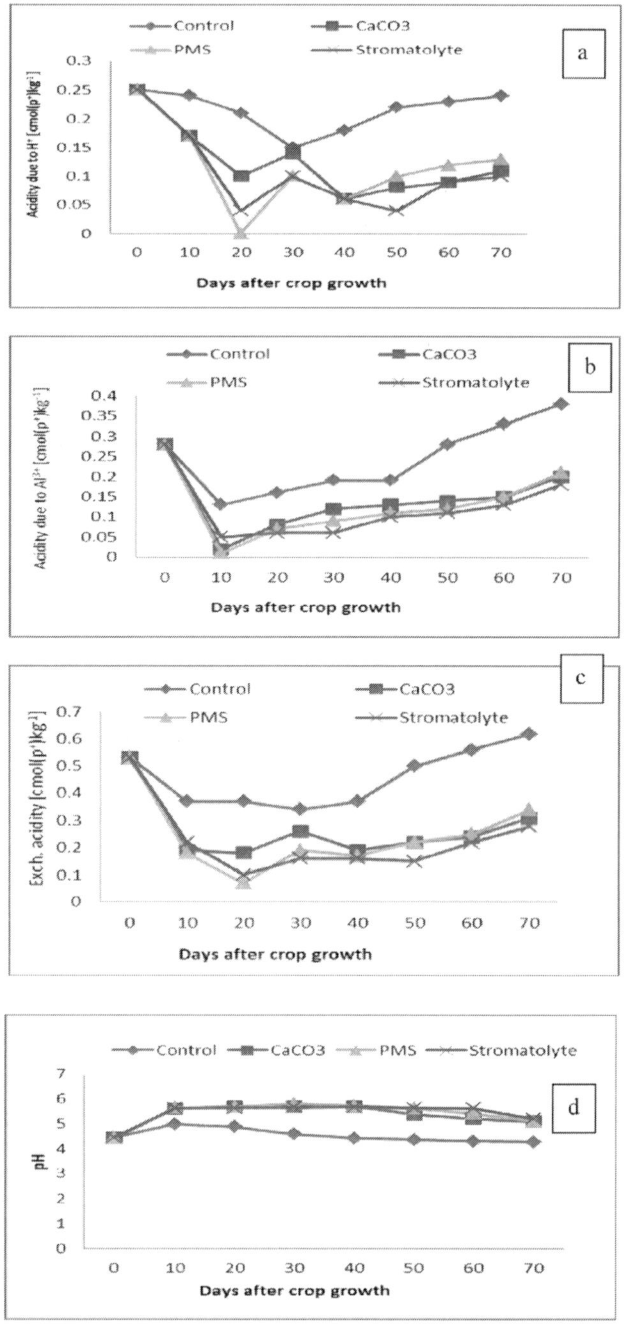

Fig.1. Influence of liming of acid soil on (a) Neutralization of acidity due to H^+, (b) Inactivating/reducing acidity due to Al^{3+}, (c) Neutralization of exchange acidity and (d) Raising pH of acid soil

Table 3. Cation composition of exchange sites as related to the soil pH value

pH	Al	Ca+ Mg + K cmol(P^+) kg^{-1} soil	Effective CEC
4.5	0.91	0.22	1.16
5.4	0.34	0.91	1.35
6.0	0.09	1.59	1.78

Pattanayak and Pandey (2012)

Table 4. Influence of liming with paper mill sludge on soil acidity and availability of major plant nutrients in soil

Treatments	Soil reaction	Soil acidity (cmol (P^+)kg^{-1} soil)			Avalable nutriments (kg ha^{-1})			Exch. Ca [(cmol(p^+)kg^{-1}]
	pH	Total	Exch.	Al	N	P	K	
Control	5.4	3.0	0.17	0.06	349	50	138	6.1
Lime (PMS)	6.9	1.9	0.15	0.00	359	66	148	8.1
FYM (F)	5.9	2.4	0.16	0.00	384	70	160	7.0
PMS + F	6.8	1.9	0.14	0.00	370	60	160	7.9

Mohanty and Pattanayak (2000)

While liming is advocated for correction of soil acidity and increasing CEC of soil, the plant factor should also be considered, because response to lime application varies with the crops. The dicots remove more Ca and Mg and respond more to lime application than monocots (Table 5). Integrating organics with lime (PMS) improves the CEC (by creation of more negative charge sites for holding cations preventing leaching losses).

Table 5. Influence of liming of soil with PMS on CEC [cmol (P^+) kg^{-1} soil] of soil subjected to cropping with monocot (maize) and dicot (cowpea) crops.

Dose of PMS (LR)	Initial	Monocot (maize)				Dicot (cowpea)			
		15 DAS		90 DAS		15 DAS		90 DAS	
	0 DAS	F_0	F	F_0	F	F_0	F	F_0	F
0	3.9	3.8	6.4	3.0	4.2	3.6	6.8	2.9	6.0
0.1	3.9	6.4	8.0	6.0	6.8	6.4	8.2	4.8	6.4
0.2	3.9	9.4	9.6	7.9	8.0	8.0	9.0	6.4	7.6
0.3	3.9	11.2	11.6	9.2	9.6	9.6	10.4	8.0	8.4

*Soil pH: 4.47, loamy sand texture, organic carbon 2.8 g kg $soil^{-1}$ (Mishra and Pattanayak, 2002)

Even though the CEC is found to decrease gradually towards the harvest of crop, but higher status of it is maintained in lime and organic manure amended soil.

Crop response to PMS application in acid soils

Crops differ in their response to lime application. High lime responsive crops include cowpea, pegeonpea, soybean and cotton. Medium responsive crops include maize, wheat, groundnut, bengal-gram, lentil, pea, cabbage, cauliflower, knolkhol, bean, ladies finger etc. Upland paddy, potato and small millets are low lime responsive crops (Mandal *et al.* 1966 and Sahu *et al.,* 1988 Pattanayak *et al.,* 2008).

Rice

Coastal acid saline soils (pH 5.10-5.15) low in calcium [exch.Ca 1.65 cmol$(P^+)kg^{-1+}$] responded significantly to lime application through PMS @ 5 q ha^{-1} integrated with Blue green algae (BGA) application @ 15 kg ha^{-1} after 7 days of rice transplanting yielding 10.5 and 18.1 per cent higher grain yield with higher uptake of major nutrients compared to the yield with soil test based application of fertilizers of 4.3 t ha^{-1} and uptake of N, P and K of 110,11 and 140 kg $ha^{-1,}$ respectively (Table 6). Even though pH is raised after submergence of rice soil, for calcium nutrition of rice crop, lime is beneficial when grown under acid soil condition having low exchangeable calcium content.

Table 6. Grain yield and nutrient uptake by rice as influenced by liming with PMS

Treatment	Mean grain yield (t ha^{-1})	Nutrient uptake (kg ha^{-1})			
		N	P	K	Ca
Soil test based dose (STD)	4.30	110	11	140	37
STD + PMS @ 5 q ha^{-1}	4.75(10.5)	120	13	163	44
STD + PNMS + BGA @ 15 kg ha^{-1}	5.08(18.1)	138	14	188	50
CD (P = 0.05)	0.09	8.8	0.8	12.1	3.1

*Data in the parenthesis indicate per cent increase over STD
(Pattanayak, 2012)

Maize

Maize crop grown in acid sandy loam soil (pH: 4.7) with lime requirement by woodruff buffer method of 3.4 t ha^{-1} responded (16. 7 %) significantly to liming through PMS application @ 0.1 LR mixed with FYM applied @ 5 t ha^{-1} at the time of planting in maize rhizosphere compared to the green maize cob yield of 14.4 t ha^{-1} due to 100 % NPK + FYM alone. Not only the yield was influenced by liming, but also, the recovery of major nutrients increased considerably, N from 44.3 to 83.9 per cent, P from 41 to 52 per cent and K from 98 to 142 per cent, respectively and the B: C ratio from a level of 2.52 for every rupee investment with full dose of NPK to 3.17 making liming a profitable practice (Table 7).

Table 7. Green maize cob yield (t ha^{-1}), nutrient recovery and B:C ratio as influenced by PMS application.

Treatment	Green cob (t ha^{-1})	Recovery (%)			B:C ratio
		N	P	K	
\Absolute control	6.48	-	-	-	1.19
100% NPK (Soil test)	12.00	44.3	41.0	98.0	2.52
100 % NPK + FYM	14.40	66.3	44.0	113.0	2.76
100 % NPK + F + PMS	16.80	83.9	52.0	142.0	3.17
CD (P = 0.05)	0.41	-	-	-	-

Pattanayak *et al.*, (2012)

Finger millet (Ragi)

The hardy, drought tolerant and calcium rich crop, finger millet responded significantly to soil amelioration with PMS in acid *Inceptisols* (pH 5.4) of Bhubaneswar yielding 31.5 per cent higher grain yield with higher mineral content particularly Ca, lower fat and carbohydrate content in grains compared to other essential input integrations. Soil amelioration had considerable influence on apparent recovery of nutrients (Table 8) by the crop.

Table 8. Influence of soil amelioration with PMS on grain yield, quality and nutrient recovery by finger millet

Treatments	Grain (tha^{-1})	Recovery of nutrients			Grain quality			Ca (mg/ 100 g)
		N	P	K	(%) Fat	Carbohydrate (%)	Mineral (%)	(%)
Absolute control	0.57	-	-	-	2.6	81	2.70	331
100% NPK (Soil test)	1.80	48.0	23	64	2.3	84	2.81	349
100 % NPK + FYM	2.22	78.3	24	92	2.2	82	2.86	360
100 % NPK + F + BF	2.82	110.2	44	134	1.8	77	2.99	370
100 %NPK + F + BF + PMS	3.71	130.3	52	165	1.6	73	3.18	401
CD (P = 0.05)	0.62	-	-	-	0.21	2.3	0.08	24

Pattanayak (2012)

Response of Pulse crops to PMS application

Greengram

The greengram, a multipurpose pulse crop of short maturity period, is preferred for multicropping systems. Being a legume crop, it fixes atmospheric N_2. The pulse crop, greengram responded significantly to PMS application as acid soil (pH 5.4) ameliorant, as one of the components of INM in *Inceptisols* of Bhubaneswar. Its integration in the INM practice, not only resulted in highest seed yield (1160 kg ha^{-1}), but also decreased half filled seed percentage (9.6 per cent) and extra N_2 gain of 55 kg ha^{-1} with recovery of P, K and S by 35.8, 74.2 and 38.8 per cent compared to the seed yield of 560 kg ha^{-1}, per cent half filled seed of 24.8 per cent, extra N_2 gain amount of 18 kg and recoveries of 12.1, 48.6 and 24.6 per cent due to inorganic fertilizers alone (Table 9).

Table 9. Influence of liming with PMS and seed yield, extra N_2 gain and nutrient recovery by green gram (cv. PDM 139)

Treatments	Seed yield (kg ha^{-1})	Half-filled Seed (%)	Extra N_2 grain (kg ha^{-1}	Apparent nutrient recovery (%)		
				P	K	S
Absolute control	320	24.8		-	-	-
Soil test NPK	560	23.1	18	12.1	48.6	24.6
Soil test NPK + FYM	670	21.5	26	16.1	50.5	32.2
NPK + F + *Rhizobium*	910	15.2	39	25.8	65.0	37.7
NPK + F + R + PMS	1160	9.6	55	35.8	74.2	38.8
CD (P=0.05)	80	1.01	5.0	-	-	-

(Pattanayak 2012)

Fertilizer dose : $N-P_2O_5 - K_2O-SO_4-S$: 20-20-40-13 kg ha^{-1}.

*Seed inoculation with *Rhizobium* and treatment with 10 g Na Molybdate per 25 kg seed * Lime requirement to raise pH to 6.5: 3.2 t PMS ha^{-1}

Blackgram

Like greengram, the blackgram is also a multipurpose pulse crop. Generally, it is grown in uplands, which are acidic in reaction. Acid soil (pH 5.3) ameliorated with PMS @ 0.2 LR brought significant improvement in plants (nodule number, unit nodule weight and nodule N) and seed yield (Table 10) compared to all other integrated (seed inoculation with *Rhizobium*, application of soil test based fertilizers with FYM) as well as sole treatments, indicating role of soil amelioration on increasing the yield potential of blackgram crop. Soil amelioration with PMS increased seed yield by 16 per cent compared to NPK + F + R treatment. Integration of

PMS, FYM and fertilizers application with seed inoculation increased the yield of blackgram by 134 per cent compared to FP (320 kg ha^{-1}).

Table 10. Acid soil amelioration with PMS as an integral component of INM practice for yield improvement of blackgram

Treatments	Nodule properties			Seed yield (kg ha^{-1})
	Nodule No.	Unit nodule weight (mg)	Nodular N (mg)	
Absolute control (FP)	40	6.1	3.02	361
Rhizobium (R)	60	8.9	8.59	501
NPK	53	7.7	5.22	544
NPK + R	57	9.7	7.19	605
NPK + F	74	7.8	7.62	630
NPK + F + R	82	8.0	9.25	728
NPK + L + F +R	95	8.2	10.60	844
CD (P = 0.05)	6.5	0.4	0.45	21

(Pattanayak, 2012)

* Absolute control reflect farmers practice

Arhar (Red gram)

The sole Arhar crop (cv. UPAS-120) responded significantly to PMS application @ 0.2 LR, resulting in 44 per cent higher seed yield compared to local farmers practice seed yield of 1050 kg ha^{-1} with extra N2 gain of 33.2 kg ha^{-1} (Table 10). The application of PMS integrated with soil test based application of fertilizers increased the seed yield of Arhar by 23.1 per cent compared to the seed yield due only to PMS i.e., 1640 kg ha^{-1}. The extra N_2 gain was to the extent of 68.8 kg ha^{-1}, with recovery of P, K and S of 36, 102.3 and 17.1 per cent, respectively (Table 11).

Table 11. Influence of PMS application on sole Arhar crop (continuously grown for 3 years)

Treatments	Seed yield (kg ha^{-1})	Extra N gain (kg ha^{-1})	Response of nutrients (%)		
			P	K	S
Farmers practice (FP)	1050	-	-	-	
FP + PMS	1510	33.2	-	-	
Soil test dose (STD)	1640	41.2	18.7	67.3	
STD + PMS	2020	68.8	36.0	102.3	17.1
CD (P = 0.05)	80	-	-		

Farmers field experiment: 18 (3 years), pH 5.5, LR for pH 6.5 = 4 t ha^{-1}. Texture: sandy – sandyloam, OC: < 5 g kg^{-1} soil, N, P, K status: Low-Medium, PMS: Paper mill sludge- CCE: 60%, FP = No fertilizer, STD = 20-40-40 kg N-P_2O_5-K_2O ha^{-1}(Misra and Pattanayak, 2003)

Oilseed crop

Oilseeds also need better growing environment for good yield. Under acid soil conditions, in addition to nutrient deficiencies, seed / pod filling is hampered leading to yield reduction.

Groundnut : The acid soils ameliorated with PMS applied @ 0.2 LR mixed with FYM in the rhizosphere at the time of sowing of groundnut , increased the pod yield by 56 and 25 per cent compared to the yield due to FP and STD of 800 and 1430 kg ha$^{-1,}$ respectively. Amelioration of acidity and Ca supplementation through liming practice increased the shelling per cent by 6.1 to 6.8 per cent compared to that without liming. Soil amelioration also helped in the extra N gain and recovery of nutrients (Table 12).

Table 12. Crop productivity, nutrient uptake and recovery by groundnut crop as influenced by PMS application.

Treatments	Pod yield (kg ha^{-1})	Shelling (%)	Extra N gain (kg ha^{-1})	Recovery of nutrients (%)		
				P	K	S
Farmers practice (FP)	800	59.3	-			-
FP + PMS	1250(56)	63.3(6.8)	52.8	-	-	-
STD	1430	62.8	63.9	37.7	204	14.2
STD + PMS	1790(25.2)	66.6(6.1)	117.1	58.5	296	24.2
CD (P = 0.05)	90	1.0	-	-	-	-

Data in the parenthesis indicate per cent increase compared to respective un-limed treatment. (Misra and Pattanayak, 2003)

Sunflower: Acid soils (pH : 4.7-5.1) ameliorated with PMS as lime source along with soil test based fertilizers application, integrated with bioinoculation (*Azotobacter*, *Azospirillum* and PSB) , significantly increased the seed yield and oil content but decreased the chaff production in farmers fields (Table 13) in coastal acid saline condition.

Table 13:.Influence of liming acid soils with PMS on seed yield and quality of sunflower crop

Treatments	Seed (kg ha^{-1})	Chaff (kg ha^{-1})	Oil (%)
Farmers practice (FP)	1680(-14)	75(+25)	33.3(-4)
Soil test based (STD)	1940	60	34.5
STD + PMS @ 0.2 LR	2280(+17)	43(-28)	35.1(+2)
STD + BI + PMS	2430(+25)	41(-32)	35.4(+3)
CD (P = 0.05)	130	10	-

* Data in the parenthesis indicate per cent decrease (-) or increase (+) compared to soil test dose of fertilizers application. (Pattanayak, 2012)

Mustard: The mustard crop under farmers field condition responded significantly to the organic (FYM) and inorganic (PMS) amelioration integrated with soil test based fertilizer application yielding 28 and 48 per cent higher seed yield compared to the yield of 640 kg ha^{-1} without amelioration. Combining the soil amelioration measure almost doubled the seed yield of the oilseed crop. Similar response (3.3 to 6.2 per cent increase) was observed in the oil content (Table 14) .

Table 14. Seed yield and oil content of mustard as influenced by organic and inorganic soil amelioration.

Treatments	Seed yield (kg ha^{-1})	Oil content (%)
Soil test based fertilizer (STD)	640	33.8
STD + FYM (5 t ha^{-1})	820(28)	34.9(3.3)
STD + PMS @ 0.2 LR	950(48)	35.4(47)
STD + FYM + PMS	1260(97)	35.9(6.2)
CD (P = 0.05)	108	0.4

(PRC Report, 2006-2010)
No.of farmers = 24, Soil pH 5.0-5.2, LR to raise pH to 6.5= 6.8 t $CaCO_3$ ha^{-1}
Exch Ca: 1;1-1.21 [cmol (P^+) kg^{-1} soil]., STD based fertilizers : 30-20-30 kg N-P_2O_5-K_2O ha^{-1}.

Arhar-Groundnut intercropping system

Generally, the productivity with appropriate intercropping systems is more than their sole crops, where the intercrops not only benefit individual crops, but also the counter crops,with better use of space, time and agro- inputs . This acts as an insurance for the farmers especially of dry land areas .

Intercropping of Arhar:groundnut (2:6) in acid upland soils, responded significantly to soil amelioration with PMS applied @ 0.2 LR, integrated both with farmers practice as well as soil test based dose of fertilizers in increasing crop yields, crop quality, extra N gain and recovery of applied nutrients compared to unlimed soils (Table 14).

Soil amelioration also improved crop yields by 46 per cent, shelling per cent of groundnut by 6 per cent, its oil content by 3.4 per cent, protein content in Arhar seed by 6 per cent, extra N gain by 70.6 kg ha^{-1} compared to the corresponding values with FP of 1484 kg ha^{-1} of crop yield, shelling % 56.2, oil content of 44.7 per cent, 21.6 per cent of protein, respectively. The crop performance improved significantly with soil test based dose of fertilizers compared to FP or FP + PMS (Table 15).

Table 15. Influence of PMS on crop productivity, crop quality, nutrient gain and apparent nutrient recovery by Arhar-groundnut intercropping system under farmers field conditions

Treatments	Crop yield (kg ha^{-1}			Shelling	Oil %	Protein	N gain	Recovery (%)		
	G.nut	Arhar	G.nut equivalent	(%) G. nut	G. nut	(%)Arhar	(kg ha^{-1})	P	K	S
Farmers practice (FP)	810	770	1484	56.2	44.7	21.6	-	-	-	-
FP + PMS	1200	1100	2162	59.7	46.2	22.4	70.6	16	70	8
	(48)	(43)		(6)	(3.4)					
STD	1530	1670	2991	60.4	46.6	23.3	89.1	25	115	15
STD + PMS	1880	2040	3665	64.0	47.9	24.4	160.6	47	163	24
	(23)	(22)	(23)	(6)	(2.8)	(4.7)				
CD (P = 0.05)	160	140	290	1.4	0.57	-	-	-	-	-

*Data in the parenthesis indicate per cent increase over their respective control treatments (Pattanayak *et al.*, 2011)

Integration of PMS application (0.2 LR) with half the recommended dose of chemical fertilizers not only produced yields at par with full dose of fertilizers but also improved quality of produce with extra N gain and better recovery of applied nutrients (Table 16). This provides better option for resource poor farmers. The results corroborate with the findings of Sudharmaidevi *et al.* (2012).

Table 16. Effect of FYM application with PMS and half of the fertilizers on performance of Arhar and groundnut crops

Treatments	Crop yield (kg ha^{-1})		Shelling %	Oil (%)	Protein (%)	Extra N gain
	G. nut	Arhar	Groundnut crop		(Arhar)	(kg ha^{-1})
100% NPK	2250	1200	65	44.9	30.0	129.4
50 % NPK + PMS	2360	1220	64	44.6	28.5	129.6
50 % NPK + FYM + PMS	2660	1460	67	45.9	29.9	195.6
CD (P=0.05)	193	142	1.3	0.8	1.7	-

(Misra and Pattanayak, 2003)

Fruit crop

Mango

Even though mango is grown in acid soils, but liming of mango rhizosphere with PMS application @ 0.5 LR (12.4 kg tree^{-1}) along with fertilizers, decreased carbohydrate and TSS content in mango panicle, C: N ratio in mango pulp and increased N (%) in panicle, TSS (%) in mango pulp along with fruit yield significantly both under unfertilized and fertilized condition in lateritic soils. Soil amelioration improved the yield as well as quality of the crop with better economic return (Table-17).

Table 17. Influence of liming of mango crop with PMS on fruit yield and quality

Treatment	Mango panicle			Mango fruit pulp			Economic
	Carbohydrate (%)	N (%)	TSS (%)	C:N	TSS (%)	Yield (kg/plant)	benefit (Rs./tree)
L_0F_0	15.1	0.54	2.15	27.4:1	10.12	28.6	143
L_0F_2	14.7	0.62	2.10	23.7:1	12.95	46.4	172
L_1F_0	14.9	0.57	1.95	26.1:1	11.50	42.8	194
L_1F_2	14.4	0.64	1.90	22.5:1	15.55	68.4	242
CD (P = 0.05)	0.12	0.03	0.04	-	0.96	9.8	-

Variety: Latsundari, 15 years old tree, soil texture clay loam, soil pH -5.0, OC -2.6 g kg^{-1}soil^{-1}, PMS cost Rs 60/50 kg bag and mango sale price Rs.50/kg), F = N-1000g, P_2O_5 – 50 kg, K_2O-1000 g/ tree. L. 12.4 kg PMS/tree (Bal *et al.*, 2008)

Vegetable crops

Vegetable crops growing in acid upland soil condition responded differently to soil amelioration with PMS, (ranging from 7 to 68 per cent, Table 18). Research results are in support of application of PMS @ 0.2 LR to acid soils growing vegetable crops mixed with FYM at least 2/3 days before planting of seedlings integrated with soil test based inorganic nutrients use. The research result on garden pea crop grown in hilly areas of Himachal Pradesh support the above findings (Sharma *et al.* ,2012).

Table 18. Response of vegetable crops to acid soil amelioration with PMS

Crops	Yield (kg ha^{-1})		Response (%)
	Soil test dose	Soil test dose + PMS	
Tomato (Hybrid)	28.9	32.9	14
Tomato (High yield)	16.4	20.3	24
Brinjal	31.8	36.6	15
Okra	5.71	6.3	11
Capsicum	40.2	43.1	7
Garden pea	10.8	12.6	17
Cucumber	14.8	16.8	14
Pointed gourd	32.5	35.4	9
Spine gourd	34.7	38.7	12
Cowpea (trailing)	18.0	30.3	68
Radish	24.2	26.5	10
Cauliflower	11.2	13.6	21
Potato(hilly tribal area)	14.0	15.2	09

(Sharma *et al.*, 2012)

Greenmanure crops

Green manuring is an age old practice for increasing soil fertility. The efficiency of green manure crop can be increased by liming. Ameliorating acid soil with PMS @ 0.2 LR dose at the time of sowing of green manure crop increases biomass production, N-gain and recycling of P, K, Ca, Mg after incorporation compared to unlimed green manuring practice (Table 19).

Table 19. Influence of liming of Dhanicha crop on green biomass, quantity of N_2 fixation and nutrient cycling.

Treatment	Biomass (t ha^{-1})	N gain (kg ha^{-1})	Recyclable nutrients (kg ha^{-1}) P	K	Ca	S	Mg
No liming	3.43	26.7	3.0	6.9	8.2	5.5	3.4
Liming with PMS @ 0.2 LR	6.30	73.1	9.4	31.8	25.1	13.3	13.5
CD (P=0.05)	0.2	2.4	1.3	2.6	2.4	5.3	2.3

(Pattanayak and Poonam Kumari, 2004)

PMS for increasing efficiency of Pressmud

The pressmud is a byproduct of sugar industry. Its calcium corborate equivalent is lower (28-29) than PMS, but it is a good source of carbon and many essential nutrients. Attempt has been made for the use of locally available pressmud mixed with PMS. Combining PMS with pressmud increased the pod yield of groundnut along with shelling per cent. Highest pod yield of 4170 kg ha^{-1} with highest shelling per cent of 81.5 was recorded with the application of 25 % pressmud with 75 % PMS (Table 20).

Table 20. Influence of combined use of PMS and pressmud (PM) on productivity of groundnut in an acid soil

Treatments	Pod yield (kg ha^{-1})	RAE (%)	Shelling (%)	pH	OC (g kg^{-1})	Exch. Acidity cmol(P^{+}) kg^{-1}
No lime (control)	2420	-	66	4.79	3.5	0.98
100 % PM + 0 PMS	3700	100	67.5	4.84	3.7	0.87
75 % PM + 25 % PMS	3799	108	69.0	4.88	3.9	0.71
50 % PM + 50 % PMS	3890	115	75.0	5.05	4.4	0.63
25 % PM + 75 % PMS	4170	137	81.5	5.40	4.8	0.47
0 PM + 100 % PMS	4030	128	80.0	5.46	4.8	0.47
CD (P=0.05)	96	-	0.9	0.19	0.12	0.08
Initial	-	-	-	4.89	3.4	0.93

(PRC Report 2006-2010)

PMS use in the state of Odisha

Quantity of PMS available in the state is presented below.

Sl.No	Place	District	Quantity (Million tonne)
1.	Rayagada	Rayagada	25,000
2.	Jaypore	Koraput	18,000
3.	Brajarajnagar	Jharsuguda	20,000
4.	Choudwar*	Cuttack	20,000

*Factory is closed; deposits are there for supply of PMS

Material cost

The material is available almost free of cost. The State Government, Department of Agriculture has signed MOU with companies for lifting the PMS free of cost. However, the Government is spending for excavating, grinding, bagging of PMS. The transport cost is charged @ Rs. 10 per 50 kg bag from the farmers.These are available at block level. Users of PMS other than farmers i.e.Institutions ,NGOs,Private parties pay Rs. 60 per 50 kg of PMS, which is processed and sold through Agro-Industries Corporation of Odisha at district level.

Specification for packing

The PMS material is supplied for farmers use under the banner of "State Plan" Department of Agriculture, Odisha with State Government Emblem. The content in the bag is written clearly in "Odia" language as:

a) Amount of Calcium-20-35%,

b) Maximum moisture-15-20%, and

c) Weight (Minimum)-50 kg

Recommendations for use

State Agricultural University

Application of PMS @0.1 LR(Lime requirement) for monocot and @0.2 LR for dicots, fruits and vegetables etc. mixed with organic manures, below the seed zone, behind the plough on the day of sowing of seeds , planting of seedlings for each crop.

For seedlings of vegetables, the PMS is to be applied at least two days before planting mixed with organic manures to each crop.

For plantation crops, the quantity of PMS to be used vary between 500g and 5 kg per tree, depending upon the age and canopy of the tree species with the onset of monsoon every year.

State Department of Agriculture

Blanket recommendation of PMS @ 500kg (10 bags) ha^{-1} irrespective of the crops after soil tests for pH.

Farmers Feed Back

- Farmers in the acid soil regions in the state have benefited from using PMS for pulses (arhar, green gram, black gram), oil seeds (groundnut, sunflower , mustard) and cereals (maize, ragi).
- The use of PMS for vegetable,fodder and fruit crops is gaining momentum.
- On an average ,10 to 35 per cent yield increase have been achieved in different crops by the farmers in the state

Problems in use of PMS

- Timely availability of PMS near farmers fields at the time of need.
- Poor quality bag used for packing is a problem for transportation and carrying to farmers fields.
- Poor quality PMS material (burnt coal,stones ,weeds etc) inside the bag some times cause dis-satisfaction among farmers.
- Demonstration programme on PMS and its use in acid soils by Government functionaries is less leading to poor visibility of impact.

Conclusion

Use of paper mill sludge, as an acid soil ameliorant is time tested and established technology. Its efficiency under field condition increases when it is applied mixed with FYM or any other organic material in the rhizosphere. The monocots respond less than dicots, hence it should be applied accordingly. It should be applied based on lime requirement of the soil for the crop to be grown. The dicot-pulse/oilseed/vegetable crops need to be limed @ 0.2 LR and cereals @ 0.1 LR. For better efficiency, it should be applied in the rhizosphere in furrows instead of broadcasting. For young seedlings, it should be applied mixed with FYM 2/3 days before transplanting. For better efficiency, PMS application should be included as a component of integrated nutrient management along with soil test based fertilizers application.Its use has no deleterious effect both on crop and soil.

References

Bal, S.C., Nayak, R.K. and Sahu, S.K. (2008). Effect of NPK and lime on C/N ratio, malformation and yield of mango grown in an acid lateritic soil. *The Orissa Journal of Horticulture* 36(2):118-120.

Mandal, S.C., Sinha, H. Prasad, C.R.a nd Ali, M.A.(1966). *Science Culture*, 32: 149.

Mishra, M. and Pattanayak, S.K. (2002). Response to graded doses of lime amended with or without FYM. Professor IFFCO Chair Report, OUAT, Bhubaneswar.

Misra, U.K. and Pattanayak, S.K. (2003). *In* Compiled Report of The ICAR Net work Adhoc Scheme"*Soil Characterization and Resource management of Acid Soil Regions for Increasing Productivity*" OUAT, Bhubaneswar.

Misra, U.K., Satapathy, S. and Panda, N. (1989) Characterization of some acid soils of Orissa: I. Nature of acidity. *Journal of the Indian Society of Soil Science*, 37: 22.

Panda, N. (2003). Acid Soils of Eastern India" In: Characterization and Sustainable Management of Acid Soils of Eastern India Compilation for Winter School. November 18 to December 8, 2003. Department of Soil Science Agricultural Chemistry, OUAT, Bhubaneswar.

Panda, N. and Koshy, M.U. (1982). In: Review of Soil Research in India. I, XII International Congress of Soil Science, 8-16th February, New Delhi.

Pattanayak, S.K. (2012) QRT Report, AINP on Biodiversity Biofertilizers. Bhubaneswar centre, IISS, Nabibagh, Bhopal.

Pattanayak, S.K. and Misra, U.K. (1989). Transformation of phosphorus in some acid soils of Orissa. *Journal of the Indian Society of Soil Science*, 37:455-460.

Pattanayak, S.K. and Pandey, R. (2012). In: "Stromatolyte as an acid Soil Ameliorant". *Journal of the Agricultural Sciences* (Communicated).

Pattanayak, S.K. and Poonam Kumari (2004). Efficacy of green manure (Sesbania cannabina) as influenced by micronutrients and liming.*In* "Development in soil Science" 69th Annual Convention of ISSS, New Delhi.

Pattanayak, S.K., Misra, U.K., Sarkar, A.K. and Majumdar, K. (2011). Integrated nutrient management for groundnut and redgram on acid soils in Odisha. *Better Crops, South Asia*, 5(1): 8-10.

Pattanayak, S.K., Mohanty, S.K., Mishra, K.N., Nayak, R.K. and Mohanty, G.P. (2008). Technical bulletin on "*Biofertilizers for tropical vegetables*"ANPN on Biofertilizers, Orissa University of Agriculture and Technology, Bhubaneswar, Orissa. *Biofertilizers, Bulletin* No.1.

Pattanayak, S.K., Satyanarayana, T., Majumdar, K. and Mandal, M. (2012). Site specific nutrient management for Maize grown in *Alfisols* of Odisha. *Journal of Agricultural Sciences* (Communicated).

Peer Review Committee Report (2006-2010). Niche area of excellence on "*Management of Acid Soils for Sustainable Crop Production*" Department of Soil Science and Agricultural Chemistry, OUAT, Bhubaneswar.

Sahu, G.C.and Patnaik, S. (1990). Influence of lime or organic matter on pH, and ion exchange phenomenon of some *Alfisols. Journal of the Indian Society of Soil Science* 38: 154-157.

Sahu, S.K. and Nanda, S.K. (1988) Paper mill sludge benefits the acid soils of Orissa *Indian Farming* 36:21-23

Sharma, A., Sharma, Munish, Sharma, R.P.a nd Sharma, S.P. (2012) Effect of boron and lime integration with organic manures and NPK fertilizers on Garden Pea productivity under acidic soils in humid sub-temperate climate. In compendium of Abstracts. *The 8th International Symposium on Plant –Soil Interactions at low pH.* October 18-22 Bengaluru, India: 274-275.

Sudharmaidevi, C.R., Monorama, Thampatty, K.C. and Venugopal V.K. (2012).Effect of liming on nutrient availability and yield of cowpea in *Ultisols* of Southern Kerala. *In Compendium of Abstracts. The 8th International Symposium on Plant-Soil Interactions at low pH.* October 18-22. Bengaluru, India,: 324-325.

11

Way Forward

A.K. Sarkar

The adjustment and maintenance of soil pH at a value suitable for optimal crop production is the first step in soil management, because soil pH affects soil properties, nutrient availability and plant nutrition. In India, about 30M ha cultivated land suffer from high soil acidity (pH < 5.5). These soils are critically degraded and are of poor health producing crop yields, which are often less than 50% of that in normal soils. Due to this, farmers use much less fertilizers for crops (50 to 80 kg NPK ha^{-1}), since this is not profitable.

Amelioration of acid soils by liming is desirable for raising crop productivity. This has been amply documented by large scale field and laboratory studies across India (Mandal et al. 1975; Panda and Koshy 1982; Mathur et al. 1985; Sarkar et al 2000; Mishra 2004; Sharma & Sarkar, 2005; Panda, 2007; NAAS, 2010; Sarkar, 2013). But, availability of liming materials of a desired specification [CaO (%) > 30%, 80 mesh size] locally at a reasonable cost (Rs. 1 to 1.50 per kg) to farmers is a problem,since good quality liming materials have industrial use. This can be solved to a considerable extent by using industrial wastes such as basic slag, paper mill sludges,pressmud, or poor grade lime with less $CaCO_3$ & high silica. Effective market chain must be there, so that lime application is cost effective. Further, we have to do away with broadcast of lime in fields,because of the problems of transportation,one-time cost,and poor efficiency and use 200 to 400 kg lime ha^{-1} in furrows each year along with 50 to 75% of the recommended fertilizer dose to atleast double the present crop productivity levels in acid soil regions.

Technology Package

The technology of lime application developed under ICAR network project on acid soils (Sharma and Sarkar,2005) is as follows:

Land situation: Rainfed/irrigated uplands and medium lands with soil pH 5.5 or less.

Dose of lime: 200 to 400 kg lime ha^{-1} in furrows ($1/10^{th}$ of lime requirement of soil). Dose will be higher for fine textured soils and for soils with high organic matter content and lower pH.

Source of lime

Basic slag, paper mill sludge, limestone/ dolomite which is locally available in ground form (about 80 mesh) and of low cost.

Method of lime application

To be applied in furrows alongwith 50 to 75% (if 100% recommended NPK is not possible) NPK fertilisers. Basal application of organic manure is desirable.

Crops & Cropping sequences

Pigeon pea, groundnut, lentil, gram, pea, cotton, maize, soybean, wheat, linseed, forages, mustard, sorghum, wheat are important crops of acid soil regions. Lime can be added to any crop/ cropping system in acid soils , especially pulses & oilseeds. Preference should be given to varieties, which are relatively tolerent to Al^{3+} toxicity.

The technology has to be assessed and refined for benefitting agro-forestry and horticulture based systems. Orientation towards new and emerging areas such as precision agriculture, farming systems, minimum tillage, micro- irrigation, conservation agriculture, residue recycling in acid soils of India with lime as one of the key inputs for crop production need to be addressed. The eastern and the north-eastern states (comprising about 80% of the acid soils of the country) suffer from very low agricultural productivity not only on account of soil acidity but also due to land degradation, poor soil organic matter and inadequate and unreliable surface irrigation systems. Thus, an integrated approach is essential to check the spread of such problems and accelerate the crop productivity and bring to smile in the face of the farmers.

Mode of Technology Transfer

Transfer of agricultural technologies has been seen as the outcome of research, extension and integration with farmers' organizations. The role of each of these, extent of initiative and interaction among them with tools and techniques of working have varied from time to time and place to place.

In order to accept and integrate into social systems, technologies must provide solutions to farmers' problems and bring about positive changes. Development of technology is not enough. The policy, support services, feed back mechanisms, market linkage, training and orientation have to be ensured for success.

Indigenous technological knowledge of farmers (ITK) on maintenance of soil fertility, soil and water conservation need to be actively considered in developing new technologies since they are tested over a period of time, low cost, locally adapted, and simple. Farmers have to be convinced that the existing ITKs are less effective than new scientifically developed technologies.

In the technology transfer mechanisms, some of the areas which need emphasis are:

- Training of farmers and participation of farmers during technology development and its dissemination.
- Awareness creation through field functionaries, NGOs, farmers groups, KVKs, ATMAs.
- A consolidated GIS database on soils, water, inputs, crops, seeds, market at Block/Panchayat level
- Ensuring the input supply system at local level such as lime, dolomite, fertilizers, seeds, plant protection chemicals etc.
- Public Private Partnership to reap the benefits of the Technology such as tie-up with steel plants for supply of basic slag.

- Hand holding at local level to promote dissemination of technology, keeping in view the farmer's resources, investment capacity and level of management.
- The toll free helpline (1551) established at Kisan Call Centre should be equipped to answer farmer's problems.
- Working on a missionary mode to solve farmer's problem with a targeted approach.

Operation of Technology and Expected outcome

1. The technology of lime + fertilizer application needs to be operationalised in 30 million hectares of cultivated soils where soil pH below 5.50. This may be done in phases i.e. taking 10 million hectares in the first phase, to ensure proper monitoring.
2. The Central and State Governments should facilitate creation of a marketing and distribution network for easy availability of liming materials. Locally available materials at a low cost in 50 to 100 kg bags with a desired quality (80 mesh size and containing > 30-35% CaO) need to be promoted.
3. Public-Private partnership in grinding, bagging and transport of materials is desirable. For instance, basic slag lumps from steel plants can be transported and ground to proper mesh size, bagged and made available in blocks to ensure ready and timely use of the material by the farmers.
4. Posters, charts & folders in local languages should be prepared & distributed among farmers for awareness creation. Field demonstrations on the benefits of liming, lime+fertilizer, compost + lime + fertilizer use for crop production in acid soils vis-à-vis their non-application are necessary to popularize the practice.
5. Soil health cards should be made available to the participating farmers. Soil tests for available N, P, K, S, B, Zn, org. C and pH after every 2 or 3 crops grown each year is advisable.
6. Active participation of KVK and ATMA will facilitate to popularize the soil health improvement programme in acid soil regions.

7. Acid soil management should not be an isolated work in the states. It should be integrated with soil and water conservation, water saving technologies, rain water conservation, quality seeds of crops, integrated plant nutrient management, farming system and conservation agriculture. This will have the desired impact and we address the issue in totality.

8. The economic benefits will depend on the action- plan drawn and its implementation in a phased manner. Additional foodgrains from the acid soil regions with the programme will definitely improve the food,nutrition and livelihood security of the masses.

Estimate of Economic gain*

i) Area : 30 million hectares under Acid soil(pH <5.5)

ii) Present crop production level : 30 million tonnes per annum.

iii) Average dose of lime : 0.3 t/ha

iv) Total requirement of lime in one year : 9.0 million tonnes .

v) Total cost of liming material @ Rs. 2000 per tonne (Approx.): Rs. 18, 000 Million

vi) Cost of Fertilisers (Rs. 20,000 /t) for 30 million hectares @ 0.15 t/ha : Rs. 90,000 Million.

vii) Yield advantage

- @ 50% response due to lime + fert. use: 15.0 Million tonnes.
- @ 100% response due to lime + fert. use: 30.0 Million tonnes.

viii) • Additional economic output in case of 50% increase in crop yields @ Rs. 20,000/ tonne : Rs. 30,0000 Million

- Additional economic output in case of 100% increase in crop yields Rs. 60,0000 Million

ix) Net gain (Rs. Million/annum)

- @ 50% increases due to lime + Fert. use: Rs. 192,000 Million

- @ 100% increases due to lime + Fert. use: Rs. 492,000 Million

**The above calculation is an estimate of expected gain by managing acid soils by lime+ fertilizer application in crops. For exact figures, one has to take into account the area, crops, cost of lime and fertilizers (N, P, K, B as per soil tests) in each state having problems of soil acidity.*

Researchable issues in Acid soils

- Effect of soil amendment x nutrient x water interactions in crops.
- Lime and fertilizer needs for conservation tillage based crop production systems.
- Exploration and utilization of microbial diversity in acid soils.
- Measures for greater utilization of subsoil water and nutrients in case of deep rooted crops such as groundnut and pigeonpea.
- Interactive effects of soil conservation, lime and nutrients for agroforestry and multi-tier horticultural systems in hilly areas.
- Constraint analysis in adoption of recommended technologies for acid soil management.
- Phosphorus x lime x Boron interactions in soils of varying pH.
- Breeding of crops to modify root characteristics for adjustments of nutrient deficiencies and toxicities in problem soils.
- Physiological and molecular interventions to explore and develop plants with higher tolerance to Al-toxicity.
- Research on crop diversification under integrated farming systems for income and employment generation

References

Mandal, S.C., Sinha, M.K. and Sinha, H. (1975). In Acid soils of India and liming, *Technical Bulletin* No. 51, ICAR, New Delhi.

Mathur, B.S., Rana, N.K., Sinha, H. and Lal, S. (1985) Lime as an effective source of fertilizer in acid red loam soils of Bihar. *Journal of the Indian Society of Soil Science* 33: 328-332.

Mishra, U.K. (2004). Acid soil and its management. *Journal of the Indian Society of Soil Science* 52; 332-343.

NAAS (2010) "Exploring untapped potential of acid soils of India. Policy paper no. 48, National Academy of Agricultural Sciences, New Delhi. pp 8.

Panda, N and Koshy, M.U. (1982). Chemistry of acid soils. Review of Soil Research in India Part I, pp 160-168. 12th International Congress of Soil Sciences, New Delhi.

Panda, N. (2007) Management of acid soils. In Nutrient Management in Acid soils (Ed. Rattan, R.K.) *Bulletin of the Indian Society of Soil Science*, New Delhi 25:1-9.

Sarkar, A.K., Singh K.P., Singh, B.P., and Singh, R.P. (2000) Long term effect of fertilizer, manure and amendment on crop production and soil fertility. SSAC (BAU) *Tech.Bull.* 21;157.

Sarkar, A.K. (2013) Soil health management in acid soils. In Soil Health Management (Edited by Dr. H.L.S. Tandon) Fertiliser Development and Consultation Organisation, New Delhi-110048 pp 142-166

Sharma, P.D. and Sarkar, A.K. (2005) Managing acid soils for enhancing productivity. *Technical Bulletin* 23; ICAR, Natural Resource Management Division, KAB II, New Delhi. pp 1-24.